Radar Meteorology

Advancing Weather and Climate Science Series

UK Series Editor: Dr Peter Michael Inness,
US Series Editor: Dr John A. Knox

Scientific advances in weather and climate science over the past decade are now making their way into the teaching literature, and climate change has become a major issue that pervades all aspects of weather and climate science.

The Royal Meteorological Society (RMetS) is a world-leading professional and learned society in the field of meteorology. Based in the UK, it encourages and facilitates collaboration with organisations that are active in earth science. It serves its professional and amateur members and the wider community by undertaking activities that support the advancement of meteorological science, its applications and understanding.

The RMetS *Advancing Weather and Climate Science Book Series*, in conjunction with Wiley-Blackwell, brings together both the underpinning principles and new developments in meteorology in a unified set of books suitable for undergraduate and postgraduate study, as well as being a useful resource for the professional meteorologist or Earth system scientist. Topics covered include; atmospheric dynamics and physics, earth observation, weather forecasting, climate variability and climate change science, impacts and adaptation. A full list of titles currently available in the series can be seen below. Click on the links for further information including table of contents and sample chapters.

Other titles in the series:

Meteorological Measurements and Instrumentation
Giles Harrison
Published: December 2014

Fluid Dynamics of the Mid-Latitude Atmosphere
Brian J. Hoskins, Ian N. James
Published: October 2014

OperationalWeather Forecasting
Peter Inness, University of Reading, UK and
Steve Dorling, University of East Anglia, UK
Published: December 2012
ISBN: 978-0-470-71159-0

Time-Series Analysis in Meteorology and Climatology: An Introduction
Claude Duchon, University of Oklahoma, USA and
Robert Hale, Colorado State University, USA
Published: January 2012
ISBN: 978-0-470-97199-4

The Atmosphere and Ocean: A Physical Introduction, 3rd Edition
Neil C. Wells, Southampton University, UK
Published: November 2011
ISBN: 978-0-470-69469-5

Thermal Physics of the Atmosphere
Maarten H.P. Ambaum, University of Reading, UK
Published: April 2010
ISBN: 978-0-470-74515-1

Mesoscale Meteorology in Midlatitudes
Paul Markowski and Yvette Richardson, Pennsylvania State University, USA
Published: February 2010
ISBN: 978-0-470-74213-6

Radar Meteorology

A First Course

Robert M. Rauber
University of Illinois, IL, US

Stephen W. Nesbitt
University of Illinois, IL, US

WILEY Blackwell

The right of Robert M. Rauber and Stephen W. Nesbitt to be identified as the authors of this work has been asserted in accordance with law.

Registered Office(s)
John Wiley & Sons, Inc., 111 River Street, Hoboken, NJ 07030, USA
John Wiley & Sons Ltd, The Atrium, Southern Gate, Chichester, West Sussex, PO19 8SQ, UK

Editorial Office
The Atrium, Southern Gate, Chichester, West Sussex, PO19 8SQ, UK

For details of our global editorial offices, customer services, and more information about Wiley products visit us at www.wiley.com.

Wiley also publishes its books in a variety of electronic formats and by print-on-demand. Some content that appears in standard print versions of this book may not be available in other formats.

Library of Congress Cataloging-in-Publication Data

Names: Rauber, Robert M., author. | Nesbitt, Stephen W., author.
Title: Radar meteorology : a first course / by Robert M. Rauber, Dr. Stephen W. Nesbitt.
Description: First edition. | Hoboken, NJ : John Wiley & Sons, 2018. | Series: Advancing weather and climate science series | Includes bibliographical references and index. |
Identifiers: LCCN 2017051770 (print) | LCCN 2017061104 (ebook) | ISBN 9781118432655 (pdf) | ISBN 9781118432631 (epub) | ISBN 9781118432624 (cloth)
Subjects: LCSH: Radar meteorology.
Classification: LCC QC973.5 (ebook) | LCC QC973.5 .R38 2018 (print) | DDC 551.63/53–dc23
LC record available at https://lccn.loc.gov/2017051770

Cover design: Wiley
Cover image: Courtesy of Josh Wurman

Set in 10/12pt and PalatinoLTStd by SPi Global, Chennai, India

Dedication

To: Ruta, my wife, Carolyn, and Stacy, my daughters, Josh and Fabian, their husbands, and my new granddaughter Molly—the rays of light illuminating my life. —Bob Rauber.

To my wife Anna and my boys Ian, Tadhg, and Leif, for all of their help in keeping life perfectly calibrated. —Steve Nesbitt.

Contents

Preface

From Past to Future: Why Study Radar Meteorology?

Radar, an acronym for *RA*dio *D*etection *A*nd *R*anging, was initially developed to remotely detect aircraft and ships. During the 1930s, military applications of radar for aircraft detection were explored by Britain, Germany, and the USA, but the devices available were limited to very low frequencies and low power output. In 1940, the British invented the *cavity magnetron*, a device that permitted radars to operate at higher frequencies and high power output. The new, secret radars employing the cavity magnetron gave the Allies a distinct advantage throughout World War II—the capability to detect aircraft and ships at long ranges. Studies of atmospheric phenomenon with radar began almost as soon as the first radars were used. These studies were initiated because weather and atmospheric echoes represented undesirable "clutter" that hampered detection of military targets.

Following the war, the first large weather-related field campaign with non-military application, the *Thunderstorm Project*, was organized to study coastal and inland thunderstorms. Data from this and other projects stimulated interest in a national network of weather radars. Enthusiasm for a national network was spurred by efforts to estimate precipitation from radar measurements. The discovery of the hook echo and its association with a tornado led to widespread optimism that tornadoes may be identified with radar. Following the war, several surplus military radars were adapted for weather observation. These were replaced beginning in 1957 by the Weather Surveillance Radar (WSR-57), which became the backbone of the U.S. National Weather Service radar network until the Weather Surveillance Radar-1988 Doppler (WSR-88D) Doppler radars were installed three decades later.

The advent of digital technology, the rapid growth in the number of scientists in the field of radar meteorology, and the availability of research radars to the general meteorological community led to dramatic advances in radar meteorological research. A fundamental change that spurred the revolution was the advance in digital technology. A basic problem facing radar scientists was the large volume of data generated by radars. A typical pulsed Doppler radar system, for example, samples data at rates as high as three million samples per second. This amount of data is impractical to store—the data must be processed in real time to reduce its volume and convert it to useful forms. Beginning in the early 1970s, advances in data storage technology, digital displays, computer hardware and software, and processing algorithms all made it possible to collect, process, store, and view data at a rate equal to the rate of data ingest. A key advance was the development of efficient software to process the data stream from Doppler radars. This development occurred at about the same time that the hardware became available to implement it, leading to rapid advances in Doppler measurements.

Doppler radars were soon developed with antennas that rotate in azimuth and elevation so that the full hemisphere around the radar could be observed. A network of Doppler radars, the WSR-88D network, was installed throughout the USA in the early 1990s to monitor severe weather. Countries of the European Union and China have also installed Doppler radar networks for storm monitoring. Mobile, airborne, spaceborne, and dual-polarization meteorological research radars were developed in the decades of the 1980s, 1990s, and 2000s for specialized applications. The WSR-88D radars were upgraded in 2011–2012 to employ polarization technology.

Early studies of "weather clutter" with the first-generation radars of the 1940s have evolved today into a complete scientific discipline called *radar meteorology*. Today, radar scientists worldwide meet regularly to learn about the latest advances in the science and to decide how best to employ these advances to better protect the public from weather hazards. With the development of the Internet and smart phone technology, displays of virtually all radar variables from the WSR-88D radars are now available in near real time to anyone in the world within range of a cell phone tower. Radar data are displayed regularly on television news, and ordinary people with no meteorological training "check the radar" every morning before venturing out into a rainy day.

So why study radar meteorology? The answer is straightforward—radar is the only tool with which we can observe the detailed structure of storms. Radar allows operational meteorologists to routinely detect tornadoes and flash floods in sufficient time to warn the public to seek shelter. Weather radars used by air traffic control provide controllers with the information to warn approaching and departing aircraft about microbursts and wind shear events. Meteorologists use radar to discriminate precipitation type in winter and determine the onset, end, and type of winter weather to be expected.

Atmospheric scientists use radar for a wide range of research, from diagnosing the circulations in hurricane eyewalls to investigating convective initiation along the Great Plains dry line. Radar data are now assimilated into numerical models used in research and forecasting. Radar is a critical tool in hydrological research.

In future decades, radar will be a key component of emerging science frontiers in the atmospheric sciences. Global climate change impacts will drive scientific research—the predicted increase and impact of extreme events, for example, will demand a greater focus on these events by the radar research community. New frontiers will continue to be forged in tropical and extra-tropical cyclone research, studies of convective, tropical and winter weather systems, measurement of rain and snowfall, and detection of floods. New technologies will be exploited to measure atmospheric properties such as water vapor and particle types in clouds. A diversity of radar technologies—and trained individuals to employ them—will be needed to accomplish this research. Capabilities will extend across the full range of current technologies—Doppler and polarization diversity, standard and phased array antennas, and multiple wavelengths—to yet-to-be-invented technologies. Radars will be deployed on a wide array of platforms such as aircraft, ships, and trucks to nearby locations as well as remote regions of the earth. The future of radar meteorology is vibrant and exciting—and studying radar meteorology gives you the tools to be part of the adventure.

Robert M. Rauber and Stephen W. Nesbitt
December 2017
Champaign-Urbana, IL, USA

Acknowledgments

Although only two authors appear on the cover, this book is really a compilation of the work of hundreds of radar meteorologists who spend their careers determining how to extract and use meteorological information from microwave signals generated by radars. Their work has saved countless lives. We could not have developed this book without the generosity of our many colleagues at universities, government laboratories and agencies, research laboratories, and in private industry who allowed us to use their photographs, images, and figures in this book. We are especially indebted to Prof. Paul Smith, who generously provided his excellent class notes to Bob to develop his radar course back in the 1990s and then permitted us to use his notes as a basis for developing Chapters 3, 5, 6, and 13 in this book. We also thank L. Jay Miller, who taught Bob radar meteorology and dual-Doppler processing during a summer sabbatical at NCAR and provided the notes that became the basis of Chapter 12. We are also indebted to Larry Di Girolamo, Scott Ellis, Peisang Tsai, Eric Loew, Jeff Trapp, Wen-Chau Lee, Jeff Waldstreicher, Michael Coniglio, and Bart Geerts who provided reviews of various chapters in this book. We particularly want to thank Eric Loew and Scott Ellis for providing a tour of the SPOL radar and taking the pictures that appear in Chapter 2, Patrick Gatlin for the large drop images in Chapter 7, and Ali Tokay for the pictures of disdrometers in Chapter 13. Open source software including Py-ART[1] (supported by the Department of Energy Atmospheric Radiation Measurement program) and CSU_Radartools[2] was used to prepare some figures in this text.

The Department of Atmospheric Sciences at the University of Illinois at Urbana Champaign has supported this effort through many years of development and revision. We collectively thank all our current and former faculty and staff, and especially our graduate students, for their help over our years together. Bob wishes to specifically acknowledge the late Lewis O. Grant, his PhD advisor, who opened his eyes to the atmosphere and launched his career in field research. Steve wishes to specifically acknowledge his mentors and colleagues in radar meteorology (Steve Rutledge, Larry Carey, Walt Petersen, Timothy Lang, and V. Chandrasekar) and those that got him dreaming of studying the atmosphere with field work (Ed Zipser, Gerry Heymsfield, Bob Houze, and Ana Barros).

Our publisher, Wiley, has been a superb partner. It has been a pleasure to work with all their staff.

---

[1]Helmus, J.J. and Collis, S.M. (2016) The Python ARM Radar Toolkit (Py-ART), a library for working with weather radar data in the Python programming language. *J. Open Res. Softw.* **4**(1), DOI: 10.5334/jors.119.

[2]See: https://github.com/CSU-Radarmet/CSU_RadarTools.

Finally, we want to thank our families. As anyone who has written a book knows, family members must sacrifice so much time, provide so much support, and have so much patience or the project will never get done. So Ruta Rauber, Carolyn, Josh and Molly Bishoff, Stacy and Fabian Trevino, and Anna, Ian, Tadhg, and Leif Nesbitt—THANKS!!! This book is as much yours as it is ours.

About the Companion Website

Instructors: Do not forget to visit the companion website for this book:

www.wiley.com\go\Rauber\RadarMeteorology

There you will find valuable material designed to enhance your teaching, including:

- Powerpoint Slides.
- Solutions.

Scan this QR code to visit the companion website. Note that the website should be available to instructors, not students.

1

Properties of Electromagnetic Waves

Objectives

By the end of this chapter, you should understand the following:

- The basic nature of electric and magnetic fields and how these fields are related through Maxwell's equations;
- Electromagnetic waves consist of oscillating electric and magnetic fields propagating at the speed of light;
- The electromagnetic spectrum, and why radars use the microwave frequency part of the spectrum;
- Why the degree of absorption of electromagnetic waves by a medium such as air, water, or ice depends on the wave frequency;
- That electromagnetic radiation can be characterized as waves or as particles called *photons*;
- How radiation interacts with matter through processes of refraction, reflection, scattering, and absorption and how each of these relate to radar meteorology;
- What is meant by polarization of an electromagnetic wave.

1.1 Introduction

A radar transmits **electromagnetic radiation**, normally microwaves, and measures the properties of the radiation scattered back to its antenna by objects in the path of its beam. Radar meteorologists face an enormous challenge because a radar, at best, measures only six pieces of information: the amplitude, phase, and polarization state of the returned electromagnetic energy, the time the radiation took to travel to and from the objects, and the azimuth and elevation of the radar antenna at the time the radiation was transmitted. From this scant data, they must deduce meteorologically relevant information such as the location of precipitation, rainfall rate, precipitation type and wind speed, and from this information quickly report to the public the location of a flash flood, hail, or a tornado. How they accomplish that is the subject of this book. To understand how radars work and, more importantly, how the energy

Radar Meteorology: A First Course, First Edition. Robert M. Rauber and Stephen W. Nesbitt.
© 2018 John Wiley & Sons Ltd. Published 2018 by John Wiley & Sons Ltd.
Companion website: www.wiley.com/go/Rauber/RadarMeteorology

radars transmit and receive can be used to determine atmospheric properties, we must first develop a basic understanding of electromagnetic energy.

1.2   Electric and magnetic fields

As you sit reading this book, you are surrounded by electric and magnetic fields. The light entering your eyes as you stare at this page reaches your retina as propagating **electromagnetic waves**. Cell phone transmissions, the energy heating food in your microwave, medical X-rays, and sunlight are all forms of electromagnetic energy—electric and magnetic fields oscillating in time and space.

1.2.1   The electric field

To help understand electric fields, let us start by imagining two infinite, parallel horizontal plates separated by 1 m (Figure 1.1). Assume for the moment that a perfect vacuum exists between the plates and an excess positive charge density of $10^{-12}\,\text{C}\,\text{m}^{-2}$ exists on one of the plates (i.e., there are fewer electrons per square meter on the positively charged plate, where a coulomb is a unit of electric charge). An **electric field** exists in the presence of a charged body, so an electric field exists between the plates. The **electric field intensity** ($\vec{E}$), a vector quantity, has a magnitude and a direction. The magnitude of $\vec{E}$ is proportional to the force acting on a unit positive charge at a point in the field. The direction of $\vec{E}$ is the direction in which that force acts. The electric field intensity is measured in units of volts per meter, or even more fundamentally in $\text{J}\,\text{C}^{-1}\,\text{m}^{-1}$. By convention, the electric field is directed away from positive and toward negative charges, so the electric field vector at any point between the plates in our experimental apparatus points from the positive to the negative plate.

We often represent $\vec{E}$ by drawing lines of force, or "flux lines," represented by a vector, $\vec{D}$, called the **electric displacement vector**. This term "displacement" comes about because there is a slight displacement of charges (atomic nuclei and electrons) that occur in a medium exposed to an electric field. The vector $\vec{D}$ is related to $\vec{E}$ by the equation:

$$\vec{D} = \varepsilon_0(1 + \chi)\vec{E} = \varepsilon_0\varepsilon_r\vec{E} \tag{1.1}$$

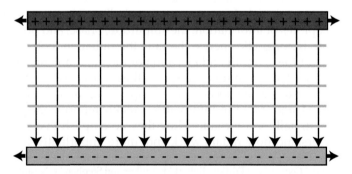

Figure 1.1   The electric field (black arrows) and lines of constant voltage (green lines) between two infinite, parallel, oppositely-charged plates

where ε_0 is called the **permittivity of free space**, χ is the **susceptibility** of the medium, and ε_r is the **relative permittivity**. The permittivity of free space is simply defined as the ratio of $\vec{D}/\vec{E}$ in a vacuum. In our apparatus in Figure 1.1, there is a vacuum, so there is no medium and $\chi = 0$. A vacuum has a permittivity of $\varepsilon_0 = 8.85 \times 10^{-12} \, C^2 \, J^{-1} \, m^{-1}$. Dry air at standard temperature and pressure has a relative permittivity of 1.000569.

The units of $\vec{D}$ are coulomb per square meter. As, by the design of our experiment, there are exactly $10^{-12} \, C$ of charge per square meter, and $\vec{D}$ and $\vec{E}$ are parallel, in our vacuum

$$\vec{E} = \varepsilon_0^{-1} \vec{D} \tag{1.2}$$

and therefore $\vec{E}$ has a magnitude of $0.112 \, V \, m^{-1}$, implying that 0.112 joules of electrical potential energy are stored in the electric field.

The susceptibility of a medium describes the degree to which a medium becomes electrically polarized (the charges within the medium align) under the influence of an electric field. Let us suppose that air (containing nitrogen, oxygen, water vapor, and trace gases), all at standard sea level pressure and temperature, is inserted between the plates of our experiment replacing the vacuum. Would the electric field between the plates be stronger or weaker? If the charge on the plates remained the same, then the $\vec{E}$ field would be weaker as $\varepsilon_r > 1$ in Eq. (1.1). Physically, the $\vec{E}$ field becomes weaker because molecules that have polarized structure (which means they have a positive and negative end), particularly water vapor, align opposite the electric field, the negative side of the molecules pointing toward the positive plate. Polarized molecules have their own electric fields which, when added to the background field, reduce the background field's overall strength.

Electric fields can be visualized by drawing lines of constant $\vec{D}$. Figure 1.2 shows a **dipole**, a separation of positive and negative charges, in this case of equal and opposite sign. The electric field is parallel to lines of $\vec{D}$ and its magnitude is

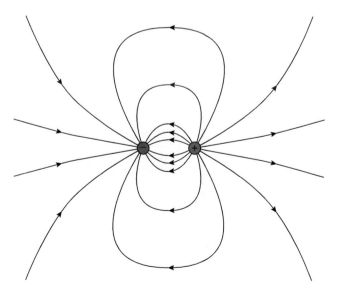

Figure 1.2   The electric field of a simple dipole

proportional to the distance between lines. Lines of force along which charges in a medium will accelerate originate on positively charged bodies and terminate on negatively charged bodies.

1.2.2   The magnetic field

Magnetic fields develop in conjunction with **electrical currents**—moving charged particles—and produce a force that acts on other moving charged particles within the field. This statement seems at odds with common experience—small refrigerator magnets, for example, appear to have no electrical current. In fact, they do. In some natural metals such as iron, magnetic fields are generated by the movement of electrons within atoms of the material—an internal electric current within the atoms themselves. Natural magnets such as the ones on a refrigerator door act on other material—iron in particular—which support internal electric currents at the atomic level.

The force exerted by a magnetic field on a charge, q, moving at velocity, $\vec{V}$, within the field is a vector at right angles to the direction of $\vec{V}$ and a vector $\vec{B}$, called the *magnetic induction* or *magnetic flux density* (i.e., force $= q\vec{V} \times \vec{B}$). Moving charges form a current, $\vec{I}$, that is measured in amperes or, more fundamentally, coulomb per second. Units of $\vec{B}$ are $JA^{-1}m^{-2}$. Lines of $\vec{B}$ are parallel to the direction a compass needle points in a magnetic field. As shown in Figure 1.3, magnetic field lines are closed lines surrounding the currents that produced them. Lines of $\vec{B}$ are everywhere parallel to lines of **magnetic field intensity**, $\vec{H}$. $\vec{H}$ has units of ampere per meter. The magnitude of $\vec{H}$ is proportional to the number of magnetic flux lines passing through a unit area perpendicular to the lines, so that

$$\vec{B} = \mu\vec{H} \tag{1.3}$$

where μ is called the **magnetic permeability**. The magnetic permeability is a measure of the ability of a material to store magnetic potential energy. A vacuum has a magnetic inductive capacity of $\mu_0 = 1.26 \times 10^{-6} J\,A^{-2}\,m^{-1}$. The ratio of the magnetic inductive capacity of air to a vacuum is unity everywhere in the atmosphere.

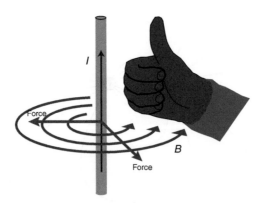

Figure 1.3   The magnetic field induction field (red arrows, B) associated with a current (I) and the force associated with the field (blue arrows). The right hand rule (force $= I \times B$) can be used to determine the relationship between the vectors

1.2.3 Relating the electric and magnetic fields—a simple dipole antenna

We have just learned that electric fields exist in the presence of charged particles, and magnetic fields exist in the presence of moving charged particles. The behavior of these fields with time is not independent—magnetic fields are generated by and influence changing electric fields and electric fields are generated by and influence changing magnetic fields. This indeed is the basis for electromagnetic waves. To understand how these fields influence each other, consider a simple **dipole antenna**, a device designed to transmit electromagnetic waves (Figure 1.4).

Suppose charges in this simple antenna are forced externally to oscillate back and forth across the antenna. At the initial time, the positive and negative charges are stationary and located at opposite sides of the antenna. As the charges are motionless, there is no current and no magnetic field. However, a voltage exists across the antenna, and the electric field is a maximum. All the energy at this time is stored in the electric field. As the charges accelerate toward the center of the antenna, the current increases and the voltage decreases until, as the charges pass the center, the current—and therefore the magnetic field—is maximum, and the electric field disappears. All the energy is now stored in the magnetic field. As the charges reach the opposite sides, the electric field is restored and the magnetic field disappears. At all other times, energy is partitioned between the electric and magnetic fields.

When the fields are caused by some external forcing to alternate rapidly in the antenna, in addition to the stored energy (which is transferred back and forth between the electric and magnetic fields), part of the energy is **radiated** as electromagnetic waves, which means the energy propagates away from the antenna, and a small amount is dissipated as heat in the antenna. The energy radiated per cycle is much

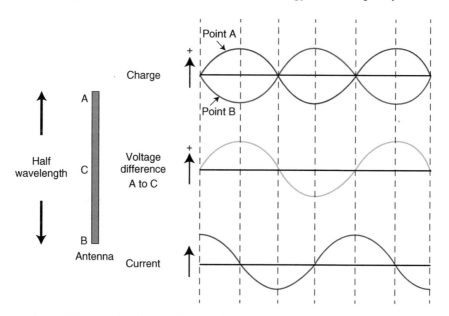

Figure 1.4 Oscillation of the charge, voltage, and current in a simple dipole antenna. (Adapted from Battan, L.J. (1973) Radar Observation of the Atmosphere. © University of Chicago Press, Used with Permission)

smaller than that stored in the fields, which are referred to as the **induction fields**. Near the antenna, beyond a distance equivalent to a few wavelengths of the radiated waves, the **radiated fields** dominate, and farther from the antenna, only the radiated fields exist.

1.2.4 Maxwell equations

Maxwell's equations, first formulated by Scottish scientist James Clerk Maxwell in the 1860s, are the foundation of classical electrodynamics and formalize the behavior of electric and magnetic fields discussed in the previous sections. The four equations

$$\nabla \cdot \vec{E} = \frac{\rho_c}{\varepsilon_0 \varepsilon_r} \tag{1.4}$$

$$\nabla \times \vec{E} = -\frac{\partial \vec{B}}{\partial t} \tag{1.5}$$

$$\nabla \cdot \vec{B} = 0 \tag{1.6}$$

$$\nabla \times \vec{B} = \mu \vec{I} + \varepsilon_r \varepsilon_0 \mu \frac{\partial \vec{E}}{\partial t} \tag{1.7}$$

in this form state that (1.4) the divergence of electric field is a function of charge density, ρ_c, (1.5) a closed loop of $\vec{E}$ field lines will exist when the magnetic field is changing with time, (1.6) the magnetic field is non-divergent and therefore consists of closed loops, and (1.7) a closed loop of $\vec{B}$ field lines will exist in the presence of a current and/or a time-varying electric field. In the vacuum of space, and in almost all locations in the troposphere (thunderstorms are exceptions), we can assume that there are no charges ($\rho_c = 0$) and no current ($\vec{I} = 0$), and the magnetic inductive capacity is that of free space, so Maxwell's equations can be reduced to

$$\nabla \cdot \vec{E} = 0 \tag{1.8}$$

$$\nabla \times \vec{E} = -\frac{\partial \vec{B}}{\partial t} \tag{1.9}$$

$$\nabla \cdot \vec{B} = 0 \tag{1.10}$$

$$\nabla \times \vec{B} = \varepsilon_r \varepsilon_0 \mu_0 \frac{\partial \vec{E}}{\partial t} \tag{1.11}$$

Solving for $\vec{E}$ and $\vec{B}$, one obtains a wave equation for which solutions are linear combinations of plane waves traveling at the speed of light in a vacuum, $\sim 3 \times 10^8$ m s^{-1}, given by

$$c = \frac{1}{\sqrt{\mu_0 \varepsilon_0}} \tag{1.12}$$

Maxwell's equations explain how electromagnetic waves physically propagate through space. A spatially varying electric field is always associated with a magnetic field that changes in time (1.9). Likewise, a spatially varying magnetic field is associated with changes in time of the electric field (1.11). In an electromagnetic wave,

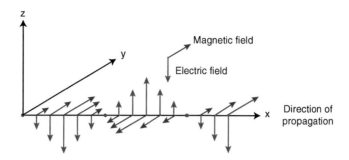

Figure 1.5   Oscillation of the electric and magnetic fields in a propagating electromagnetic wave

changes in the electric field are accompanied by a wave in the magnetic field in one direction, and changes in the magnetic field are accompanied by a wave in the electric field in the other direction. The character of both fields is described through (1.8) and (1.10). The relationship between the oscillating electric and magnetic fields can be understood by examining Figure 1.5. Note that the magnetic field is maximum when the electric field is minimum and *vice versa*. In general, we can describe electromagnetic waves by specifying the properties of either the electric field or the magnetic field. In radar meteorology, we customarily choose the electric field. Note that $\vec{E}$ and $\vec{B}$ are perpendicular to each other and the direction of wave propagation. We should note that the speed at which electromagnetic waves propagate through transparent or semitransparent materials, such as water, ice, or air, is less than c because electromagnetic waves interact with matter. The ratio between c and the speed v_c at which light travels within a material is called the **refractive index**, n, of the material ($n = c/v_c$). At frequencies used by radars, the refractive index of air at standard sea-level temperature and pressure is about 1.0003 and decreases with altitude.

One special solution of Maxwell's equations is a sinusoidal, **monochromatic** (single frequency, f) plane wave, the type of waveform used in many weather radars. Here, the electric field oscillates in time (t) according to

$$\vec{E}(r, \alpha, \beta, t) = \frac{\vec{A}(\alpha, \beta)}{r} \cos\left[2\pi f\left(t - \frac{r}{c}\right) + \phi\right] \tag{1.13}$$

where (r, α, β) are the range from the radar antenna, and the antenna's elevation, and azimuth, $\frac{\vec{A}(\alpha,\beta)}{r}$ is the amplitude of the electric field far from the antenna, and ϕ represents a constant transmitter phase angle. It is common in radar meteorology to write Eq. (1.13) and other equations in complex notation using the identity:

$$e^{ix} = \cos x + i \sin x \tag{1.14}$$

so that

$$\vec{E}(r, \alpha, \beta, t) = \frac{\vec{A}(\alpha, \beta)}{r} \exp\left[i\left(2\pi f\left(t - \frac{r}{c}\right) + \phi\right)\right] \tag{1.15}$$

where $i = \sqrt{-1}$. We keep in mind that we mean "the real part of" when using this notation. Complex notation becomes especially useful when discussing attenuation of radar energy by gases, clouds, or precipitation. We can understand why by considering the behavior of the simple mathematical wave function $Ae^{-t}\cos(\omega t)$.

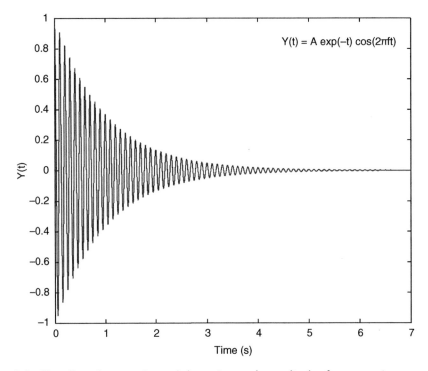

Figure 1.6   The effect of propagation and absorption on the amplitude of a propagating wave

In this expression, A is the amplitude of a wave at an initial time $t=0$ and ω is the angular frequency of the wave, where $\omega = 2\pi f$. Figure 1.6 shows a plot of this function. The cosine term represents the oscillating wave. The exponential term represents the rate at which the amplitude of the wave decreases with time. In radar meteorology, the exponential term might represent the rate of **absorption** of wave energy by a medium through which the wave was passing. Mathematical operations such as taking a derivative of a function like $Ae^{-t}\cos(\omega t)$ are cumbersome. However, if we use complex notation and write $\cos(\omega t)$ as Re $(e^{i\omega t})$, where Re denotes the real part, then the wave function can be written as $Ae^{(-1+i\omega)t}$. Here, we have dropped the Re notation and assume that we are considering only the real part. Mathematical operations with equations containing only exponents are generally much easier to perform. In radar applications, when we state that a medium has a "complex refractive index" what is meant is that the medium both absorbs wave energy (which can be represented by a real exponential function) and transmits wave energy (which can be represented by an oscillatory function such as a cosine wave, but expressed as the real part of a complex exponential function).

1.3   The nature of electromagnetic radiation

1.3.1   The electromagnetic spectrum

Electromagnetic waves are electric and magnetic force fields that propagate through a vacuum at the speed of light, c, or through a medium such as air, water, or ice at a speed slightly less than c. Electromagnetic waves are characterized by a **wavelength**, λ, and a **frequency**, f, where in free space, $c = \lambda f$, and in a medium, $v_c = \lambda f$, where v_c is the speed of light in the medium. Frequency has basic units of s^{-1}, which are commonly expressed in hertz (Hz). We will typically use megahertz (MHz) and gigahertz (GHz), respectively, for 10^6 and 10^9 Hz. More often in radar meteorology, we refer to wavelength because wavelength, particularly when expressed in units of centimeters or millimeters, can be easily compared to the dimensions of **hydrometeors**—the generic term for cloud droplets, raindrops, ice crystals, snowflakes, and hail.

The **electromagnetic spectrum**, shown in Figure 1.7, consists of all frequencies of electromagnetic waves from long-wavelength radio waves to short-wavelength gamma rays. Figure 1.7 also shows the atmosphere's opacity to electromagnetic radiation across the electromagnetic spectrum. Parts of the atmosphere are highly absorbing across parts of the spectrum, meaning that electromagnetic energy cannot propagate far through the atmosphere at those wavelengths. On the short-wavelength end, gamma rays, X-rays, and ultraviolet energy, which have natural sources outside earth's atmosphere, are nearly all absorbed by gases such as ozone and oxygen in the earth's upper atmosphere. Long-wavelength radio wave radiation, such as AM and FM radio waves, propagate in the lower atmosphere, with some absorption in earth's ionosphere. Infrared radiation and high-frequency microwaves are selectively absorbed by atmospheric gases throughout earth's atmosphere. There are two "windows" in the spectrum where the atmosphere is nearly non-absorbing through its depth, the first at visible wavelengths (permitting sunlight to reach the earth and for us to see beyond our noses) and the second in the microwave/radio wave portion of the spectrum (allowing us to have electronic communications). Weather radars operate in the short-wavelength, high-frequency end of the microwave region, and even into the range of shorter wavelengths within the radio window where the atmosphere is partially absorbing, primarily due to water vapor. Operational weather radars in the USA and China use wavelengths near 10 cm ($f = 3$ GHz), whereas in Europe, operational radars have wavelengths near 5 cm. Mobile radars typically use shorter wavelengths, typically 3 cm, due to the need for small, portable antennas. Cloud radars use even shorter wavelengths (about 0.3–1.2 cm) where, from Figure 1.7, atmospheric absorption is clearly an effect that must be considered. Radar wind profilers employ longer wavelengths corresponding to the VHF (very high frequency, $\lambda = 1$–10 m, $f = 30$–300 MHz) and UHF (ultra high frequency, $\lambda = 1$–0.1 m, $f = 300$–3000 MHz) range.

1.3.2   Electromagnetic wave interactions

The phase relationship among electromagnetic waves scattered by objects along a radar beam must be considered for many aspects of radar meteorology. Phase differences arise because of the difference in range (and therefore in round-trip transit time) from the radar to the various scattering objects. The integral number of wavelengths in the round-trip distance is unimportant in determining the phase, but the residual fraction of a wavelength is critical. The total scattered electric field depends on the mutual interference effects among the waves scattered by individual objects in the beam.

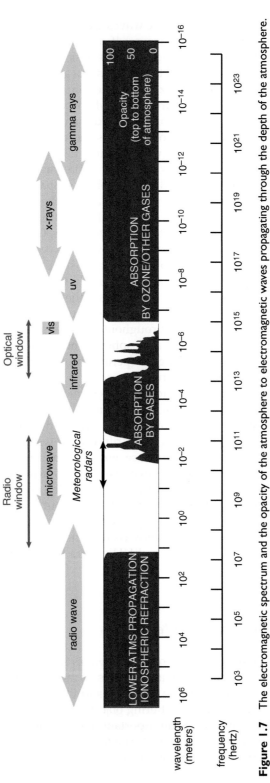

Figure 1.7 The electromagnetic spectrum and the opacity of the atmosphere to electromagnetic waves propagating through the depth of the atmosphere. Meteorological radars operate in the microwave part of the spectrum

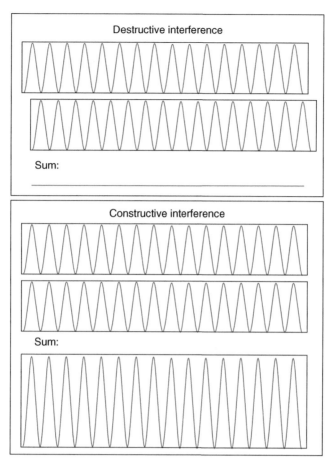

Figure 1.8   An example of destructive and constructive interference on two waves with the same amplitude and frequency

Figure 1.8 shows the electric field scattered by two objects of equal size, shape, and composition. In the upper panel, the objects are located at an integral number of quarter wavelengths apart. The back and forth distance is therefore one half wavelength, so the returned waves are 180° out of phase. When the waves combine, they undergo **destructive interference**, and the total amplitude of the electric field is zero. In the lower panel, the objects are located at an integral number of half wavelengths apart. The back and forth distance is therefore one wavelength, so the returned waves are in phase. When the waves combine, they undergo **constructive interference**, and the total amplitude of the electric field is double the amplitude of a single wave. In a situation where there are many objects, the amplitude of the scattered electric field is the "phasor sum" of the electric field from each of the objects.

1.4   Interactions of electromagnetic waves with matter

The interaction of electromagnetic waves with matter depends both on the frequency of the radiation and on the size, shape, composition, and distribution of the material.

It is easiest to visualize these interactions by considering electromagnetic waves in the visible part of the spectrum, but the various types of interaction have important applications in the microwave part of the spectrum where radars operate.

All interactions of electromagnetic waves ultimately occur on the molecular scale, with atoms or molecules interacting with quanta of electromagnetic energy called **photons**. The nature of these interactions is quantitatively described within the discipline of physics called **quantum mechanics**. Here, we will limit our discussion to describing the outcome of these interactions. There are two possibilities on the molecular scale. The first is **absorption**. In this case, the electromagnetic energy is removed—absorbed and realized as heat in the absorbing substance (an increase in the kinetic energy of the molecules of the substance). A simple demonstration is heating food in a microwave oven, where electromagnetic energy (microwaves) is converted to heat. The second is that the electromagnetic energy is absorbed but immediately re-radiated at the same frequency in a different (or possibly the same) direction. This process is called **scattering**. Both processes are important in radar meteorology.

Interactions of electromagnetic waves with matter can also be considered on a larger scale, where we no longer consider the interaction of a photon with a molecule, but rather the interaction of electromagnetic energy with a large body of matter. Interactions of particular importance in radar meteorology are refraction, reflection, Mie scattering, and Bragg scattering.

1.4.1   Refraction

Refraction is the change in direction of propagation of an electromagnetic wave that occurs as the wave crosses from one medium into another. Figure 1.9 shows an example of refraction for optical wavelengths where light passes from water to air. Refraction is governed by Snell's law

$$\frac{n_2}{n_1} = \frac{\sin(\theta_1)}{\sin(\theta_2)} = \frac{v_{c1}}{v_{c2}} \tag{1.16}$$

where θ_1 is the angle of incidence and θ_2 is the angle of refraction, both *measured with respect to vertical*, and v_{c1} and v_{c2} are the respective speeds of light in each layer. Snell's law is illustrated graphically in Figure 1.10 where a radar beam is propagating upward across the boundary between two layers with different indices of refraction, the lower layer having a great n than the higher layer. Microwaves undergo continuous refraction as they propagate through the atmosphere because of changes in atmospheric density and water vapor content. We will examine this process in Chapter 4 when we consider propagation of radar beams.

1.4.2   Reflection

Reflection is the change in direction of an electromagnetic wave at an interface between two different substances, where the interface extends over a distance much greater than the incident wavelength. With reflection, the incident wave returns into

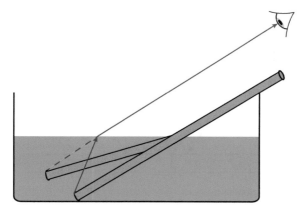

Figure 1.9 An example of refraction of light at an optical wavelength. The straw appears to be bent to the observer because the light undergoes refraction at the water/air interface

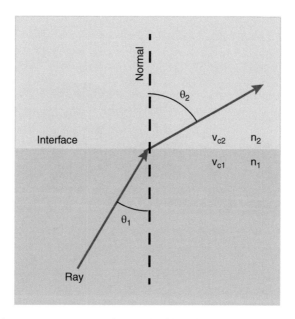

Figure 1.10 A radar ray crossing an interface marked by a change in the index of refraction will bend according to Snell's law

the medium where it originated. **Specular reflection** is mirror-like—electromagnetic energy arrives from a single direction and is reflected in a single direction (Figure 1.11). **Diffuse reflection** occurs when the electromagnetic energy arrives from one direction, but departs in multiple directions, as might occur with a rough surface. Reflection from the earth's surface is a major concern for spaceborne radars, as well as airborne radars with downward pointing beams. Reflection can also occur when large objects such as buildings intercept the beams of ground-based radars. We will address issues related to reflection of radar energy in later chapters.

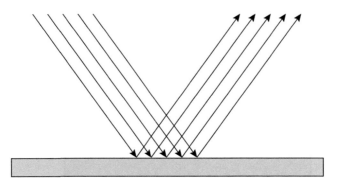

Figure 1.11   Specular reflection

Figure 1.12   The blue sky above the three Doppler on Wheels radar trucks results from scattering of sunlight by molecules in the atmosphere

1.4.3   Mie scattering

The term **scattering** is generally used to describe a process by which an electromagnetic wave deviates from a straight path as the result of an encounter with an object. For visible light, scattering occurs as light waves encounter air molecules, aerosol, and cloud droplets and is the process that makes the sky appear blue and clouds appear white (Figure 1.12). For weather radars, the primary objects of concern are cloud droplets, raindrops, ice crystals, snowflakes, and hailstones and non-meteorological targets, particularly insects.

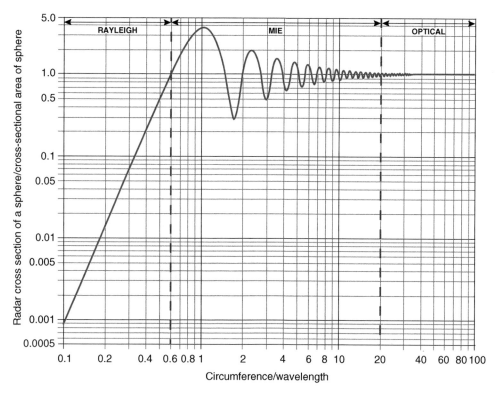

Figure 1.13   The radar cross section of a sphere as a function of the ratio of the sphere circumference to the radar wavelength

Mie scattering theory describes scattering of radiation by a spherical particle impinged upon by a plane wave of electromagnetic radiation. The theory, a comprehensive solution of Maxwell's equations for the interaction of radiation with a sphere, was first derived by German Physicist Gustav Mie.

The general solution for Mie scattering for a dielectric sphere such as a water droplet illuminated by electromagnetic waves generated by a radar located at a distance r from the droplet is shown in Figure 1.13. On the abscissa is the ratio of the circumference of a sphere of radius a, $2\pi a$, to the wavelength of the radar, λ. On the ordinate is the ratio of the **radar cross section**, σ, to the cross-sectional area of the droplet, πa^2. The radar cross section of a sphere at range, r, has the dimensions of area and is defined by the equation

$$\sigma = 4\pi r^2 \left(\frac{S_r}{S_{inc}} \right) \tag{1.17}$$

where S_{inc} is the power flux density (power per unit area) intercepted by the sphere and S_r is the back-scattered power flux density at the receiving antenna. Although the factor r^2 appears in the defining equation, the radar cross section is not a function of range. For a linear scattering process, S_r is proportional to S_{inc} but S_r decreases with the distance r back to the receiving antenna according to the inverse-square law governing the propagation of electromagnetic waves. For this reason, the radar cross section is independent of range.

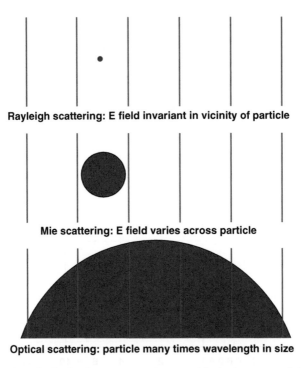

Figure 1.14   Phase lines (red) of a plane wave passing a particle. In the top panel the particle diameter is much smaller than the wavelength, in the middle, the diameter is comparable to the wavelength, and in the bottom, the diameter significantly exceeds the wavelength, leading to different types of scattering

The radar cross section of a dielectric sphere in Figure 1.13 exhibits distinct behaviors that depend on $2\pi a/\lambda$: on the left side of the figure, the increase with $2\pi a/\lambda$ appears linear on a log–log graph, in the middle it is oscillatory, and on the right it is a constant. Note the areas on the graph denoted **Rayleigh scattering** and **geometric optics**. These denote the regions where Rayleigh scattering theory and geometric optic principles for refraction and reflection can be applied. Figure 1.14 illustrates the electric field from a radar-transmitted wave impinging on the sphere in each of the three regimes. The sphere is assumed to be far enough from the radar that the wave fronts are plain waves with no curvature.

In the Rayleigh regime, the sphere is much smaller than the wavelength, sufficiently small in fact that we can assume that the electric field experienced by the sphere is constant across the sphere. This is the criteria for **Rayleigh scattering**. The theoretical predictions for Rayleigh scattering, first derived by British Physicist John William Strutt (in later life, Lord Rayleigh), form the basis for the weather radar equation that we will derive in Chapter 5. **Rayleigh scattering theory** applies when electromagnetic radiation is scattered by small spherical particles with radii less than a factor of $1/2\pi$ times the wavelength of the radiation.

Outside the Rayleigh regime, the sphere is too large for the approximation that the electric field is constant across the sphere to be true. This leads to resonances in the induced electric field within the sphere that can cause the radar cross section to increase or decrease as a raindrop grows. The term **Mie scattering** is generally used to describe scattering associated with spherical particles with diameters larger than

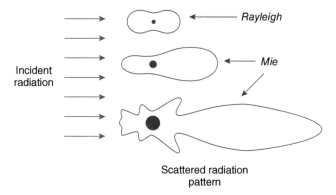

Incident radiation

Scattered radiation pattern

Figure 1.15 Scattered radiation pattern for progressively larger spherical particles. The distance is proportional to the amplitude of the scattered radiation

about an eighth or less of the wavelength of the radiation. With Mie scattering, it is possible for the radar reflectivity to increase or decrease as raindrops get larger. For this reason, it is preferable to avoid Mie scattering by using a radar with a sufficiently large wavelength.

In the optical regime, the sphere's diameter exceeds many wavelengths. The radar cross section in this case is simply the geometric cross section of the sphere. When the sphere is so large that its surface is essentially flat, the scattering becomes equivalent to specular reflection.

The geometry of the scattering function for Rayleigh and Mie scattering is shown in Figure 1.15. The pattern of scattered energy for Rayleigh scattering is peanut shaped, with equal scattering in the forward and return direction. The pattern for Mie scattering depends on the drop's size and can take on many shapes as illustrated in Figure 1.15.

Scattering occurs with non-spherical objects such as large raindrops (which assume a "hamburger" shape as they fall due to frictional drag) and ice crystals. Theoretical work and numerical calculations have been employed to estimate the scattering functions of large deformed drops and a small number of idealized ice crystal shapes. However, no comprehensive theory exists for the wide range of irregular ice particles that are common in snow and cold clouds.

1.4.4 Bragg scattering

A special type of scattering, called **Bragg scattering,** is named after co-discoverers William H. Bragg and William L. Bragg. In radar applications, Bragg scattering occurs when electromagnetic waves impinge on regularly spaced objects or regions of air with different indices of refraction leading to constructive interference between the scattered waves (Figure 1.16). The objects or regions of air must be spaced at distances of half the radar wavelength for Bragg scattering to occur. The half-wavelength criteria is important because waves scattered by the objects will constructively interfere—the backscattered wave from each object arrives in phase with waves scattered by other objects so that the amplitude of the backscattered energy adds together, rather than undergoing phase cancellation (Figure 1.16).

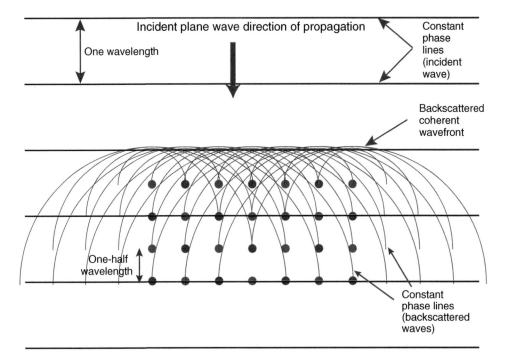

Figure 1.16 Bragg scattering occurs when electromagnetic waves impinge on regularly-spaced objects or regions of air with different indices of refraction leading to constructive interference between the scattered waves and a coherent wave front

Conditions for Bragg scattering can develop in the atmosphere when turbulence mixes air with different moisture and temperature characteristics (and therefore different indices of refraction). Turbulence creates all scales of eddies, one of which corresponds to half the radar wavelength. Bragg scattering can also occur in clouds where some droplets, by random chance, have the appropriate half-wavelength spacing along the direction of the beam.

1.4.5 Absorption

Absorption of electromagnetic radiation occurs when photons of a specific frequency of radiation are absorbed by an atom or molecule and that energy is converted to thermal energy, adding heat to the medium. Absorption of microwaves, for example, occurs regularly in your microwave oven as radiation is used to heat your food. In the atmosphere, some or all of microwave energy transmitted by a radar can be absorbed by gases, water droplets, and ice particles. The amount of absorption that occurs is greater for shorter radar wavelengths and less so for longer radar wavelengths.

1.5   Polarization of electromagnetic waves

The orientation of an electromagnetic wave is described conventionally by specifying the orientation of its electric field vector. The **plane of polarization** is the

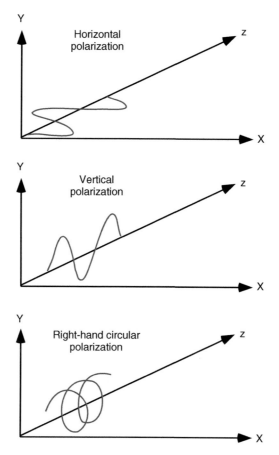

Figure 1.17 Oscillation and propagation of the electric field vector in a (a) horizontally polarized, (b) vertical polarized, and (c) right hand circularly polarized wave

plane in which the electric field oscillates. Figure 1.17 shows three orientations of an electric field propagating through space that are particularly important in radar meteorology.

Consider a monochromatic electromagnetic wave propagating along a direction such that one principal axis (x) is normal to the direction of propagation and parallel to a tangent to the earth's surface (i.e., the axis is horizontal). The second principal axis is perpendicular to (x) and the propagation direction similar to that shown in Figure 1.17. The electric field will oscillate in the x–y plane with amplitudes E_x, and E_y with a phase difference, ϕ. The two components of the electric field for this wave can be written as

$$E_x = E_{xm} \cos(2\pi f t) \tag{1.18}$$

$$E_y = E_{ym} \cos(2\pi f t + \phi) \tag{1.19}$$

where the subscript m stands for maximum. Both E_x and E_y oscillate between values of 0 and E_{xm} and E_{ym}, respectively, with the phase angle, ϕ, specifying the difference between the times that E_x and E_y reach their maximum value. The following conditions specify states of polarization particularly important for radar meteorology:

1. $E_{ym} = 0$, the electric field oscillates in the x direction and the wave is said to be **horizontally polarized**.
2. $E_{xm} = 0$, the electric field oscillates in the y direction and the wave is said to be **vertically polarized**.
3. $E_{xm} = E_{ym}$ and $\phi = 0$, or $\phi = \pi$, the electric field will oscillate back and forth along a 45° axis between the x and y direction. Such a wave is **linearly polarized**.

Most **conventional radars** are horizontally polarized. However, more and more radars, including the National Weather Service radars, have the capability of controlling the polarization state of the transmitted electromagnetic waves to determine information about precipitation and weather characteristics. Radars with this capability are called **polarization diversity radars** if they transmit at a fixed polarization and receive at multiple polarizations (e.g., WSR-88D radars), or **polarization-agile radars** if they transmit at multiple polarization states. Both of these radar types fall in the class of **polarimetric radars**. Polarization-agile radars used for weather transmit electromagnetic waves by alternating between (1) and (2), whereas linearly polarized polarization diversity radars transmit using (3) and examining the horizontal and vertical components of the returned electromagnetic waves. We will explore polarimetric radar hardware in Chapter 2 and polarimetric radar measurables in Chapter 7.

A special case for polarization occurs if $E_{xm} = E_{ym}$ and $\phi = \pi/2$ or $\phi = -\pi/2$. Under these conditions, the electric field vector will rotate in a clockwise ($\pi/2$) or counterclockwise ($-\pi/2$) circle as the wave propagates (Figure 1.17). This type of polarization is called **circular polarization**. This type of polarization diversity radar has not seen much use in radar meteorology but is important in aircraft tracking radars, for which the technology is used for "weather clutter" suppression so that an aircraft can be detected in adverse weather conditions.

Important terms

absorption	electrical current
Bragg scattering	electromagnetic radiation
circular polarization	electromagnetic spectrum
constructive interference	electromagnetic wave
conventional radar	frequency
destructive interference	geometric optics
diffuse reflection	horizontally polarization
dipole	hydrometeor
dipole antenna	induction field
electric displacement vector	linear polarization
electric field	magnetic field
electric field intensity	magnetic field intensity

magnetic permeability	radiate
Maxwell's equations	radiated fields
Mie scattering	reflection
Mie scattering theory	Rayleigh scattering
monochromatic wave	Rayleigh scattering theory
permittivity of free space	relative permittivity
photon	refraction
plane of polarization	refractive index
polarization-agile radar	scattering
polarization diversity radar	specular reflection
polarimetric radar	susceptibility
quantum mechanics	vertically polarization
radar cross section	wavelength

Review questions

1. What four characteristics of an electromagnetic wave are used by radar scientists to deduce properties of the atmosphere?
2. Electric fields exist in the presence of charged particles, and magnetic fields exist in the presence of currents, yet electromagnetic fields can be found propagating across vast reaches of empty space where no charges exist. Why?
3. If a simple dipole antenna was oriented vertically, would the resulting electromagnetic wave be horizontally or vertically polarized?
4. Explain why complex notation is useful to describe the behavior of an electromagnetic wave propagating through a medium.
5. What range of wavelengths is commonly used for weather radars? Based on Figure 1.7, are shorter or longer wavelengths more likely to be absorbed when passing through the atmosphere?
6. Describe different ways that radiation interacts with matter. How are each of these important in radar meteorology?
7. What do we mean when we say that a medium such as air or water has a "complex index of refraction?"
8. Why is it preferable to have a radar operate in the Raleigh scattering regime?
9. What do we mean when we say that an electromagnetic beam is horizontally polarized?
10. Large raindrops, when they fall, take on a shape like a hamburger, whereas smaller raindrops are more spherical. If a microwave beam was passing through a field of large raindrops, would a horizontally polarized wave or a vertically polarized wave travel faster? Conceptually, how might we use this information to learn something about the intensity of the rain?

Challenge problems

11. A lightning strike can occur when the electric field in air at mean sea-level pressure and temperature reaches about 3 million $V\,m^{-1}$. Assume that air in the apparatus in Figure 1.1 is at mean sea-level pressure and temperature. What must the excess charge density be on one of the plates for a spark to jump between the plates? What is the excess electron density?

12. Common wavelengths for meteorological radars are 10.7, 5.6, and 3.2 cm, and 8.6 and 3.2 mm. What are the corresponding frequencies in GHz? Which will undergo Rayleigh scattering when encountering raindrops with maximum diameters of 3 mm?

13. Based on Figure 1.13, would a 1 cm diameter hailstone fall within the Rayleigh, Mie, or Optical scattering regimes for radar wavelengths of 10, 5, 3, 1, and 0.3 cm?

14. Suppose a radar sends out a microwave pulse that illuminates a field of raindrops within a volume of $106\,m^3$. A fraction of a second later, the same radar sends another pulse that illuminates the exact same drops in the exact same volume, except the drops have all moved ever so slightly between pulses due to wind and their falling toward the earth. Would the amount of energy scattered back to the radar and received at the antenna be the same from each pulse? (Hint: Consider the effects of constructive and destructive interference.)

15. To develop an appreciation of the processes of reflection, refraction, scattering, and absorption, use the Internet to find examples of how these processes create optical effects associated with visible light propagating through the atmosphere and clouds. List at least one optical effect (e.g., a rainbow) associated with each process, and which process causes the optical effect.

2

Radar Hardware

Objectives

By the end of this chapter, you should understand the following:

- The basic components of a radar system.
- How the microwave signal is generated by a radar.
- The path that the signal takes from transmitter to antenna and back to the receiver.
- How frequency, phase, and polarization are controlled within a radar system.
- The concept of in-phase and quadrature channels and their use in recovering amplitude and phase.
- Special types of radar systems, including phased-array, mobile, airborne, and spaceborne radar systems.

2.1 Introduction

Radars are used in a wide variety of civilian and military applications. Every day, airport radars routinely track aircraft in flight, police radars are out in force on highways to determine the speed of cars, and sophisticated radars are used in specialized applications, such as ground-penetrating radars that investigate structures beneath the Earth's surface. In this chapter, we limit our focus to equipment comprising weather radar systems, those used to monitor and study the atmosphere. The goal of this chapter is to provide you with sufficient information to understand the components of a radar system—how and where radar microwave signals are generated, how these signals are transmitted from their source to the antenna, how they are directed to a location in space, and how the returned signals are collected and processed. We will not be concerned with details of electronics or engineering design but limit ourselves to basic functionality. Engineering aspects of radar technology can be found in many technical books on radar.

2.2 Frequency and wavelength

Meteorological radars transmit at frequencies, f_t, ranging from 2.7 to 95 GHz, corresponding to wavelengths between 11.1 cm and 3.2 mm. Phased-array profiling

Radar Meteorology: A First Course, First Edition. Robert M. Rauber and Stephen W. Nesbitt.
© 2018 John Wiley & Sons Ltd. Published 2018 by John Wiley & Sons Ltd.
Companion website: www.wiley.com/go/Rauber/RadarMeteorology

radars operate at lower frequencies, for example, 404 MHz ($\lambda = 74$ cm) and 915 MHz ($\lambda = 32$ cm). The use of frequency and wavelength is interchangeable because their product is the speed of light:

$$c = \lambda f_t \tag{2.1}$$

Frequency is most commonly used by radar engineers, but wavelength is often preferred by meteorologists because wavelength can be conveniently compared to the size of meteorological targets such as raindrops, hailstones, and snowflakes. The choice of wavelength depends on the mission of the radar, the platform on which the radar antenna is mounted, and the available power to run the radar system. Several factors dictate the choice. The most important consideration in weather surveillance radar is avoiding signal attenuation. As we will see in Chapter 9, signal attenuation is a strong function of wavelength, with large-wavelength radars experiencing minimal attenuation and small-wavelength radars suffering severe attenuation. Radars such as the WSR-88Ds ($\lambda = 10.0$–11.1 cm), whose primary mission is severe weather detection and rainfall estimation, must operate at wavelengths that minimize the effects of attenuation. Unfortunately, large-wavelengths radars require large antennas to maintain a narrow beamwidth, have much greater power requirements, and are generally much more expensive to operate. For these reasons, longer wavelength radars make poor choices for mobile research radars mounted on trucks, aircraft, or satellite. These mobile radars typically operate at wavelengths 3 cm or smaller, which, in some cases, may mean that the researcher may have to apply attenuation corrections to the data in post-analysis.

During World War II, radar operations were secret, and to avoid having the actual frequencies discovered by the enemy, letter **band designations** were used to identify the frequencies of radar systems. This practice continues to this day and is common in radar meteorology. Table 2.1 lists band letter designations used with meteorological radars. There are other bands not listed in the table that are commonly used for non-meteorological applications. Note that only a very limited number of frequencies within each band are allocated for meteorological radar usage; nearly, all other frequencies are allocated by the Federal Communications Commission for applications such as cell phones, aircraft communications, and emergency operations. The last two columns in Table 2.1 list a specific frequency and wavelength used for existing meteorological radars.

Table 2.1   Letter band designations and frequencies used with meteorological radars

Band	Frequency range	Wavelength range (cm)	Met. frequency (GHz)	Met. wavelength (cm)
VHF	30–300 MHz	90–600	0.037	800
UHF	300–1000 MHz	30–90	0.915	35
S	2.0–4.0 GHz	7.5–15	2.8	10.7
C	4.0–8.0 GHz	3.75–7.5	5.5	5.5
X	8.0–12.0 GHz	2.5–3.75	9.4	3.2
K_u	12.0–18.0 GHz	1.67–2.5	15.5	1.94
K	18.0–27.0 GHz	1.11–1.67	24	1.25
K_a	27–40 GHz	0.75–1.11	35	0.86
W	75–110 GHz	0.40–0.27	94	0.32

The returned signal arriving at the antenna from targets also has a frequency, but in most cases, it is not the transmitted frequency. The arriving signal has a frequency that is Doppler shifted slightly from the original frequency as a result of motion of the targets toward or away from the radar. The frequency *shift* is very slight, at most a few kilohertz, so a transmitted frequency might be 3,000,000,000 Hz and the returned signal might be 3,000,001,000 Hz. Doppler radars use this information, specifically the phase shift that accompanies this frequency shift, to retrieve the radial velocity of the targets toward or away from the radar.

2.3   Components of a weather radar system

Figure 2.1 shows a block diagram of major radar components involved in the transmission, reception, and processing of radar signals for a **polarization-agile**, pulsed Doppler radar system employing a klystron transmitter. Except for the radome, the diagram in Figure 2.1 is based on the National Science Foundation/National Center for Atmospheric Research (NSF/NCAR) S-Pol Radar, a national research facility. S-Pol, a 10 cm wavelength dual-polarization Doppler S-band radar, is pictured in Figure 2.2 at its home base in Colorado. The diagram in Figure 2.1 is not

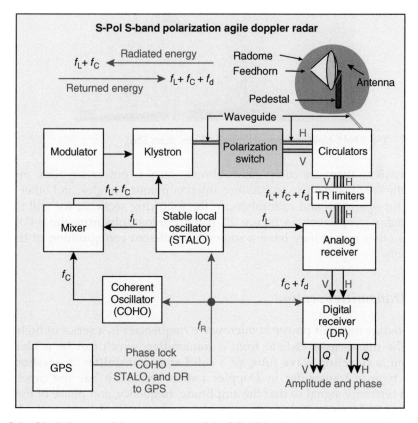

Figure 2.1   Block diagram of the components of the S-Pol S-band radar system, which has a klystron transmitter

Figure 2.2   The S-Pol S-band antenna and trailer (Photo: Scott Ellis)

comprehensive—there are components absent, such as power supplies, motors to position the antenna, devices to measure antenna-pointing angles, and other devices required for operation and calibration. In the following sections, we will illustrate the key radar components in Figure 2.1 using photographs from the S-Pol Radar. Note that other radars may have a somewhat different configuration of the radar components.

2.3.1   Transmitter section

Doppler radars transmit power at microwave frequencies in a series of high-energy pulses. The microwaves radiate from a **transmitter**, which can be a **klystron**, a **magnetron**, a **traveling wave tube,** or a **solid state transmitter**. A klystron is one common type of transmitter in Doppler radars because it has the capability to amplify a reference signal so that the amplitude, frequency, and phase of the output signal is precisely controlled, a necessary step to accurately determine the **Doppler radial velocity**—the velocity of objects (such as raindrops) moving along a radar beam toward or away from the radar.

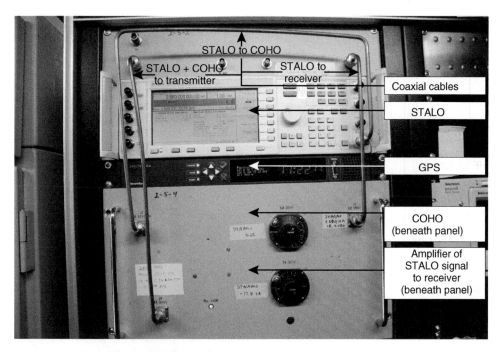

Figure 2.3   The STALO and COHO component housing in the S-Pol radar (Photo: Scott Ellis)

A klystron is a **power amplifier**. The microwave frequency and phase of the input signal to the klystron is determined by two devices, a **stable local oscillator** called the **STALO** and a **coherent oscillator** called the **COHO** (Figure 2.3). Both oscillate at frequencies other than the radar frequency. The signals from the STALO and COHO in the S-Pol system are "phase locked" with a signal from a global positioning system (GPS) clocking device, a requirement so that the phase of the signal delivered to the klystron is precisely known and can be compared to the return signal. This is necessary to determine the Doppler radial velocity. The output of the STALO (frequency $=f_L$) and the COHO (f_C) is passed through a mixer, the output of which is a signal oscillating at the sum of the STALO and COHO frequencies ($f_L + f_C$), the operating frequency of the radar.

A klystron delivers power pulses at regular intervals controlled by a **modulator**. The S-Pol klystron, a type of linear-beam tube, is shown in Figure 2.4, along with a diagram illustrating a klystron's typical inner structure. The klystron consists of an electron gun, a **cathode** that emits electrons, a modulating **anode** that focuses the electrons into a fine beam, and accelerates the beam due to the strong electric field. The electron beam travels past a series of cavities that are part of the anode structure. A weak microwave signal is introduced in the first of the cavities. This signal is delivered to the klystron as a low-power radio frequency (RF) wave[1] from the mixer. The

---

[1]The term RF refers generically to signals that have a rate of oscillation in the range of 3 kHz to 300 GHz, which corresponds to the frequency of radio waves and alternating currents that carry radio signals. This range includes microwaves, which have frequencies between 0.3 and 300 GHz.

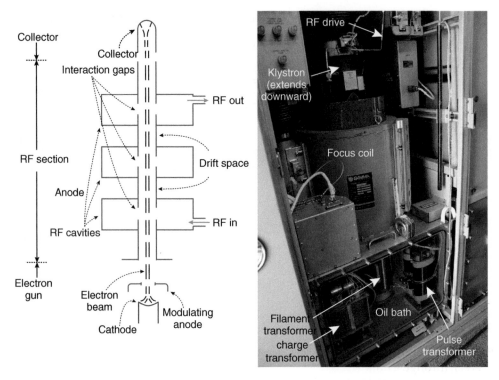

Figure 2.4 (a) Diagram of the key components of a klystron transmitter. (b) The klystron transmitter in the S-Pol radar (Note the tape measure is extended 6 in. to give an idea of scale.) (Photo: Scott Ellis)

cavity structure causes the electron beam velocity to modulate at the frequency of the input microwave signal, amplifying the signal and creating high-power microwaves. These microwaves exit the tube through an opening in the final cavity and enter the waveguide. The electron beam continues past the cavities and reaches a collector at the end of the klystron tube. Klystrons operate at high voltage and generate considerable heat. In the picture of the S-Pol radar transmitter, the top of the klystron is visible. The remaining part is encased within the cylinder labeled "focus coil." The **focus coil** generates a magnetic field that keeps the electron beam centered in the klystron to avoid arcing within the klystron tube. Power to the klystron is delivered through several **transformers**. The components on the left, center, and right below the klystron are called *charge, filament*, and *pulse* transformers, respectively. The function of a transformer is to multiply (or divide) an alternating current voltage. The transformers are mounted in an oil bath to prevent arcing due to the high voltages.

2.3.2 Waveguides, rotary joints, polarization switching devices, and circulators

A radar's transmitter and antenna are located at some distance from each other. In small, shorter wavelength radars, this distance can be a few meters or less, whereas in long-wavelength S-band radars, the distance can be tens of meters. Electromagnetic

Figure 2.5   Waveguides for S, C, X, K$_a$, and W-band radars (Photo: Scott Ellis and Jonathan Emmett)

waves travel from the transmitter to the antenna, and back from the antenna to the receiver, through a hollow conductive rectangular metal pipe called a **waveguide**. The dimensions of a waveguide are normally half the wavelength of the electromagnetic waves passing through it. Figure 2.5 shows sections of waveguides for (left to right) S (10 cm), C (5 cm), X (3 cm), K$_a$,(0.86 cm), and W (0.32 cm) band wavelengths. Figure 2.6 shows long stretches of waveguide connecting to the antenna of the S-Pol radar.

The size and material of a waveguide are designed to support the dominant modes of electromagnetic waves and suppress the unwanted modes from propagation. Propagation through waveguide structures minimizes interference and losses due to absorption. Scanning radars generally have **rotary joints** (Figure 2.7) that allow the waveguide attached to the rotating antenna to connect to the fixed waveguide attached to the transmitter and receiver. The peak power that can be transmitted through a waveguide is limited by dielectric breakdown and arcing that can occur if too much power is supplied. Some radars increase the usable peak power by either pressurizing the waveguide or filling it with certain gases that inhibit breakdown, such as sulfur hexafluoride.

In some dual-polarization radars, **waveguide switches** are incorporated into the waveguide system so that alternating horizontal (H) and vertical (V) polarization states can be transmitted from the radar. Other radars transmit and receive both H and V polarization states simultaneously and still other radars transmit one polarization state (e.g., H) and receive both. The details of how this is done depends on the particular radar system, specifically whether the radar changes the radiated polarization state on a pulse-to-pulse basis (typically between horizontal and vertical orthogonal states), and whether it has a single- or dual-transmitter or receiver system. Some early polarization radar systems transmitted circularly polarized waveforms, and these employ a still different design. The polarization switching system in the S-Pol radar includes a dual-waveguide system, with switching devices that direct signals into the

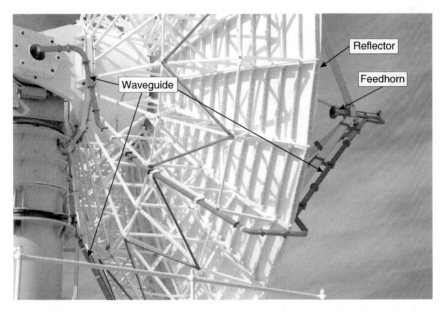

Figure 2.6   Close up of the S-Pol antenna system showing the waveguides for both polarizations, the feedhorn, and the reflector (Photo: Scott Ellis)

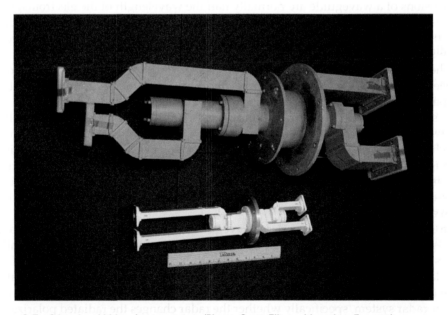

Figure 2.7   S-band and X-band rotary joints (Photo: Scott Ellis and Jonathan Emmett)

Dummy loads-
prevents reflection
at waveguide
junctions

Slow waveguide
switches-
set for choice of H, V,
simultaneous HV,
or H–V–H–V operation

Fast waveguide switch synced to PRF,
switches signal from H to V waveguides
during H–V–H–V operating mode

Figure 2.8 Waveguide polarization switches and dummy loads to prevent reflection at waveguide junctions (Photo: Scott Ellis)

appropriate waveguide. The switching mechanisms of S-Pol were specially adapted to allow transmit signals at rapidly alternating polarization states (horizontal and vertical, HVHV, etc.) or a single polarization state (H or V) or H and V simultaneously, effectively creating a 45° polarization transmission. The WSR-88D radars transmit and receive H and V simultaneously, which does not require a switch, but rather power splitters. A photograph of the polarization-switching mechanism in the S-Pol radar is shown in Figure 2.8. The choice of how dual-polarization capabilities are implemented impacts the sensitivity, data quality, and availability of dual-polarization variables, as will be discussed in Chapter 7.

The transmitted power exceeds the received power by many orders of magnitude. It is critical in a radar that the receiver, which must be sensitive to very low power levels, be shielded from the transmitter. As the transmitted and received signals normally pass through a portion of the same waveguide, a device must be located within the path of the signal that acts as a switch, alternately directing the outgoing signal from the transmitter to the antenna, while shielding the receiver, and then directing the incoming signal away from the transmitter and to the receiver. The devices that perform this function are called **circulators**. The S-Pol radar has two circulators, one within the horizontal polarization waveguide line and one within the vertical polarization waveguide line (Figure 2.9). Although these devices alternately direct the signal from transmitter to antenna, or antenna to receiver, there is still sufficient leakage of the transmitted signal into the receiver waveguide that the receiver components can be damaged. To protect these components, devices called **transmit–receive (T-R) limiters** are placed along the waveguide (Figure 2.9). These

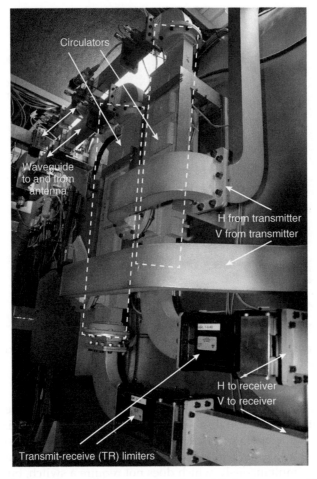

Figure 2.9   Circulators and transmit–receive limiters in the S-Pol radar system. (Photo: Scott Ellis)

block the transmitted pulse, but permit the much weaker received signal to pass through to the receiver.

2.3.3   The antenna section

A radar's **antenna** section consists of a **reflector**, typically dish-shaped in the form of a circular paraboloid, and a waveguide **feedhorn** placed at the focal point of the reflector (Figure 2.6). Larger dishes, such as those used in S, C, and X-band radars, are supported on a pedestal, their motion controlled by drive motors and devices to measure the antenna pointing angles. Antennas of smaller wavelength radars, such as K- or W-band, are small and often, but not always, fixed in place (e.g., vertically pointing), so that the extensive support structures common with larger radar systems are unnecessary. Larger antennas are often protected by a **radome**, a structurally weather- and wind-resistant dome that minimally attenuates the microwave signals passing through it (Figure 2.10).

Figure 2.10   The radome covering the KILX WSR-88D radar antenna. The building on the left is for rawinsonde launches

The purpose of the antenna is to create a narrow, focused beam. Most radar beamwidths are nominally about 1° for meteorological radars. To achieve a narrow beamwidth, an antenna must be large compared to the wavelength. The diameter of radar reflectors are typically 50–100 times the wavelength. For example, the reflector of the S-Pol radar is 8.5 m, and the radar wavelength is 10 cm, so the ratio of diameter to wavelength of 85 achieves a beamwidth of 0.92°.

Diffraction effects at the antenna boundary, as well as the waveguide, feedhorn, and their supporting structures lead to a complicated distribution of radiation over a spherical surface surrounding the radar, rather than just a simple narrow beam. A way to visualize the **antenna beam pattern** is to first consider a radar with an isotropic antenna, that is, one that radiates power equally in all directions. Figure 2.11 shows the power radiated in a vertical plane passing through the center of the antenna relative to the power that would be radiated by an isotropic antenna. The radiation pattern consists of a **main lobe**, the primary beam, and **sidelobes**. Sidelobes are not necessarily symmetric about the main beam in all directions. For example, Figure 2.12 shows a 3-D view of the radiation pattern from a radar with a very narrow (0.33°) beamwidth. For this antenna, the lobes are obviously not symmetric about the center of the main lobe. However, for most antennas, a 2-D graph is sufficient to characterize the beam pattern.

Figures 2.11 and 2.12 map out the antenna's **gain function**, where gain, $G_{\theta,\phi}$, is defined as the ratio of the power flux density at angular coordinates (θ,ϕ) to the power flux density that would be incident at angular coordinates (θ,ϕ) if the antenna were a lossless isotropic antenna radiating the same power. The angular width of the main

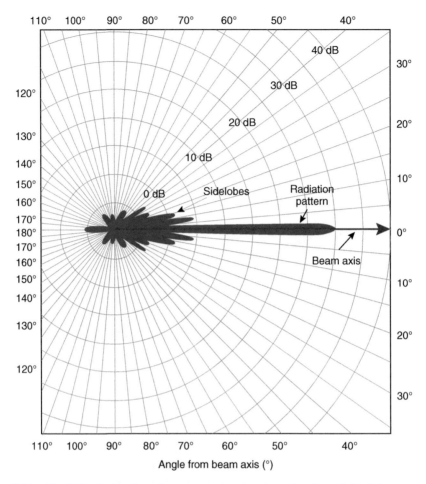

Figure 2.11   The 2-D gain function of an antenna showing the main lobe and side lobes

lobe is nominally defined by the points where the power is one-half, or 3 dB down, from its maximum value. In operations, all power returned to the radar is assumed to come from targets located within the main lobe. Power transmitted from the sidelobes can create false echoes on radar displays, particularly from ground clutter contamination, or extend the vertical or horizontal dimensions of a storm (e.g., Figure 2.13).

   Antennas are not designed to illuminate the entire reflector with uniform power. The reason uniform power illumination is undesirable is because it produces sidelobe power levels that are too large for meteorological applications. Normally, the feedhorn is designed to taper illumination toward the outer radius of the antenna. Tapered illumination has three effects: a reduction of sidelobe energy, a reduction in maximum power gain, and an increase in beamwidth. The latter two are considered an acceptable trade-off for the reduction in sidelobe energy. Radar antennas routinely achieve maximum sidelobe levels at least 20–25 dB below the gain along the beam axis. The first sidelobe of the S-Pol radar antenna in Figure 2.6, for example, is 27 dB below (or 1/1000th) the power along its beam axis.

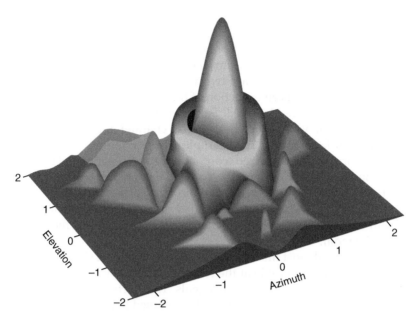

Figure 2.12   A "three-dimensional" image of the gain function for an antenna with very narrow beamwidth (about 0.33°). The height of the contour above the base plane is proportional to the power gain in decibels. Numerous sidelobes are evident (Adapted from Howard, D.D. (1967) Analysis of the 29 Foot Monopulse Cassegrainian Antenna of the AN/FPQ-6 and AN/TPQ-18 Precision Tracking Radars. Memorandum report 1776, Naval Research Lab., Washington, D.C., 24 pp)

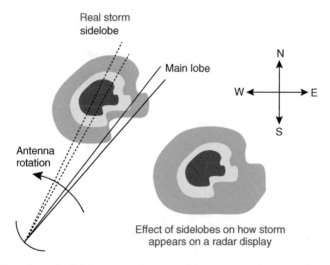

Figure 2.13   The effect of sidelobes on the horizontal appearance of a radar echo

2.3.4   *The receiver section*

Energy scattered back to the antenna from objects illuminated by the radar beam is gathered by the antenna and returned through the waveguide and circulator, arriving at the **receiver**. In S-Pol, there are two waveguides, one for each polarization state. Figure 2.14 shows the receiver section of the S-Pol radar system with the major components described below identified. For reference, the transmitted polarization state, whether it is H or V, is called the **co-polar channel** and the orthogonal state is called the **cross-polar channel**. For example, if a horizontally polarized pulse is transmitted, the horizontally polarized (co-polar) returned energy will be transmitted to the receiver in one waveguide, and the vertically polarized (cross-polar) returned energy will be transmitted to the receiver in the other waveguide. The frequency of the returned signal $(f_L + f_C + f_d)$ is the transmitted frequency, plus or minus a Doppler frequency shift induced by the motion of the objects illuminated by the radar toward or away from the radar.

The microwave signal must first be converted to an electrical signal. This is accomplished by inserting a coaxial cable at the terminus of the waveguide that acts much like a car antenna does in receiving radio signals. The electrical signal is then amplified using a **low-noise amplifier** (LNA) and then passed to a mixer, which mixes the signal from the STALO (f_L) with the received signal $(f_L + f_C + f_d)$, the output of which is an intermediate frequency $(f_C + f_d)$ (Figure 2.14). In S-Pol, this signal then passes to a switch that directs the signal into one of the two chains of amplifiers, depending on whether the signal is from the co-polar or cross-polar channel (Figure 2.14). The signal is then passed to the **digital receiver**.

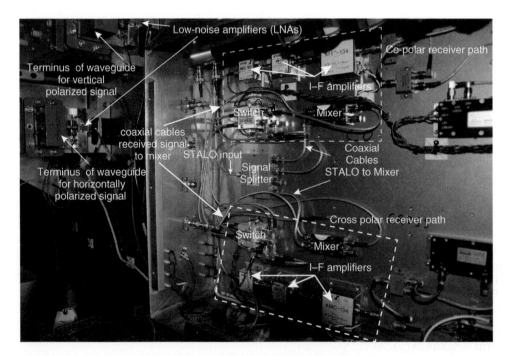

Figure 2.14   The receiver system in the S-Pol radar (Photo: Scott Ellis)

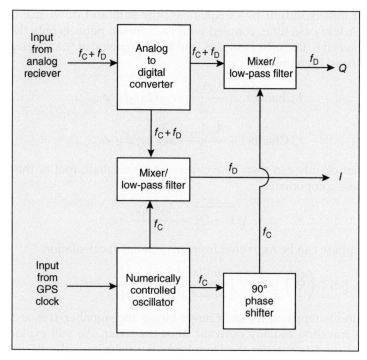

Figure 2.15   Box diagram showing the retrieval of the *I* and *Q* signals from the input Doppler frequency

The amplitude and phase of the returned signal are derived from output from the digital receiver. The amplitude is used to recover the radar reflectivity factor, which will be discussed in Chapter 5, and the phase is used to recover the Doppler radial velocity, which will be discussed in Chapter 6. The digital receiver uses a method called *quadrature demodulation*. Figure 2.15 shows how the *I* and *Q* signal components are generated within the digital receiver. Engineers use complex representation of the signals, associating the *I* component with the real part of the complex signal and the *Q* component with the imaginary part. We will simplify this here by not resorting to complex notation at this point in the text.

The signal to the digital receiver from the intermediate frequency amplifier originating from pulse *m* at range gate *n* can be represented as

$$E_r = A_{(m,n)} \cos[2\pi(f_C + f_d)t + \phi_{(m,n)}] \tag{2.2}$$

where $A_{(m,n)}$ is the amplitude and $[2\pi(f_C + f_d) + \phi_{(m,n)}]$ is the phase of the signal. This signal is first digitized by passing it through an **analog to digital converter**. The digital receiver also contains a numerically controlled oscillator (NCO), which is driven by the GPS clock and oscillates at the COHO frequency.

The signal from the NCO, $A_0 \cos(2\pi f_c t)$, passes to one mixer without alteration. The signal is also sent to a second mixer, but before it arrives, it passes through a 90° phase shifter so that it is in **quadrature** with the original NCO signal, so that it has the form $-A_0 \sin(2\pi f_c t)$. At each mixer, the signal is mixed with the output of the analog to digital converter.

Recall that mixers output two frequencies, the sum and difference of the input frequencies. A low-pass filter, coupled with each mixer, permits only the difference frequency to survive. The mixer outputs lead to the removal of f_c, so the signal emerging from the I and Q channels is

$$I \text{ Channel} \quad \frac{A_0 A_{(m,n)}}{2} \cos(2\pi f_d t + \phi_{(m,n)}) \tag{2.3}$$

$$Q \text{ Channel} - \frac{A_0 A_{(m,n)}}{2} \sin(2\pi f_d t + \phi_{(m,n)}) \tag{2.4}$$

The echo amplitude can then be recovered as the square root of the sum of the squares of these components:

$$\sqrt{I^2 + Q^2} = \frac{A_0 A_{(m,n)}}{2} \tag{2.5}$$

and the echo phase can be recovered from an arctangent calculation

$$\tan^{-1}\left(\frac{Q}{I}\right) = \tan^{-1}\left[\frac{\sin(2\pi f_d t + \phi_{(m,n)})}{\cos(2\pi f_d t + \phi_{(m,n)})}\right] = 2\pi f_d t + \phi_{(m,n)} \tag{2.6}$$

This calculation is repeated for each range bin as the signal arrives at the receiver from a pulse traveling radially outward from the radar. We will explore how this information is used to recover the Doppler radial velocity and reflectivity in later chapters.

2.3.5 *Magnetron transmitters*

Magnetron transmitters are generally smaller than klystron transmitters with similar power output. Magnetrons are often used in situations where small size is important, for example, in airborne and truck-mounted radars. Modern magnetrons can perform satisfactorily for many Doppler radar applications. Figure 2.16 shows a picture of the magnetron used in the K_a-band radar that is sometimes deployed with the S-Pol radar system. A diagram of the inner structure of a magnetron is shown in Figure 2.17. The magnetron's anode is a positively charged circular outer structure that has a series of evenly spaced circular cavities that open through slots into a central open circular channel. At the very center is the magnetron's cathode, or negatively charged center. A very powerful magnet surrounds the structure and creates a magnetic field that is oriented parallel to the cathode. Because of interaction of the magnetic field and the electric field created by anode and cathode and modulated by the cavities, electrons emitted from the cathode orbit in the central channel in a manner that resonates at a frequency determined by the structure and spacing of the cavities and the interaction space. Microwaves are generated as a result of the resonant behavior of the electrons. The microwaves pass out an opening (actually through an iris made of glass or ceramic) in one of the cavities and into a waveguide where they propagate to the antenna.

For Doppler radar applications, it is critical to retain the frequency and phase of the transmitted signal from the magnetron. In the magnetron system used for the K_a-band at S-Pol (Figure 2.18), a sample of the signal (f_T) is fed to a mixer, where it is mixed with

Figure 2.16   A K$_a$-band magnetron (Photo: Scott Ellis and Jonathan Emmett)

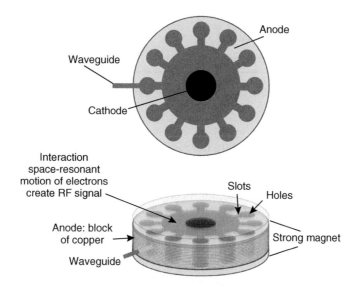

Figure 2.17   Internal structure of a magnetron transmitter

the signal from the STALO (f_L). The STALO and the digital receiver are phase-locked to a reference oscillator. At S-Pol, the GPS serves as this oscillator for the K$_a$-band radar. Magnetron frequencies have a tendency to drift due to temperature. The frequency drift causes the output signal of the mixer to oscillate at a lower frequency, which is the difference between the two input frequencies ($f_T - f_L$). To maintain the received signal within the receiver bandwidth, the frequency of the STALO is tuned so that the output signal of the mixer arriving to the digital receiver maintains a constant value. The phase of the signal is measured by the digital receiver for each pulse

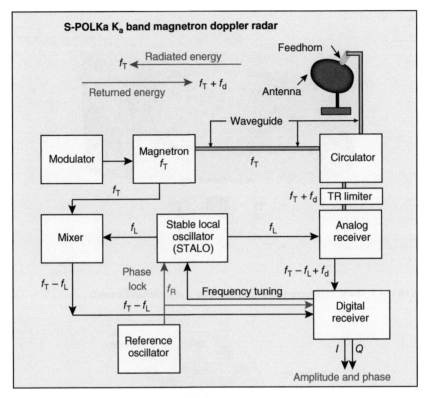

Figure 2.18   Block diagram of the components of the S-PolKa K$_a$-band radar system, which has a magnetron transmitter system

and the result is used to compensate the received signal. When the signal returns to the antenna, it arrives with a frequency ($f_T + f_D$). The signal then passes through a circulator and T-R limiter and into an analog receiver where it is mixed with the STALO frequency and then input to the digital receiver. The output of the digital receiver is the I and Q signal, from which the amplitude and phase are derived.

2.4   Specialized radar systems

Meteorological radars are either operational (perform weather monitoring for the purpose of providing hazardous weather warnings and public information about precipitation location and intensity) or used for research. In Chapter 10, we explore operational radar network operations in the USA and other countries. Here, we limit the discussion to research radar systems. A number of specialized radars have been developed for meteorological research.

2.4.1   Phased-array radars

Phased-array radars were first used for weather applications as wind-profiling radar systems. The technology more recently has been applied for rapid scanning of severe storms and is under consideration as a possible future replacement or enhancement

Figure 2.19   A phased-array antenna on a mobile Doppler radar (Photo: Joshua Wurman)

of National Weather Service WSR-88D radars. Two research radars that employ phased-array technology for severe storm research are the National Severe Storms Laboratory Multi-Function Phased-Array Radar (MPAR) and the Center for Severe Weather Research Rapid-Scan Doppler on Wheels (Figure 2.19).

Figure 2.20 shows a wind profiler phased-array antenna. A phased-array antenna is composed of an array of radiating elements. With this type of antenna, a phase shifter is employed that routes the signal to each radiating element through cables of varying length. Beams are created by shifting the phase of each successive element in the array so that constructive interference between the waves occurs in the desired direction of the beam. Figure 2.21 illustrates this basic principle for beam steering with a phased-array antenna containing several radiating elements. Phased-array antennas have the advantage that the beam is electronically steered. This means that a storm can be scanned much more rapidly than with conventional antennas. However, as the array is fixed in position, the coverage is limited to a 120° sector in both azimuth and elevation. As with all antennas, phased-array antennas have sidelobes that must be considered in their design. There are other technological challenges facing implementation of phased-array technology. For example, phased-array antennas do not preserve horizontal/vertical orthogonality when the beam is steered at an angle to its principal plane (i.e., orthogonal to the antenna), so that dual-polarization capability is compromised. As these challenges are addressed, the application of phased-array technology will become more common in atmospheric research and operations.

2.4.2   Mobile and deployable radars

Severe thunderstorms, hurricanes, and other types of hazardous weather are relatively rare events. As a result, geographically fixed research radar systems are often

Figure 2.20   A wind profiler phased-array antenna. The devices on the corners are part of a radio-acoustic sounding system and are not part of the radar itself (Courtesy NOAA)

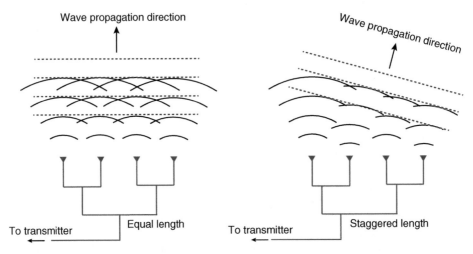

Figure 2.21   A beam can be steered with a phased-array radar by staggering the length of the transmission lines between the transmitter and the radiators (Courtesy NOAA, with changes)

not ideally located to sample storms, particularly during short research field campaigns targeted at collecting research quality data. To address this issue, a fleet of mobile W, K_a, X, and C-band truck-mounted radars, operated by universities, government agencies, and private companies, are now commonly used in atmospheric research.[2] The rapid-scan Doppler on Wheels, a phased-array mobile radar depicted in Figure 2.19 and operated by the Center for Severe Weather Research, is one member of this fleet.

Mobile radars are often deployed in hazardous environments that require a quick setup and departure. For data to be of research quality, it is critical that the antenna-pointing angles are well known. To achieve this, the truck must be accurately leveled, and the geographic pointing direction of the truck must be measured. This is particularly important if the collected data are to be used in multiple Doppler analyses (see Chapter 12), where geographic co-location of data from other radars (often also mobile) is done in post-analysis. Mobile radars must also be self-sufficient and carry electrical generators. The antenna must not exceed size limits imposed by vertical and horizontal road clearance tolerances.

Some research radars, such as the S-Pol radar featured in Section 2.2, are deployable to locations around the world, but operate at a fixed location where they are deployed. Radars such as S-Pol must be designed to be disassembled, packed into transportable containers, and then reassembled at a remote location. With S-Pol, the entire transmitter and receiver section rests in one container and the operator computer and research scientist stations in a second container. The radar can be packaged for transport without too much disruption of equipment during disassembly (although the entire radar system and all equipment fills 10 containers during shipping).

2.4.3   Airborne radars

Although great progress has been made using ground-based Doppler radars to study storm structure, remote storms such as oceanic cyclones and hurricanes cannot be observed with these systems. Also, some phenomena are remote, such as orographic storms over mountains or clouds over the arctic, or hard to observe at high resolution from ground-based radar systems. Airborne meteorological radars provide a means to make measurements of the structure and dynamics of these difficult-to-observe weather systems. Currently, meteorological radar systems are deployed on a number of research aircraft. Because of size and weight considerations, these radars normally use X, K, or W-band wavelengths. The scanning strategies differ from platform to platform, with some radars having only upward and downward pointing beams, whereas others scan at angles along or normal to the direction of flight. The sophisticated scanning strategy employed by the NOAA P-3 X-band tail Doppler radar system is shown in Figures 2.22 and 2.23. The radar performs alternating conical scans fore and aft of the aircraft as the aircraft moves forward, as shown in Figure 2.22. The effect of this strategy is to map out a cylindrical grid of data (Figure 2.23). Because of the different beam orientations, Doppler radial velocities are sampled from different directions, permitting full three-dimensional wind fields to be derived using dual-Doppler techniques similar to those described in Chapter 12. Before data can be processed, the location of individual range gates, and the values of Doppler radial velocity data at

---

[2] A summary of mobile radar facilities available within the atmospheric sciences community as of 2014 is given in Bluestein, H. *et al.* (2014) Radar in atmospheric sciences and related research: current systems, emerging technology, and future needs. *Bull. Amer. Meteor. Soc.*, **95**, 1850–1861.

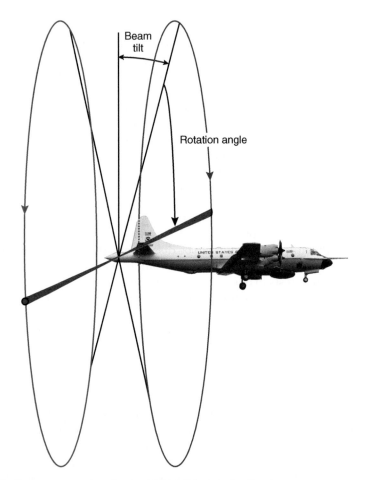

Figure 2.22   Scanning procedure for the NOAA P-3 aircraft tail radar system

those range gates, must be corrected for the forward motion of the aircraft, as well as aircraft's roll, yaw, and pitch angles.

2.4.4   Spaceborne radars

Space radars include altimeters, scatterometers, imaging radars, and precipitation radars with measurement capabilities similar to other radars described above. Altimeters, scatterometers, and imaging radars are used primarily to determine properties of the earth's surface, such as surface wave height, the location of ocean currents, eddies and other circulation features, soil moisture, snow cover, and sea ice distribution.

The first precipitation radar flown in space was launched aboard the Tropical Rainfall Measuring Mission (TRMM) satellite in November 1997. The TRMM precipitation

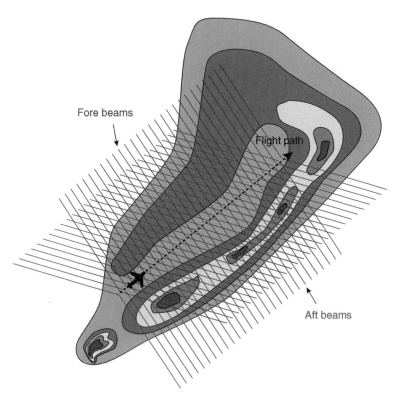

Figure 2.23   The beam pattern produced by the P-3 radar scanning as the aircraft moves along a weather system. The cross beam pattern permits dual-Doppler retrieval of the wind field

radar was designed to obtain data concerning the three-dimensional structure of rainfall over the tropics where ground-based and ocean-based radar measurements of precipitation are almost nonexistent. The TRMM K_u-band radar used a phased-array antenna that operated at a frequency of 13.8 GHz. It had a horizontal resolution at the ground of about 4 km and a swath width of 220 km. TRMM is no longer operational. A second radar, the cloud profiling radar (CPR), was launched on the CloudSat satellite on April 28, 2006. The CPR is a W-band 94-GHz nadir-looking radar. The CPR was designed to provide observations to advance understanding of cloud abundance, distribution, structure, and radiative properties. The CPR uses a 1.85 m antenna and has an along-track resolution of 1.7 km and a cross-track resolution of 1.4 km. The latest satellite launched for precipitation measurements is the Global Precipitation Mission satellite. This satellite carries the Dual-frequency Precipitation Radar (DPR). The two frequencies of the DPR allow the radar to infer the sizes of precipitation particles and provide additional information about storm characteristics. The K_a-band radar scans across a region of 125 km and is nested within the 254 km wider scan of the K_u-band radar. The DPR phased-array antennas are shown during their testing phase before launch in Figure 2.24.

Figure 2.24   The dual-frequency precipitation radar antenna on the Global Precipitation Mission satellite, shown here in its testing phase before launch (Courtesy NASA)

Important terms

analog to digital converter
anode
antenna
antenna beam pattern
band designations
cathode
circulator
coherent oscillator
COHO
co-polar channel
cross-polar channel
digital receiver
Doppler radial velocity
feedhorn
focus coil
gain function
in-phase
klystron
low-noise amplifier
magnetron
main lobe

modulator
Phased-array radars
polarization agile
power amplifier
quadrature
radome
receiver
reflector
rotary joints
sidelobe
solid state transmitter
stable local oscillator
STALO
transformer
transmit–receive limiter
transmitter
traveling wave tube
waveguide
waveguide switch

Review questions

1. In Figures 2.1 and 2.18, various components of the radar systems are connected with either black arrows or green bars. What do these represent? What role do they play in a radar system?
2. Describe fundamental differences between magnetron and klystron transmitters.
3. What is the role of a circulator and a T-R limiter in a radar system?
4. What is the role of the STALO and COHO in a Doppler radar system?
5. Which radar will have a larger waveguide, an X-band radar, or a C-band radar?
6. Why is the antenna of an S-band radar typically quite large, and a K-band radar much smaller?
7. If a polarization-agile radar transmits a vertically polarized pulse and receives both polarization states, which state (horizontal or vertical) is cross-polar? Which is co-polar?
8. What is the frequency difference in hertz, and the phase difference in degrees between the in-phase and quadrature components of the radar signal?
9. How is the direction of the radar beam steered in a phased-array antenna?
10. What fundamental constraints limit the choice of wavelength used for airborne radar systems? How about ground-based mobile radars?

Challenge problems

11. Suppose a peak power of 100 kW was transmitted along the beam axis of the radar with an antenna beam pattern similar to that depicted in Figure 2.11. What is the power, expressed in watts, in the direction of the first, second, and third sidelobes and in the backlobe. (Hint: See Eq. (3.4).)
12. Imagine you were tasked with designing a radar to be placed in orbit on a satellite, with specifications that it had to operate for 5 years and observe precipitation in weather systems over which it passes. What major challenges might you face in designing such a radar?
13. Examine Figure 8.2 in a later chapter in the book. Explain the tall curved echo on this range–height indicator (RHI) scan in terms of Figure 2.11.
14. Suppose a radar is performing an RHI scan through the core of a supercell thunderstorm that contains giant hail. How might the radar sidelobes impact a measurement of the altitude of the storm top?

(continued)

15. This chapter used the NSF/NCAR S-Pol radar as an example to illustrate the basic components of a radar. The basic components of a number of other radar systems used for research are available to various degrees on the internet. Choose one radar system and examine the documentation provided for it. Note the similarities and differences between S-Pol and the radar you choose to investigate.

3

Radar Characteristics

Objectives

By the end of this chapter, you should understand the following:

- How radars use trains of pulses to determine range to a target.
- Why the maximum range that a radar can detect a target unambiguously is related to the pulse repetition frequency.
- The technique WSR-88D radars use to avoid range ambiguities.
- The wavelengths and frequencies used by radars and their advantages and disadvantages.
- Transmitted and received power, and how it is measured.
- The various ways that meteorological radar data are displayed.

3.1 Introduction

A radar is a device capable of **remote sensing**, that is, gathering information about objects far from the instrument by intercepting and interpreting electromagnetic energy coming from those objects. Radars are **active remote sensors**—first generating and transmitting electromagnetic energy and then gathering and interpreting its properties after the radiation has been returned from objects through processes of reflection or scattering. Radars differ substantially from **passive remote sensors**, such as our eyes and many instruments on satellites. These depend on natural sources of radiation such as thermal emission or reflection of solar energy from objects.

Radars measure the power returned from objects as well as the range, azimuth, and elevation of the objects to determine their location in space. Meteorological radars convert the returned power to a quantity called the **radar reflectivity factor**, which, although technically incorrect, is called radar reflectivity in common usage. The radar reflectivity factor is the quantity commonly displayed on radar images. We will defer discussion of the radar reflectivity factor to Chapter 5. Doppler radars, in addition, measure the phase of the returned signal, and from that deduce the **radial velocity**, the velocity that objects are moving toward or away from the radar. Polarization diversity radars measure the polarization state of the returned signal, and from that deduce additional properties of objects that relate to their size, composition,

Radar Meteorology: A First Course, First Edition. Robert M. Rauber and Stephen W. Nesbitt.
© 2018 John Wiley & Sons Ltd. Published 2018 by John Wiley & Sons Ltd.
Companion website: www.wiley.com/go/Rauber/RadarMeteorology

and shape. Although our focus here is on meteorological objects such as raindrops or snowflakes, it is important to remember that meteorological radars do not discriminate and also receive returns from airborne insects, birds, and bats, as well as airplanes, buildings, and mountains in the path of the radar beam. These objects are generically called **targets**, a term that came about because of radars' initial military applications. The scattered signals received from targets are called **echoes**.

Radars operate in the microwave frequency range, which, unlike visible light, can penetrate deep into storm systems and reveal their interior structure. Meteorological radars employ a range of microwave frequencies, with different frequencies having different advantages and disadvantages for probing weather systems. A meteorological radar antenna focuses energy transmitted by the radar into a narrow beam with a width on the order of 1°. By continuously changing the azimuth and elevation of the antenna, a radar can scan the environment, mapping out either part or the entire hemisphere surrounding the radar. During the scanning procedure, radars transmit energy pulses, separated by much longer periods of quiet during which the radar senses echoes from the environment.

In this chapter, we focus on understanding basic properties of the electromagnetic energy pulses transmitted by radar and scattered back by the targets in the path of the energy pulses. We also examine different ways that data collected by radar are displayed so that the information can be interpreted in the context of meteorological phenomena.

3.2   Range and range ambiguity

The **slant range**, the distance along a radar ray from the radar to the target, is the most fundamental measurement made with radar. Meteorological radars accomplish this measurement by transmitting short pulses, on the order of a microsecond, separated by much longer listening times, on the order of a millisecond (Figure 3.1). The range is determined from

$$r = \frac{c\Delta t}{2} \tag{3.1}$$

where Δt is the time between the beginning of two adjacent pulses and c is the speed of light. The factor of 2 comes about because the microwave pulse must round trip out and back from the target. For example, if Δt were 1 ms, the target would be located 150 km from the radar.

The **pulse period**, T_r, is the time between successive pulses, whereas the **pulse repetition frequency** (PRF), the inverse of T_r, is the rate at which successive pulses are transmitted. The pulse repetition frequency determines the maximum range, r_{max}, that radars can observe targets without ambiguity. Microwaves traveling out to a target beyond r_{max} will arrive back in a time interval exceeding T_r, so the received echo will be interpreted to be from the next pulse in the pulse train rather than the pulse that struck the target. Targets must therefore be within

$$r_{max} < \frac{c}{2(\text{PRF})} \tag{3.2}$$

for the range to the target to be unambiguous. For example, for a PRF of 400 s^{-1}, r_{max} would be 375 km, and for a PRF $= 1000$ s^{-1}, r_{max} would be 150 km. Figure 3.2 shows

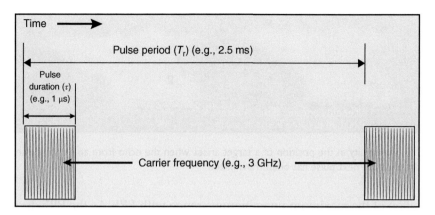

Figure 3.1 The relationship between the pulse duration and the pulse period

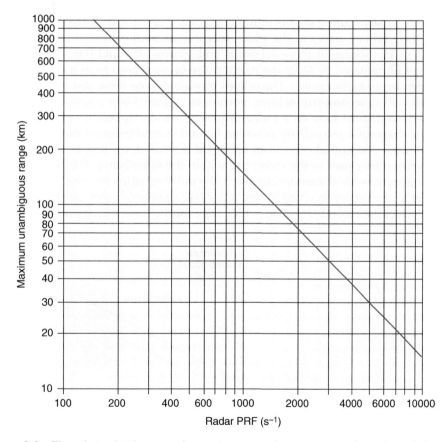

Figure 3.2 The relationship between the maximum unambiguous range of a radar and the pulse repetition frequency

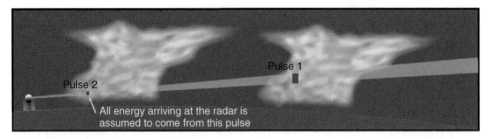

Figure 3.3   Ambiguity in the position of a target arises when the echo from an earlier pulse arrives at the radar after the next pulse has been transmitted

the variation of the maximum unambiguous range with PRF. As the PRF decreases, the maximum range increases. The curvature of the earth helps in limiting echoes from targets beyond r_{max} because many targets far from the radar will be below the radar beam, particularly for rays launched at higher elevation angles.

Problems arise for those targets illuminated by the radar and located beyond r_{max} because their echoes arrive at the radar antenna after a new pulse is transmitted (Figure 3.3). The source of these echoes is ambiguous—it could have been from a close target illuminated by the most recent pulse or have been from a distant target illuminated by an earlier pulse. The echoes produced from these distant targets are commonly called **second-trip echoes**. Fortunately, second-trip echoes from distant targets have a distinct look on a radar display, generally appearing elongated and weak. The shape of a second-trip echo can be understood by examining Figure 3.4, which shows a conceptual echo from an actual distant storm and its projection as a second-trip echo closer to the radar, but within the same beam. The echo appears weak because, as we will learn in Chapter 5, the radar equation relating the transmitted power to the power received by the radar has a range (r^2) correction. The range used to correct the second-trip echo is the range for the most recent pulse instead of

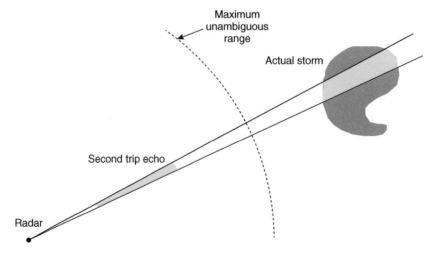

Figure 3.4   The long, narrow shape of a second-trip echo arises because the echo is constrained into the solid angle of the beam, but much closer to the radar

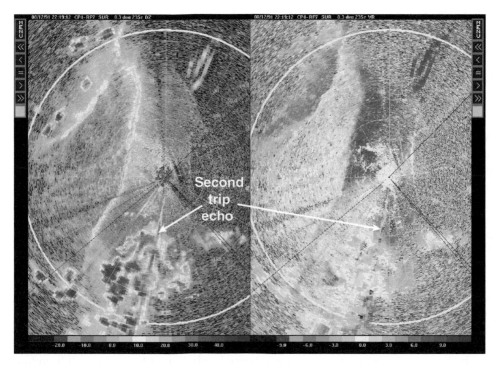

Figure 3.5 Second-trip echoes appearing in the radar reflectivity field (left) and the radial velocity field (right) (Image courtesy of Nolan Atkins)

the pulse that produced the echo, effectively reducing the intensity of the second-trip echo. Figure 3.5 shows a real second-trip echo. Note the elongated character of the echo. The Doppler radial velocity is shown as well. Note that the echoes have the radial velocity of the distant storm, making the echoes stand out more clearly.

For reasons we will discuss in Chapter 6, Doppler radars, such as the WSR-88Ds, require high PRFs to unambiguously determine Doppler radial velocity. With high PRFs, the potential for second-trip echoes is significant. Modern radars such as the WSR-88Ds get around this problem by transmitting pulses at two PRFs, one high (e.g., 1282 Hz) and one low (e.g., 322 Hz). The second-trip echoes appearing with the high PRF will not be present at the low PRF. By comparing the data from the two PRFs, second-trip echoes can be easily digitally removed, allowing recovery of both a long-range view of the weather echoes while retaining the required data to recover the Doppler radial velocity within a shorter range. This comes at a cost, as each location must be sampled twice, doubling the time required to complete a scan.

3.3   The transmitted and received signal

A radar transmits electromagnetic energy at a specific frequency in bursts of power contained in pulses of a specific duration and spacing. The returned echo arrives at the antenna with a slightly different frequency and a vastly different power level. In this section, we focus on key quantities that describe the signal and echo. Many of these quantities are closely related to one another.

3.3.1   *Pulse duration and pulse length*

A radar pulse consists of a wave packet of finite length, l, with the microwaves composing the packet oscillating at the frequency of the radar. The pulse duration, τ, is simply the inverse of l divided by the speed of light

$$\tau = \frac{l}{c} \tag{3.3}$$

A radar's capability to detect a target increases with the amount of radiation returned to the radar by the target, which is directly proportional to the pulse length. However, the radar's ability to resolve nearby targets along the path of a propagating pulse depends inversely on the pulse length. The choice of pulse length therefore becomes a trade-off between along-beam resolution and sensitivity. Typical pulse lengths used for meteorological radars are of the order of hundreds of meters. For example, the pulse duration of the WSR-88D in its short and long pulse modes is 1.57 and 4.7 μs. Multiplying the pulse duration by the speed of light produces a pulse whose length $l \sim 500$ m for the short mode and $\sim$1500 m for the long mode.

The along-beam resolution of a radar is half the pulse length, $l/2$. To understand why, consider Figure 3.6. Assume that the locations, a, b, and c in Figure 3.6 are fixed in space as the pulse travels by the points. Energy scattered back toward the radar from the leading edge of the pulse by a target located along line c will coincide exactly with energy scattered back from the trailing edge of the pulse from a target located along line b. An instantaneous sample of the returned radiation at the radar will therefore contain information about targets between b and c. The distance between b and c is exactly half the pulse length. For the WSR-88D radars, this implies that the along beam resolution is 250 and 750 m for the short and long pulse modes, respectively.

3.3.2   *Power and the duty cycle*

The fundamental unit of **power** is the Watt. Meteorological radars can operate with peak power approaching a megawatt (10^6 W). Very sensitive radar receivers, on the other hand, can detect echoes with powers as small as 10^{-15} W. Because of the enormous range of power that must be considered in radar operations, scientists and engineers often use logarithmic power ratios

$$dB = 10 \, \log \left(\frac{P_{w2}}{P_{w1}} \right) \tag{3.4}$$

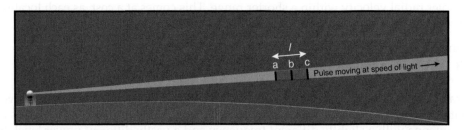

Figure 3.6   The red area denotes a radar pulse. The along-beam resolution is equal to half the pulse length

in which case the power ratio is expressed in decibels (dB). It is often desirable to express power in comparison to a standard power. If a standard power of 1 mW is used, the decibel power unit is expressed as

$$dBm = 10 \log\left(\frac{P_w}{1 \text{ mW}}\right) \tag{3.5}$$

where P_w is any value of power. For example, a megawatt would be 90 dBm, a gigawatt, 120 dBm, and a nanowatt, −30 dBm.

The **peak transmitted power**, P_t, is the power, averaged over a radar pulse, that is *transmitted into the atmosphere from the antenna*. The transmitted power must be known accurately in order for quantitative measurements to be made using radar. Ideally, the transmitted power should be measured at the antenna, or just outside the radome covering the antenna for radars which have one. However, measuring the power at this location is impractical—instead the power, P_m, is measured at the directional coupler. From this point, radiation must travel through the waveguide and feedhorn, reflect off the antenna's dish, and, depending on the radar, pass through a radome before propagating into the atmosphere. Power losses due to each hardware component, L_x, downstream of the directional coupler during the transit must be accounted for in determining the true transmitted power. Additional power losses, L_y, also occur between the directional coupler and the power meter used to measure power, and these must also be accounted for in the calibration. The losses are accounted for during regular radar calibrations, which is easy using logarithmic units (dBm) because the losses from various components sum together so that

$$P_t = P_m - \sum L_x + \sum L_y \tag{3.6}$$

The L_x and L_y are determined as part of a radar's initial calibration.

The transmitted power is limited by the size of the waveguide, which in turn is determined by the radar wavelength. High power levels lead to arcing within the waveguide. The power level at which arcing occurs can be raised by pressurizing the waveguide and/or filling it with a dielectric gas such as sulfur hexafluoride, but it will eventually arc with a high-enough power level. Large-wavelength radars have larger waveguides. For this reason, long-wavelength radars can transmit more power than short-wavelength radars. Table 3.1 shows examples of the maximum peak power that can be accommodated in waveguides for three common wavelengths.

A radar transmits energy for a short time and then pauses a much longer time during which no energy is transmitted. A radar's **duty cycle** is defined as the fraction of the time that the transmitter is on, and is given by

$$\delta = \tau(\text{PRF}) \tag{3.7}$$

The maximum duty cycle of the WSR-88D, for example, is 0.002, implying that the transmitter is on at most only 0.2% of the time the radar is operating. The **average**

Table 3.1   Maximum peak power in unpressurized waveguide

Band	Wavelength (cm)	Waveguide size (in.)	Max design P_t (MW)
S	10	1.5 × 3	3
C	5.6	1 × 2	1.3
X	3.2	0.62 × 1.25	0.4

transmitted power, P_{av}, the power averaged over the pulse period, is defined as

$$P_{av} = P_t \delta \tag{3.8}$$

For example, if the peak power of a WSR-88D is 1 MW, then the average power is 2000 W for a duty cycle of 0.002. The average power is important because it is the power actually measured when calibrating a radar system.

The power received at the antenna from targets in the path of the propagating pulse is many orders of magnitude smaller than the transmitted power. Echo power ranges widely, from about 10^{-5} to 10^{-15} W or less. Although radar receivers can be designed to detect echoes with power levels as small as 10^{-15} W, they are ultimately limited by the problem of background radiation. Microwave radiation from blackbody emissions from radar components, as well from microwave sources not associated with the radar system, forms an ever-present background that constitutes **noise**. The **minimum detectable signal** is the weakest echo power that can be discriminated from this noise background. The typical minimum detectable signals for many radars fall in the range of 10^{-13} to 10^{-14} W.

A receiver must be calibrated to determine the received power. The calibration is done by measuring the output voltage at the receiver generated from test signals of known power from a calibrated signal generator. The signal is fed into the receiver through the directional coupler in the waveguide. The test signals allow a radar engineer to develop a calibration scale that relates the voltage output at the receiver to input power to the receiver. The received power during radar operations is then determined by comparing the voltage output generated by an echo with the scale determined from the calibration.

3.4 Radar geometry and types of displays

Radar data are collected in a spherical coordinate system with coordinates given by range (r), elevation (α), and azimuth (β). The radar beam, because of atmospheric refraction and earth curvature, increases in elevation with range. In research and in some operational applications, the data in spherical coordinates are interpolated to Cartesian coordinates given, for example, by distance along the east–west (x), north–south (y), and up–down (z) directions, where east, north, and up are positive. Data may be displayed from one radar, or data from more than one radar can be combined into a composite image. When composites are created, the manner in which the data are combined is of importance, as the data collocated in x and y may differ substantially in z as the distance between the radars and the target are generally not the same.

3.4.1 Common radar displays in spherical coordinates

There are two common ways that radar data are displayed using the original spherical coordinate system in which they were collected. The name for these displays came from the early days of radar when the display was on a cathode ray tube (see Figure 3.7). The military was interested in displaying the data in a *plan* view to *indicate*

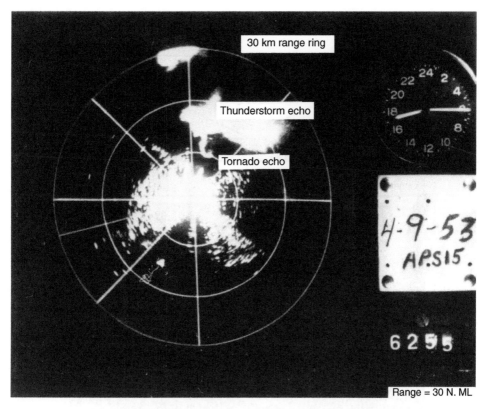

Figure 3.7 A cathode ray tube display from an early radar system showing the first thunderstorm hook echo reported in history (Photo by Illinois State Water Survey, Prairie Research Institute, University of Illinois)

the *position* of ships and aircraft, hence they used the name **plan position indicator** or **PPI** to describe the display.

The term PPI survives today in radar meteorology as the name of the display where the radar is swept through a sector or full 360° of azimuth while keeping the elevation angle constant (Figure 3.8). The data are collected on a cone, projected onto the spherical earth, and then displayed on the flat plane of a computer screen. An example of a PPI display from a WSR-88D is also shown in Figure 3.8. When viewing PPIs, it is important to remember that as one proceeds outward in range from the radar, the echoes appearing on the screen are from progressively higher elevation.

The second type of display that uses spherical coordinates also had its origin with military applications, in this case to *indicate* the *height* and *range* of a target, hence, the name **range–height indicator** or **RHI** scan. Again, radar meteorologists continue to employ this terminology. With an RHI scan, the radar is kept at a fixed azimuth while scanning in elevation (Figure 3.9). An example of a RHI scan through a stratiform cloud system is shown in Figure 3.10.

A type of display used by radar engineers, and sometimes by meteorologists, is the **A scope display**, which shows data, typically power or voltage, or other variables such as the radar reflectivity factor shown in Figure 3.11, along a single beam as a

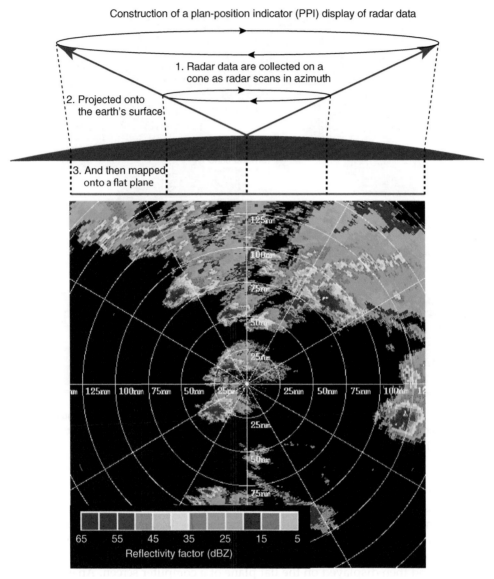

Figure 3.8 A plan position indicator (PPI) display from the WSR-88D showing a group of supercell thunderstorms

function of range. The "scope" refers to an oscilloscope, a device commonly used during radar calibrations.

Short-wavelength, ground-based radars such as K-band radars, radar profilers, radars used on research aircraft, and some spaceborne radars operate in the vertically pointing mode, continually pointing either straight up or down. In this mode, radar data are plotted as a function of height and time, producing a **time–height cross section**. Figure 3.12 shows an example of a time–height section from an airborne W-band radar as the aircraft flew across a winter cyclone.

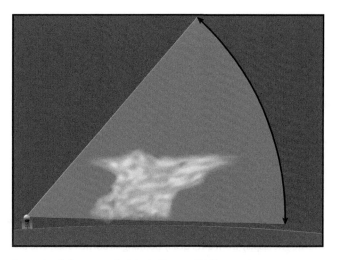

Figure 3.9   A schematic of the range–height indicator (RHI) scanning strategy

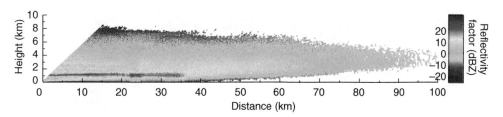

Figure 3.10   An RHI display of a stratiform winter storm. The bright red line is the melting level where snowflakes are melting into raindrops

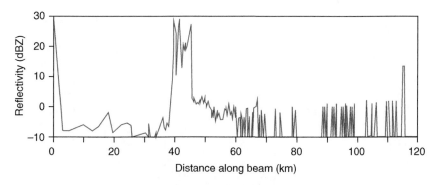

Figure 3.11   An A-scope display of data along a single beam

Radar data are normally collected in PPI mode, but sometimes it is desirable to view the data in a vertical or horizontal plane. The projection of radar data collected on a series of cones at different elevation angles in PPI mode onto a vertical plane is called a **reconstructed RHI**. Reconstructed RHIs generally have much poorer resolution than an RHI. Nevertheless, they can give some idea of storm structure. Figure 3.13 shows a PPI of a precipitation band in a winter cyclone over western

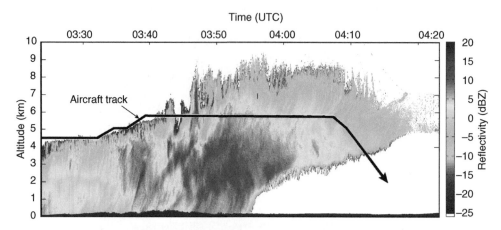

Figure 3.12　A time–height display from a vertically pointing airborne radar. The black line is the flight track

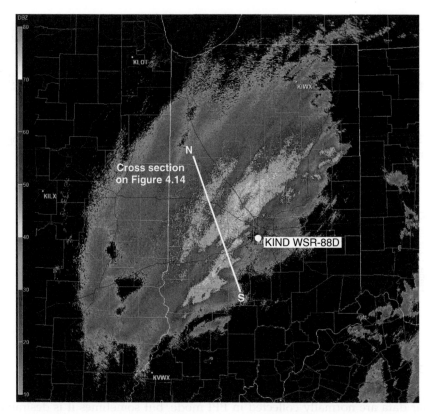

Figure 3.13　A PPI at 0.5° elevation of a winter storm over Indiana. The white line shows the location of the flight track of a research aircraft flying a vertically pointing radar

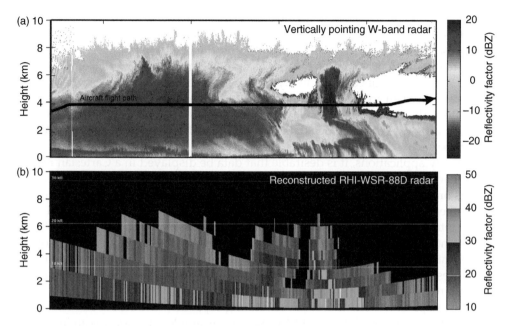

Figure 3.14   (a) A vertical cross section along the white line in Figure 3.13 taken by a high-resolution W-band radar on board an aircraft. (b) The same cloud system, but shown as a reconstructed RHI from the WSR-88D located in Indianapolis, IN

Indiana. The white line shows the location of the reconstructed RHI in Figure 3.14 from the Indianapolis WSR-88D. The resolution can be directly compared in this case with measurements made in the same cross section at approximately the same time by the University of Wyoming W-Band Cloud Radar on board the NSF/NCAR C-130 aircraft during a winter storm study. The reconstructed RHI captures general

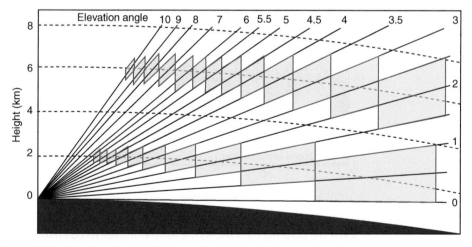

Figure 3.15   Strategy for constructing a Constant Altitude Plan Position Indicator (CAPPI) display (Adapted from Marshall, J.S. (1957) The constant-altitude presentation of radar weather patterns. Proceedings of the 6th Weather Radar Conference, American Meteorological Society)

features, but the details of the storm structure are lost because of low resolution and the reduced sensitivity at the WSR-88D's S-band wavelength.

It is also possible to project segments of radar data collected on a series of cones at different elevation angles in PPI mode onto a horizontal plane at one altitude (Figure 3.15). This is uncommon today because it is computationally easy to simply interpolate the data to a constant altitude, but in the early days of weather radar, it was the only way to operationally examine storms at a constant altitude. Radar meteorologists use the term **Constant Altitude Plan Position Indicator**, or **CAPPI**, to describe such a display.

For many applications, it is desirable that data be displayed in a Cartesian rather than spherical coordinate system. This can be accomplished by interpolating the data from its original distribution in spherical space to a regular grid in Cartesian space. The mathematics of the interpolation will influence which features in the original data will be retained and which will be damped or removed through smoothing. When data are interpolated into a Cartesian system, **horizontal cross sections** can be plotted at constant altitude, or **vertical cross sections** can be plotted along or parallel to the x or y axes of the coordinate system. An example of a horizontal and vertical cross section through a squall line is shown in Figure 3.16.

Some radars flown on aircraft have scanning capability. On the NOAA P-3s, and on aircraft that flew NCAR's ELDORA radar, for example, the radars scan in a cone aft and forward of the aircraft. Because of the aircraft motion, this scanning procedure

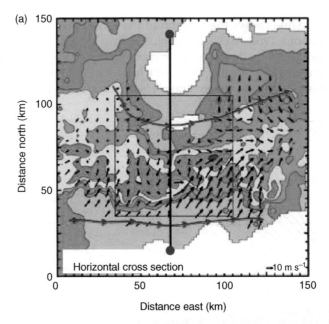

Figure 3.16   (a) Horizontal cross section through a squall line at 2 km elevation. (b) Vertical cross section through the same squall line along the black line on the horizontal cross section (From: Grim, J.A., Rauber, R.M., McFarquhar, G.M., Jewett, B.F., and Jorgensen, D.P. (2009) Development and forcing of the rear inflow jet in a rapidly developing and decaying squall line during BAMEX. *Mon. Wea. Rev.*, **137**, 1206–1229. © American Meteorological Society, used with permission)

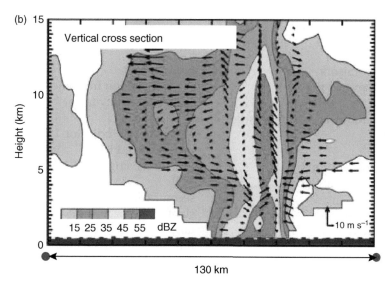

Figure 3.16   (*Continued*)

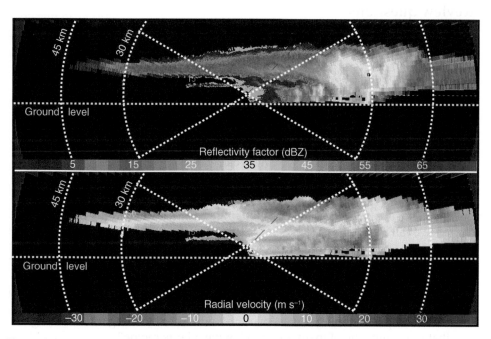

Figure 3.17   Reflectivity factor and radial velocity measured by the ELDORA radar on board the Navy Research Laboratory P-3 aircraft as the aircraft flew along a squall line

effectively maps the data onto a spiral as the aircraft moves forward. The data from the spiral can be projected on a single plane to produce a full-circle plot about the aircraft that appears similar to a ground-based radar's RHI. Figure 3.17 shows an example taken from the ELDORA radar as the aircraft flew along a mesoscale convective system.

Important terms

A scope display

active remote sensors

average transmitted power

CAPPI

Constant Altitude Plan Position Indicator

duty cycle

echoes

horizontal cross sections

minimum detectable signal

noise

passive remote sensors

peak transmitted power

plan position indicator

power

PPI

pulse period

pulse repetition frequency

radar reflectivity factor

radial velocity

range–height indicator

reconstructed RHI

remote sensing

RHI

second-trip echo

slant range

target

time–height cross section

vertical cross sections

Review questions

1. What is the difference between an active remote sensor and a passive remote sensor?

2. What properties of a radar echo on a PPI scan would lead you to believe that it is a second-trip echo?

3. What technique do the WSR-88D radars use to eliminate second-trip echoes?

4. What causes noise in a radar system and ultimately limits the capability of a radar to detect targets producing weak echoes?

5. Where in a radar system is the transmitted power actually measured? Why must corrections be applied to the measured value?

6. Why must one be cautious when looking for low-level features toward the edge of a PPI radar display?

7. If you wanted to determine the height of a thunderstorm, would you be better off using a PPI, an RHI, or a CAPPI display?

8. What is the difference between a CAPPI and a horizontal cross section?

9. Which would have a higher resolution, an RHI or a reconstructed RHI?

10. What complication might one encounter when constructing a radar image that is a composite of many radars viewing a large region?

Challenge problems

11. The WSR 88D radars at Lincoln, IL (KILX), and Chicago, IL (KLOT), are exactly 193 km apart.
    a. If both radars operated with the same PRF, what PRF would be required so the radars would cover the maximum area possible without any overlap?
    b. What PRF would be required for one radar to be at the maximum unambiguous range of the other?
    c. Under the conditions of (b), what is the duty cycle of the radar if the pulse duration is 0.5 μs.
    d. If the peak power is 100,000 W, what is the average power?
12. A weather radar is transmitting vertically:
    a. If the radar transmits a megawatt peak power and receives a nanowatt, what is the difference in power between the transmitted signal and returned signal expressed in decibels?
    b. Suppose the international space station (ISS) passes directly over the radar at an altitude of 400 km. What is the maximum PRF that the radar can use to unambiguously determine the altitude of the ISS?
    c. If the radar stays at one azimuth angle and sweeps the antenna to the horizon and back to the zenith, what type of scan did it perform?
13. The maximum unambiguous range of a radar, calculated from the speed of light, c, is 150 km. Assume that the index of refraction is everywhere 1.0003 along the path of a ray from this radar. What is the true maximum unambiguous range?
14. Suppose the width of a tornado emerging from a thunderstorm is 1 km, and as a researcher, you would like to obtain at least 20 separate measurements across the tornado's width along a single beam. What is the maximum pulse duration that you must use to ensure that you satisfy this requirement?
15. If you had a very limited amount of money and wanted to buy a cheap radar system to view weather, what wavelength radar would you likely purchase, and in what direction would you point the radar? What type of data display would you most likely generate? Justify your answer.

4

The Path of a Radar Ray

Objectives

By the end of this chapter, you should understand the following:

- How the index of refraction is related to the vertical gradient of atmospheric density and moisture.
- Why the vertical distribution of water vapor is important to ray propagation.
- Why the curvature of the earth ultimately limits the useful range at which a radar can operate.
- What factors influence the propagation of a radar ray through the earth's atmosphere.
- The equations that are used to relate altitude and distance across the earth's surface to the range and elevation angle measured by a radar.
- The conditions under which severe anomalous propagation of a radar ray will occur.

4.1 Introduction

The first step in interpreting the returned signal received by a radar is to determine the geographic position, altitude, and size of the volume in space where the **scattering elements** were located.

For clarity, we will describe electromagnetic energy transmitted by a radar using two terms: we will use the term **ray** to describe the path the point of maximum intensity of electromagnetic radiation emitted by the radar takes as the radiation propagates through space; we will use the term **beam** to refer to an area or volume of space illuminated by the radar, similar to a flashlight. This chapter focuses specifically on determining position and altitude of a **radar ray** and the potential errors that can result in this determination. We first consider what factors influence a **radar ray path**, that is, the path a pulse of energy will take as it propagates away from the radar. We then will derive equations describing the path that a pulse of microwave energy would take if it propagated through an idealized, stratified atmosphere. Finally, we will consider how radar rays may deviate from this path in an atmosphere with variable thermal and moisture stratification.

Radar Meteorology: A First Course, First Edition. Robert M. Rauber and Stephen W. Nesbitt.
© 2018 John Wiley & Sons Ltd. Published 2018 by John Wiley & Sons Ltd.
Companion website: www.wiley.com/go/Rauber/RadarMeteorology

4.2   Ray propagation in an idealized atmosphere

A radar ray path over the earth's surface is influenced by the characteristics of the atmosphere along the path of the ray and the shape of the earth itself. In this section, we consider how these effects combine to determine the distinct path a radar ray will take in an atmosphere with a simplified vertical profile of density and moisture.

4.2.1   Factors influencing radar ray paths

In the absence of an atmosphere, radar rays would be straight and would point off into space as the earth curved downward beneath the ray (Figure 4.1). The effect of **Earth curvature** is to cause a ray to become higher above the earth's surface with distance. Were it not for the atmosphere, the path of such a ray would be easy to predict.

In the near-vacuum of free space, the speed of propagation of an electromagnetic wave is given by

$$c = \frac{1}{\sqrt{\varepsilon_0 \mu_0}} \tag{4.1}$$

where c is the speed of light, a constant, and equal to 3×10^8 m s^{-1}. However, when electromagnetic waves propagate through a different medium, such as air or water drops, they no longer propagate at c, but at a slower velocity that is related to the permittivity, ε_1, and magnetic inductive capacity, μ_1, of the medium. The velocity of an electromagnetic wave through a medium other than a vacuum is given by

$$v_c = \frac{1}{\sqrt{\varepsilon_1 \mu_1}} \tag{4.2}$$

Figure 4.1   A ray launched at the horizon, because of earth's curvature, will rise higher above the earth's surface and eventually into space (Photo Courtesy NASA)

and the **refractive index** of a medium is defined as

$$n = \frac{c}{v_c} \tag{4.3}$$

For air at sea-level pressure, the refractive index is approximately 1.0003. The refractive index, as expressed here, accounts only for propagation of electromagnetic waves. In its most general form, the refractive index is expressed as a complex number:

$$m = n + ik_a \tag{4.4}$$

where $i = \sqrt{-1}$ and k_a is related to the absorptive properties of the medium. This usage will become important when we consider attenuation, but for now we will just consider the real part, n, associated with propagation.

The refractive index is related to atmospheric density and the polarization state of molecules composing the atmosphere. Molecules are said to be **polarized** if they produce their own electric field in the absence of external forces. An electric field will exist in the vicinity of a molecule if the electrons on a molecule, from a statistical point of view, spend more time on one end of the molecule than the other. Each polar molecule represents a dipole. The water molecule is such a molecule (Figure 4.2), and water vapor is an important and highly variable component of the atmosphere. In the absence of an external electric field, polar molecules align randomly in the atmosphere. An external force, such as that induced by propagating microwaves from a radar, causes the molecules to reorient and the dipole fields to align, adding constructively to the net electrical force acting on each molecule. The electrical force acting on each molecule is therefore the sum of the external electrical force and the electrical force produced by each of the polar molecules. Gradients of water vapor molecules act to bend a propagating electromagnetic wave, much as a prism bends visible light.

The relationship between n, temperature, T, the partial pressure of dry air, P_d, and the water vapor pressure, e, varies with the frequency of the transmitted electromagnetic energy. The general equation relating the index of refraction to pressure and vapor pressure is

$$n = 1 + C_1 \frac{P_d}{T} + C_2 \frac{e}{T} + C_3 \frac{e}{T^2} \tag{4.5}$$

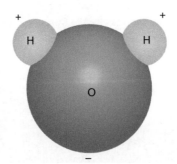

Figure 4.2 The water molecule is a polar molecule because the average distribution of electrons creates a dipole with a positive charged side at the location of the hydrogen atoms

For microwave frequencies used in meteorological radar applications, the following constants can be used:

$$n = 1 + 7.76 \times 10^{-5} \text{ K hPa}^{-1} \frac{P_d}{T} - 5.6 \times 10^{-6} \text{ K hPa}^{-1} \frac{e}{T} + 0.373 \text{ K}^2 \text{ hPa}^{-1} \frac{e}{T^2} \quad (4.6)$$

The vapor term containing e/T is small relative to the pressure term and is often combined with the pressure term with little error. Furthermore, the index of refraction is sometimes referred to in units of "radio refractivity", N, where

$$N = (n - 1) \times 10^6 \quad (4.7)$$

Combining Eqs. (4.6) and (4.7) and incorporating the second term into the first in Eq. (4.6) gives

$$N = 7.76 \text{ K hPa}^{-1} \frac{P}{T} + 3.73 \times 10^5 \text{ K}^2 \text{ hPa}^{-1} \frac{e}{T^2} \quad (4.8)$$

where P is the total pressure ($P_d + e$) in hectopascals (hPa), e is the partial pressure of water vapor in hPa, and T is the temperature in kelvin (K).

Atmospheric density decreases exponentially with height, and so n decreases exponentially from about 1.0003 at sea level to 1.000 at the top of the atmosphere. From Snell's law, this implies that a radar ray will be refracted downward toward the earth's surface as it propagates away from the radar at any angle below the zenith. The exact path of a ray, however, is determined by the local vertical profile of density and water vapor. Unfortunately, the vertical distributions of density and water vapor are variable in time and space and, at any given time, are unknown over the volume of space scanned by a radar. The values used for these variables to determine ray propagation are based on **International Standard Atmosphere** (ISA) values published by the International Organization for Standardization. Refraction that occurs in an atmosphere with this standard profile is called **standard atmospheric refraction**. The specific contributions to N from atmospheric density and water vapor, based on the ISA Standard Atmosphere temperature profile and the assumption that the atmosphere is saturated (100% relative humidity) at all altitudes, are shown in Figure 4.3. Water vapor contributes about 27% to N near the surface, decreasing to essentially zero at the tropopause.

The vertical gradient of the index of refraction, $\frac{dn}{dz}$, in the standard atmosphere is approximately -4×10^{-8} m^{-1}. The actual propagation path of a radar ray depends on $\frac{dn}{dz}$. The rate at which dry air density varies with height is influenced by the presence of stable layers, particularly **inversions**, layers where temperature increases with altitude. More importantly, the vertical gradient of n can become very large when a dry air layer overrides moist air. Changes in n in the fifth and sixth decimal place can significantly affect the propagation of electromagnetic waves in the atmosphere.

4.2.2   The path of a ray in an idealized atmosphere

The exact differential equation for a ray path through a spherical stratified atmosphere follows from application of a principle of optics called Fermat's principle and is given by

$$\frac{d^2h}{ds^2} - \left(\frac{2}{a_e + h} + \frac{1}{n} \frac{dn}{dh} \right) \left(\frac{dh}{ds} \right)^2 - \left(\frac{a_e + h}{a_e} \right)^2 \left(\frac{1}{a_e + h} + \frac{1}{n} \frac{dn}{dh} \right) = 0 \quad (4.9)$$

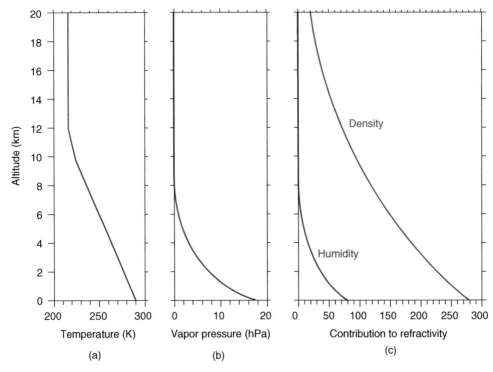

Figure 4.3   Standard atmosphere profile of temperature (a) and vapor pressure (b). The relative contribution of density and humidity to radio refractivity (c) in the standard atmosphere of (a, b) (Courtesy of David A. Hooper)

The derivation of this equation is provided in Appendix B for students who might be interested in its origin.[1] Figure 4.4 shows the geometry for this equation. The relevant parameters are the height of the radar antenna above the surface of the earth, h_0, the height of a ray above the surface of the earth, h, the distance along the surface of the earth underneath the radar ray, s, the elevation angle of the ray relative to a circle concentric with the earth's surface at the altitude of the ray, α_e, and the radius of the earth, a_e.

Under typical conditions, the elevation angle in Figure 4.4 is small (a few degrees). Since

$$\frac{dh}{ds} \approx \tan \alpha_e \tag{4.10}$$

then

$$\left(\frac{dh}{ds}\right)^2 \ll 1 \tag{4.11}$$

and the second term must be very small compared with the first and third. Furthermore,

$$h \ll a_e \tag{4.12}$$

---

[1]The equation was first derived by Hartree, D.R., Michel, J.G.L., and Nicolson, P. (1946) Practical methods for the solution of the equations of tropospheric refraction, in *Meteorological Factors in Radio Wave Propagation*, Physical Society: London, pp. 159–161.

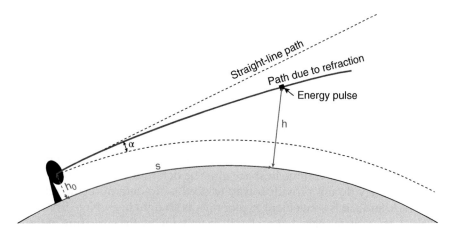

Figure 4.4 A radar ray (blue) follows a curved path through the atmosphere due to the combined effects of earth's curvature and atmospheric refraction

so we can approximate $a_e + h = a_e$. Finally, n is close to unity and can be considered unity in this equation with negligible error. Applying these "small angle" and "large earth" approximations, the path of a ray in the atmosphere can be approximated as

$$\frac{d^2h}{ds^2} = \frac{1}{a_e} + \frac{dn}{dh} \tag{4.13}$$

The first term on the right side is the effect of curvature of the earth on the height of a ray and the second is the effect of refraction. If one assumes that the vertical gradient of the index of refraction above a radar is constant, the right side of this equation is a constant and the equation can be easily integrated once to obtain an expression for the elevation angle of the ray relative to a circle concentric with the earth's surface at the altitude of the ray, and twice to obtain the height of the ray at any distance from a radar. The resulting equations are

$$\alpha_e = \tan^{-1}\left[\tan\alpha + s\left(\frac{1}{a_e} + \frac{dn}{dh}\right)\right] \tag{4.14}$$

$$h = \frac{1}{2}s^2\left(\frac{1}{a_e} + \frac{dn}{dh}\right) + s\,\tan\alpha + h_0 \tag{4.15}$$

where α is the elevation angle of the antenna.

The National Weather Service WSR-88D radars use a different form of the equation[2]

$$h' = \frac{s^2}{2ia_e} + s\sin\alpha \tag{4.16}$$

Here, h' is the height above the altitude of the antenna and $i = 1.21$. Eqs. (4.15) and (4.16) are numerically equivalent if the vertical gradient of the index of refraction is given by

$$\frac{dn}{dh} = -2.724 \times 10^{-5}\ \text{km}^{-1} \tag{4.17}$$

---

[2]http://www.wdtb.noaa.gov/courses/dloc/topic3/lesson1/Section5/Section5-4.html.

Note that $\sin \alpha$ is used in place of $\tan \alpha$ in the WSR-88D equation. $\sin \alpha$ can replace $\tan \alpha$ with an error of less than 0.01% at $\alpha < 4°$, 0.1% at $\alpha < 8°$, and 1% at $\alpha < 16°$. The sine function is used instead of the tangent to save computation time during operations.

4.2.3   The range and height of a pulse volume in space

A radar records the elevation (α) and azimuth (β) angles of the antenna, and the time it takes for a pulse to travel round trip from its antenna to target and back, from which the range, r, to the target is determined. From these measurements, the position of the center of the sampled volume in space must be determined. Equations (4.14) and (4.15) unfortunately do not provide a way to do this. To determine the relationship between α, r, h, and s, it is convenient to plot rays as straight lines. Note from 4.14 that if $dn/dh = -1/a_e$, a ray launched at $\alpha = 0$ will exactly follow the curvature of the earth. The curvature of the earth (assuming sphericity) is defined by

$$C_e = \frac{1}{a_e} \tag{4.18}$$

which implies that the ray's curvature is given by

$$C_{ray} = -\frac{dn}{dh} \tag{4.19}$$

These equations imply that the curvature of a ray *relative to the earth's surface* is given by

$$C_{r.earth} = -\frac{1}{a_e} - \frac{dn}{dh} \tag{4.20}$$

If we choose to plot rays as straight lines, then the radius of curvature of a fictitious earth relative to these straight rays must be

$$a'_e = \frac{1}{\left(\frac{1}{a_e} + \frac{dn}{dh}\right)} = ka_e \tag{4.21}$$

where

$$k = \frac{1}{1 + a_e \left(\frac{dn}{dh}\right)} \tag{4.22}$$

As an example, for a value of $dn/dh = -4 \times 10^{-5}$ km^{-1}, the fictitious earth's radius, a'_e, is about $4/3a_e$. The geometry of the fictitious earth is convenient for determining the relationship between α, r, h, and s. The "4/3 earth approximation" is commonly used to calculate ray paths in radar meteorology. Consider the geometry in Figure 4.5.

Applying the law of cosines to the triangle with sides $[a'_e, r, a'_e + h]$, we have

$$(a'_e + h)^2 = r^2 + a'^2_e - 2ra'_e \cos \kappa \tag{4.23}$$

Using $a'_e = ka_e$, (4.23) can be written as

$$h = (r^2 + (ka_e)^2 - 2rka_e \cos \kappa)^{1/2} - ka_e \tag{4.24}$$

Noting from Figure 4.5 that

$$\cos \kappa = \cos\left(\frac{\pi}{2} + \alpha\right) = \cos \frac{\pi}{2} \cos \alpha - \sin \frac{\pi}{2} \sin \alpha = -\sin \alpha \tag{4.25}$$

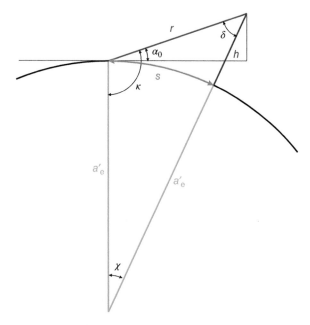

Figure 4.5   Geometry used to calculate the distance across the earth's surface and the height of a point along a ray as a function of slant range and radar elevation angle

Eq. (4.24) can be written as

$$h = (r^2 + (ka_e)^2 + 2rka_e \sin \alpha)^{1/2} - ka_e \tag{4.26}$$

Equation (4.26) relates the range from the radar to the center of the sampled volume, and the elevation angle of the antenna, to the height of the sampled volume above the surface of the earth.

Next, we apply the law of sines to the triangle with sides $[a'_e, r, a'_e + h]$ to get

$$\sin \chi = \frac{r \sin \kappa}{a'_e + h} \tag{4.27}$$

Noting that

$$s = \chi a'_e = \chi ka_e \tag{4.28}$$

where χ is expressed in radians, we can write

$$s = ka_e \sin^{-1} \left(\frac{r \sin \kappa}{ka_e + h} \right) \tag{4.29}$$

From Figure 4.5

$$\sin \kappa = \sin \left(\frac{\pi}{2} + \alpha \right) = \sin \frac{\pi}{2} \cos \alpha + \cos \frac{\pi}{2} \sin \alpha = \cos \alpha \tag{4.30}$$

Substituting into (4.29)

$$s = ka_e \sin^{-1} \left(\frac{r \cos \alpha}{ka_e + h} \right) \tag{4.31}$$

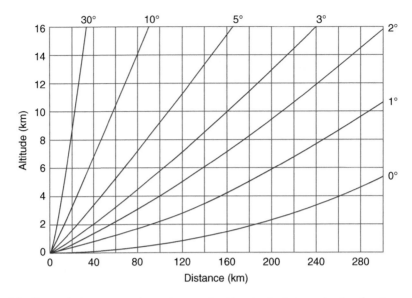

Figure 4.6   Ray paths in an atmosphere characterized by standard atmospheric refraction

Equation (4.31) relates the range, elevation angle of the antenna, and height of the sampled volume to its position along the earth's surface from the radar.

Figure 4.6 shows ray paths of several elevation angles under conditions of standard atmospheric refraction. A ray launched at the horizon will be centered 1 km above the earth's surface at about 90 km from the radar, and over 5 km above the surface at 300 km from the radar. A ray launched at 1° elevation will be over 10 km above the surface at 300 km from the radar. This figure clearly demonstrates the range limitations that the curvature of the earth imposes when using a single radar to study atmospheric storms.

4.3   Anomalous propagation

During radar operations, the vertical gradient of the index of refraction over the region scanned by a radar is unknown and is likely to vary in space and time, particularly around the weather systems that are of most interest to radar meteorologists. Some assumption must be made about the vertical gradient of the index of refraction so that radar data can be geonavigated and displayed, or possibly used in later scientific analyses. Any deviation from the assumed path of a ray resulting from an erroneous assumption about the vertical gradient of the index of refraction is called **anomalous propagation**. Anomalous propagation almost always occurs, but the positioning errors of the ray path are so small that errors in the geo-location of the sampled volume in space are of no concern. There are conditions, however, where the deviation of a ray path can be quite large, so much so that low elevation rays can bend downward toward the earth. This may cause significant error in location of the sampled volume in space, or the beam may even strike the ground or objects near the ground, leading to **ground clutter** displayed quite far from the radar.

Conditions are also possible that the vertical gradient of the index of refraction causes a ray to become trapped within a low-level layer of the atmosphere, a condition known as **ducting**. Such conditions are most prevalent when the index of refraction in the layer right above the ground decreases very rapidly with altitude, conditions that prevail when a low-level temperature inversion occurs in a layer characterized by very moist air below the inversion and dry air aloft. Unfortunately, these are precisely the conditions that often characterize the marine layer just off the west

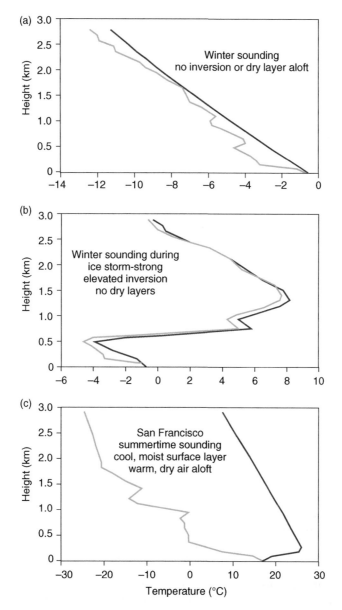

Figure 4.7   Three soundings used to calculate the ray paths in Figure 4.8

coasts of continents, often leading to **sea clutter**. Similar conditions are also common on clear nights over land when a nocturnal inversion develops and air near the surface saturates to its dew point temperature. In stormy regions, these conditions sometimes occur when a moist outflow from a thunderstorm undercuts drier air in the storm's vicinity, and the radar is within the moist flow.

To understand the effects of various weather conditions on ray paths, consider the three soundings in Figure 4.7. These are actual soundings taken on 3 days, the first a winter day with no inversion or dry layers, the second during a major ice storm, where a sharp temperature inversion was present aloft, but the atmosphere was near saturation, and the third a sounding taken in the early morning at San Francisco, CA, when a shallow marine layer was present with dry air aloft. Figure 4.8 shows sample ray paths calculated using a piecewise (layer by layer) integration of 4.15 with the actual values of the index of refraction from the soundings. The red rays are for the winter sounding with no inversion or dry layers. Compare those with the blue rays from the ice storm sounding. The change in the position of the 0.2° and 0.3° rays is small in the lower atmosphere, increasing to a horizontal distance of about a kilometer about 100 km downrange of the radar at altitudes about 1.5 km. Inside of 100 km range, the difference between rays is insignificant. The same cannot be said for the San Francisco sounding. The 0.2° and 0.3° rays undergo severe anomalous propagation, bending downward and striking the ground. In fact, all rays below an elevation angle of 0.42° return to the surface. The 0.42° ray gets trapped for a long distance before finally turning upward 140 km downrange. The horizontal difference in the position of the 0.5° ray from the winter sounding at an altitude of 2.5 km is almost 55 km.

To understand the potential impact of anomalous propagation on radar echoes, consider Figure 4.9, a composite radar image from the Lincoln, IL, WSR-88D radar

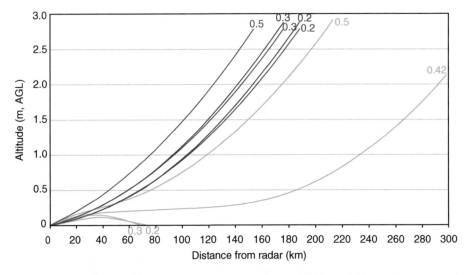

Figure 4.8   Ray paths for each of the three soundings in Figure 4.7. The red, blue, and green soundings are for the soundings in panels a, b, and c, respectively. The elevation angle of the radar is noted at the end of each ray path

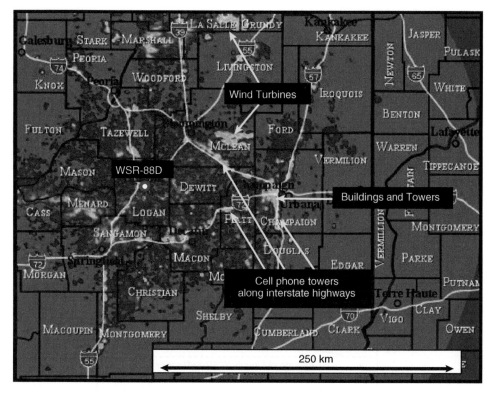

Figure 4.9 Radar image from the KILX radar in Lincoln, Illinois, on August 27, 2010 at 12:00 UTC showing effects of anomalous propagation

taken on a summer day, August 27, 2010, just before sunrise. The sky on this day was completely clear across the state of IL. The corresponding temperature and relative humidity profile, taken from the Lincoln, IL, rawinsonde launched at the same time is shown in Figure 4.10. A nocturnal inversion was present at the time, with humid air near the surface and dry air above it. The radar image shows strong radar returns from buildings and towers in towns as far away as 100 km from the radar, radar returns along the major interstates associated with cell towers, and returns from clusters of wind turbines. These echoes all disappeared shortly after sunrise as the nocturnal inversion mixed out vertically by thermals.

In summary, the position of a radar ray in space is normally calculated assuming that the vertical gradient of the index of refraction is constant. In most meteorological situations, this assumption is reasonable and the deviation of rays because of variations in the vertical gradient of the index of refraction results in errors that are small and generally of no consequence. Applications that require precise positioning of radar data in space should consider calculating rays using ambient soundings, if they are available. Under certain conditions, particularly a surface-based inversion within which relative humidity decreases rapidly with altitude, severe anomalous propagation of a radar ray can occur and the positioning of the ray should be questioned if standard refraction was assumed.

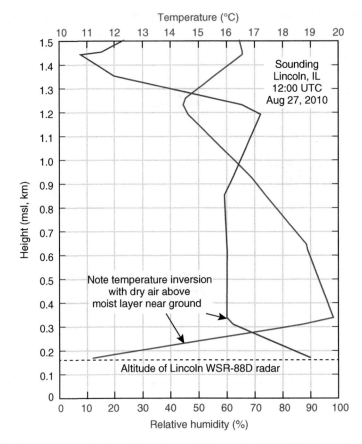

Figure 4.10   Temperature and relative humidity profile from the 12:00 UTC August 27, 2010 Lincoln, IL, rawinsonde that was taken at the time the radar data in Figure 4.9 was collected

Important terms

anomalous propagation
beam
ducting
Earth curvature
ground clutter
International Standard Atmosphere
inversion
polarized

radar ray
radar ray path
refractive index
ray
sea clutter
scattering elements
standard atmospheric refraction

Review questions

1. A radar transmits a pulse with its antenna pointed vertically. Does the speed at which the pulse travels upward away from the radar increase, decrease, or remain the same as the pulse moves from the surface to the top of the atmosphere?

2. What atmospheric properties influence the index of refraction in the troposphere?

3. What is meant by a "polar" molecule?

4. What assumption is made about the index of refraction to determine ray paths for meteorological radar operations?

5. Under standard atmospheric conditions, which effect dominates the path a radar ray will take through the atmosphere, atmospheric refraction, or earth's curvature? Justify your answer.

6. A lawyer calls you and asks you if the rainfall at a particular location exceeded 1 in. between noon and 7 p.m. on a particular day. Another person claimed (correctly) that the nearest WSR-88D measured 0.7 in. rainfall at that location during the time period. After some investigation, you determine that the location of interest was 200 km from the nearest WSR-88D and that the lowest elevation angle used by the WSR-88D that day was 0.5°. Based on information in this chapter, what might you advise this lawyer to help her with her case.

7. The outflows from microbursts are typically very shallow. Why do many microbursts go undetected by radar?

8. Would you expect anomalous propagation to occur more commonly in the early morning or mid-afternoon? Why?

9. You are a radar operator using a WSR-88D located near an urban area with lots of highways. What clues might you use to determine if anomalous propagation might be influencing your radar data on a particular day?

10. On a particular morning just before sunrise, you look at a radar screen and see radar echoes in thin intersecting lines all over the screen. You run off to tell your colleagues, but get waylaid talking about last night's football game at the water cooler. By the time you get back to your desk an hour later with your colleagues, the screen is clear. What happened?

Challenge problems

11. In summertime along the west coast of California, the marine layer is often very cool (temperature about 20 °C), while just above the marine layer, the air is hot and quite dry. The boundary between these layers can be quite distinct. Suppose on a particular day, the pressure, vapor pressure, and temperature just below and just above the boundary between the marine layer and the dry air were as follows:

	Dry air pressure (hPa)	Vapor pressure (hPa)	Temperature (°C)
Just above layer boundary	995	1	31
Just below air boundary	1000	20	19

    a. Calculate the index of refraction for each layer.

    b. Calculate the radio refractivity for each layer.

    c. Suppose a radar ray is propagating upward toward the boundary at an elevation angle α. What would α have to be for the ray to point exactly along the boundary after refraction?

12. A tornado is located halfway between two WSR-88D radars. The top of the tornado circulation is 2 km above the ground. Both radars are the same elevation above sea level, and the height of each antenna above the ground is 25 m. The radars are spaced 150 km apart. The table below shows conditions at the ground and 5722 m above sea level. Assume that the index of refraction varies linearly with height between these two altitudes and that the radius of the earth is 6371 km.

	Dry air pressure (hPa)	Vapor pressure (hPa)	Temperature (°C)
Top level (5722 m AGL)	500	0.1	−6.9
Bottom level (0 m AGL)	996	17	16.4

    a. Calculate the index of refraction at each level.

b. Calculate the vertical gradient of the index of refraction.

c. Starting with the equation: $\dfrac{d^2h}{ds^2} = \dfrac{1}{a_e} + \dfrac{dn}{dh}$, derive Eqs. (4.14) and (4.15) showing all steps.

d. Use these equations to calculate the elevation angle each radar must scan for the center of the beam to just touch the top of the circulation of the tornado.

13. Suppose that the exact distance from the nearest WSR-88D radar to your house is 95 km. Assume that both locations are at the same altitude, sea level, and that the height of the WSR-88D antenna above the ground is 25 m. Further assume that the surface pressure at both locations is 1000 hPa, the temperature is 0 °C, and the air is saturated from the surface to the 500 hPa level, where the temperature is −25 °C and the altitude is 5400 m. Calculate the height of the center of the WSR-88D 0.5° beam over your house. Hint: When air is saturated, the vapor pressure equals the saturation vapor pressure, which is only a function of temperature. An equation relating saturation vapor pressure (hPa) and temperature (°C) is

$$e_s = a + T(b + T(c + T(d + T(f + T(g + hT)))))$$

where

$$
\begin{array}{lll}
a = 6.107799961 & b = 4.436518521 \times 10^{-1} & c = 1.428945805 \times 10^{-2} \\
d = 2.650648471 \times 10^{-4} & f = 3.031240396 \times 10^{-6} & g = 2.0340880948 \times 10^{-8} \\
h = 6.136820929 \times 10^{-11} & &
\end{array}
$$

14. Suppose that a WSR-88D radar antenna is on a pedestal 50 m above the ground and is pointed directly at the horizon (elevation angle of 0°). On a particular day, engineers are amazed to find that a radar ray remains exactly 50 m above the ground out to a range of 150 km! If the value of the index of refraction at the ground was 1.000300, what was the index of refraction 1 km above the ground?

15. A ship's radar is located 20 m above the ocean surface. The captain wants to detect an enemy ship of the same size that he believes is located well beyond the horizon. A clever sailor who read this book runs to starboard where the weather center is located and sends up a weather balloon to measure weather data above the ship. He reports to the captain that from his measurements and calculations, he is certain that if they point the ship's radar at the horizon, the radar will detect the enemy ship beyond the horizon if it is there.

a. What atmospheric properties did the sailor have to measure to arrive at this conclusion?

b. Draw an example vertical temperature and moisture profile between the ocean surface and 1 km that would support the sailor's conclusion. Assume that the surface temperature was 15°C and the relative humidity was 80%.

5

Power and the Radar Reflectivity Factor

Objectives

By the end of this chapter, you should understand the following:

- The radar equation for solitary and distributed targets.
- The concept of a contributing volume.
- The steps involved and assumptions required in the derivation of the weather radar equation.
- The difference between radar reflectivity and the radar reflectivity factor.
- The definition of the radar reflectivity factor, the quantity shown on typical radar displays.
- The relationship between the radar reflectivity factor and the raindrop size distribution.
- How the weather radar equation relates the radar reflectivity factor to the returned power measured by a radar.
- Caveats one must be aware of in applying the weather radar equation.

5.1 Introduction

A radar transmits pulses of microwave energy of known power. The microwaves strike objects in their path, all of which scatter energy back toward the radar. The radar antenna intercepts this energy and the radar receiver measures the power returned to the radar. The objects intercepting radar energy are generically called **targets**. Our first challenge in this chapter is to develop a generic **radar equation**, which relates the parameters of a radar system to a single target's microwave scattering characteristics and its location with respect to the radar. Our second and more important task is to develop the **weather radar equation**, which relates the parameters of a radar system to the scattering properties of an *ensemble* of spherical raindrops and their location with respect to the radar. The latter equation contains a term called the **radar reflectivity factor**, the most important quantity in radar meteorology and the parameter

Radar Meteorology: A First Course, First Edition. Robert M. Rauber and Stephen W. Nesbitt.
© 2018 John Wiley & Sons Ltd. Published 2018 by John Wiley & Sons Ltd.
Companion website: www.wiley.com/go/Rauber/RadarMeteorology

appearing on standard radar displays used by meteorologists, the media, and the public throughout the world.

5.2   Radar equation for a solitary target

Deriving the radar equation for a radar illuminating a solitary target, whether it is a raindrop, a bird, or a bumblebee, involves three steps:

1. Determining the **power flux density** (the radiated power per unit area) incident on a target;
2. Determining the power flux density scattered back toward the radar;
3. Determining the amount of backscattered power collected by the radar antenna.

5.2.1   Power flux density incident on a target

Consider a radar that transmits energy in a rectangular pulse of length, l, with peak power, P_t, and illuminates a target centered in space at a point (α, β, r) from the antenna, where α and β are the radar antenna's elevation and azimuth angles and r is the radial distance to the target. We will implicitly assume that (1) the dimensions of the target are much smaller than the pulse length; (2) the target is sufficiently far from the radar that electromagnetic waves passing the target appear as plane waves; and (3) the waves incident on the target are linearly polarized. These three assumptions hold exceptionally well for meteorological radars. We will also assume that absorption of microwaves and/or scattering of microwaves out of the beam (collectively called **attenuation**) does not occur in the medium between the radar and the target. This assumption is valid in nearly all situations at S-band and is invalid in nearly all situations for short-wavelength radars such as K- and W-band. We will consider attenuation effects, and how we might take advantage of them to deduce more meteorological information, in a later chapter.

We will first consider an **isotropic antenna** that is **lossless**; that is, the antenna radiates the power generated at the transmitter *uniformly in all directions* without loss of power due to absorption of energy in the waveguide or other radar components between the transmitter and the antenna (Figure 5.1). For such an isotropic antenna, the power flux density, S_{iso}, incident on the target would be

$$S_{iso} = \frac{P_t}{4\pi r^2} \tag{5.1}$$

where power flux density has standard SI units of watts per square meter. Isotropic antennas can provide no information about the location of a target. To determine the angular coordinates of a target (α, β) relative to the radar position, the energy must be focused into a narrow beam, a task accomplished by the radar's antenna.

Recall from Chapter 3 that a radar antenna focuses power in the direction the antenna points. However, even with the best antennas, some power is still distributed over the spherical volume surrounding the antenna. An antenna's **gain function**, $G_{\theta,\phi}$, is defined as the ratio of the power flux density at angular coordinates (θ, ϕ) to the power flux density that would be incident at angular coordinates (θ, ϕ) if the antenna

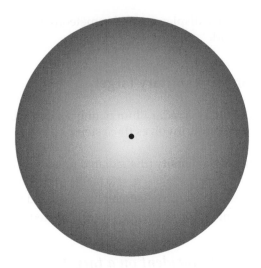

Figure 5.1   An isotropic antenna radiates energy equally in all directions

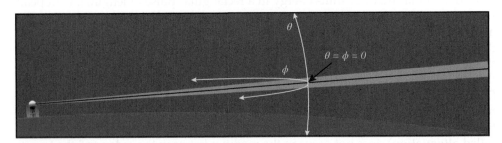

Figure 5.2   The angular coordinates θ and ϕ are measured in the azimuthal and elevation directions, respectively, relative to the beam axis, which itself is defined as the direction the antenna has maximum gain

were a lossless isotropic antenna radiating the same power. The gain function maps the power distribution for a specific antenna. As illustrated in Figure 5.2, the angular coordinates θ and ϕ are measured in the azimuthal and elevation directions relative to the **beam axis**, which itself is defined as the direction the antenna has maximum gain. The **gain** of an antenna, G, refers to the value of $G_{\theta,\phi}$ at $\theta = \phi = 0$, that is, along the beam axis.

The power flux density incident on a target, S_i, at range, r, for a radar with maximum antenna gain, G, can be calculated as

$$S_i = \frac{GP_t}{4\pi r^2} \tag{5.2}$$

We have implicitly assumed here that the target is sufficiently small that the power flux density does not vary across it. This is certainly true for any meteorological radar illuminating a single raindrop. The magnitudes of the quantities in (5.2) can be illustrated by considering a WSR-88D radar. Suppose a WSR-88D illuminates a single raindrop with diameter 0.001 m (1 mm) at 50 km range. The WSR-88D transmits $P_t = 7.5 \times 10^5$ W (750 kW) and has a gain, $G = 45.5$ dB $= 10^{4.5} = 31662.7$. From (5.2),

$S_{inc} = 0.84\,\text{W m}^{-2}$. The raindrop, which has cross-sectional area $\left(\frac{\pi D^2}{4}\right)$ much smaller than a square meter, would only intercept $5.3 \times 10^{-7}\,\text{W}$.

5.2.2 Power flux density scattered back to the radar

A target's **radar cross section** is defined as the ratio of the power flux density scattered by the target to the power flux density incident on the target. However, as targets are located at remote distances from the radar, it is not possible to measure the power flux density scattered by a target at the target's location. We can, however, measure the power flux density scattered by a target at the radar's antenna, provided that there is no absorption or scattering of the energy in the atmosphere between the target and the antenna. The radar cross section is therefore defined mathematically as

$$\sigma = 4\pi r^2 \left(\frac{S_r}{S_i}\right) \tag{5.3}$$

where S_r is the power flux density *at the receiving antenna* and the factor $4\pi r^2$ accounts for the fact that the power scattered by the target is spread over a sphere of area $4\pi r^2$ as it propagates from the target back to the radar antenna. Implicit in this definition is that the scattering processes are linear; that is, the backscattered power flux density from a target is proportional to the incident power flux density, also at the target. For scattering processes important to radar meteorology, linearity is a good assumption.

In general, the microwave scattering properties of radar targets are complicated and depend on a target's size relative to the radar wavelength, shape, viewing aspect, composition, and conductivity. The radar cross section of a very complicated object such as an aircraft, for example, changes substantially with viewing angle. The strange angular shapes of the B2 Stealth Bomber (Figure 5.3) were designed to minimize the plane's radar cross section, and thus detectability with radar.

Figure 5.3 The Northrop Grumman B-2 Spirit is an American strategic bomber, designed to avoid radar detection and be able to penetrate dense anti-aircraft defenses (U.S. Air Force)

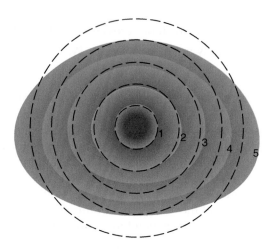

Figure 5.4 Distortion of a raindrop as a function of diameter (mm) for a raindrop falling in still air. (Adapted from Beard, K.V. and Chuang, C. (1987) A new model for the equilibrium shape of raindrops. *J. Atmos. Sci.*, **44**, 1509–1524. © the American Meteorological Society, used with permission)

Meteorological targets are composed of water, ice, or both and include cloud droplets and raindrops, ice crystals, snowflakes composed of ice crystal aggregates, ice pellets, frozen raindrops, graupel, and hailstones. Some of the ice particles may be partially melted. Water droplets with diameters smaller than about 1 mm remain spherical because of surface tension. The drag force on larger raindrops causes them to distort into oblate spheroids as they fall (Figure 5.4). Ice crystals have a wide variety of shapes that depend on the temperature and saturation conditions in which they grow (Figure 5.5). Frozen raindrops, graupel, and some hailstones tend to be quasi-spherical, although other shapes are common (Figure 5.6). Many of these meteorological objects have complex shapes and therefore complicated radar cross sections.

5.2.3   Backscattered power collected by the radar antenna

The power flux density arriving at the radar antenna can be obtained by solving (5.3) for S_i, substituting that into (5.2), and solving for S_r to give

$$S_r = \frac{\sigma G P_t}{16\pi^2 r^4} \tag{5.4}$$

The **effective area** of the radar antenna, A_e, is defined as the area, which when multiplied by the incoming power flux density, gives the total received power. The effective area of an antenna, essentially the area that is effective in collecting power from an incoming plane wave, is always less than the projected area of the antenna aperture for meteorological radars. Typically, A_e is about half the aperture area. Antennas with substantial losses have even smaller A_e.

The total received power, P_r, can be obtained by multiplying (5.4) by the effective area to give

$$P_r = S_r A_e = \frac{P_t G \sigma A_e}{16\pi^2 r^4} \tag{5.5}$$

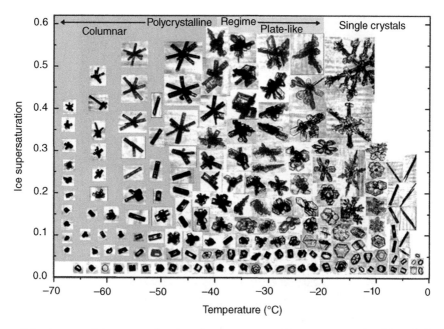

Figure 5.5   Ice crystal shapes as a function of temperature and ice supersaturation (From Bailey, M.P. and Hallett, J. (2009): A comprehensive habit diagram for atmospheric ice crystals: confirmation from the laboratory, AIRS II, and other field studies. *J. Atmos. Sci.*, **66**, 2888–2899, © the American Meteorological Society, used with permission)

Furthermore, antenna theory shows that the antenna gain and effective area are related by

$$G = \frac{4\pi A_e}{\lambda^2} \tag{5.6}$$

where λ is the radar wavelength. Solving for A_e and substituting (5.6) into (5.5), we obtain the radar equation for a solitary target

$$P_r = \frac{P_t G^2 \lambda^2 \sigma}{64\pi^3 r^4} \tag{5.7}$$

5.2.4   Implications of the radar equation

To understand this equation, it is helpful to write it by first collecting terms together that represent characteristics of the radar and the target, as shown below:

Radar   Target

$$P_r = \frac{1}{64\pi^3}[P_t G^2 \lambda^2]\left[\frac{\sigma}{r^4}\right] \tag{5.8}$$

We can also use (5.6) to write (5.7) in a different form:

Radar   Target

$$P_r = \frac{1}{4\pi}\left[\frac{P_t A_e^2}{\lambda^2}\right]\left[\frac{\sigma}{r^4}\right] \tag{5.9}$$

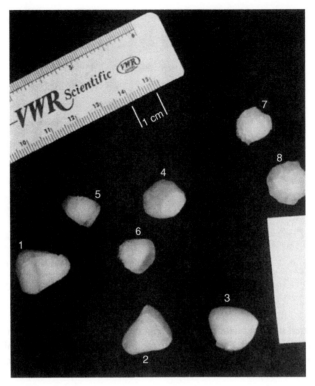

Figure 5.6 Conical and spherical hailstones (From Matson, R.J., Huggins, A.W. (1980) The direct measurement of the sizes, shapes and kinematics of falling hailstones. *J. Atmos. Sci.*, **37**, 1107–1125, © the American Meteorological Society, used with permission)

The last term in parentheses in (5.8) and (5.9) represents properties of the target and show that the power received at the radar falls off rapidly as an isolated target moves further in range. Because of the fourth power dependence, doubling the range decreases the received power by a factor of 16. Characteristics of the target, such as shape and orientation, influence the radar cross section and therefore the received power.

The quantities in the first bracket in each equation are characteristics of the radar. Increasing the transmitted power, or using a larger antenna with correspondingly higher gain, will increase the power returned to the radar. The wavelength dependence is not as obvious—one form of the equation has the square of the wavelength in the numerator and the other in the denominator. The wavelength dependence is indeed complicated because P_t, G, and σ are all wavelength-dependent quantities. From (5.9), for a fixed antenna size, decreasing the wavelength increases the gain, and therefore the returned power (provided a corresponding decrease does not occur in the radar cross section). Continuing with our example of the WSR-88D, recall that the WSR-88D transmits $P_t = 7.5 \times 10^5$ W (750 kW) and has a gain $G = 45.5$ dB $= 10^{4.5} = 35{,}481$. The WSR-88D operates at wavelength $\lambda = 0.1$ m (10 cm). Let our target this time be a single large hailstone with $\sigma = 0.001$ m^2 located at $r = 100$ km. Then

$$P_r = \frac{1}{64\pi^3}[P_t G^2 \lambda^2]\left[\frac{\sigma}{r^4}\right] = \frac{1}{64\pi^3}[(7.5 \times 10^5\,\text{W}) \times (35481)^2(10^{-1}\,\text{m})^2]\left[\frac{10^{-3}\,\text{m}^2}{(10^5\,\text{m})^4}\right] \quad (5.10)$$

$$P_r = 4.75 \times 10^{-14}\,\text{W} \qquad (5.11)$$

This amount of power is far below the noise threshold of the WSR-88D and the hailstone would go undetected. Fortunately, precipitation particles occur in large numbers rather than singly. In fact, the radar equation for solitary targets has little applicability in radar meteorology with one exception—radar calibration. One specific calibration technique involves placing metal spheres of known cross section at fixed distances from a radar. The transmitted and received power is measured by a device inserted into the directional coupler to determine the effective antenna gain. The measurements provide a way to account for both power losses due to absorption of radiation within the hardware beyond the transmitter and any defects that might exist in the antenna that impact its gain.

5.3   Radar equation for a distributed target

Raindrops, snowflakes, and hailstones (also insects) occur in enormous numbers. These airborne objects normally populate volumes of atmosphere much larger than the smallest volume of space observable with a meteorological radar. When considering how much power is returned to the radar from a region of the atmosphere at an instant in time, we must take into account the contributions from each of the targets within the contributing volume of space. The objects, collectively, are called a **distributed target** because they are distributed throughout a volume of space.

5.3.1   The contributing volume for distributed targets

The **contributing volume** for a distributed target is defined as the volume containing all of the objects from which scattered microwaves arrive back at the radar simultaneously. The length of the contributing volume in the radial direction can be determined by considering a rectangular pulse transmitted by a radar for a duration of $t = \tau$ microseconds. The pulse travels outward at the speed of light, c. At some later time, $t = \Delta t$, wave energy scattered by an ensemble of objects somewhere along the beam path arrives instantaneously back at the radar. Objects contributing to the wave energy from the leading edge of the pulse must have been located at range $r_2 = c\Delta t/2$, whereas objects contributing to wave energy from the trailing end of the pulse must have been located at $r_1 = c(\Delta t - \tau)/2$. The radial dimension of the contributing volume is determined by subtracting r_1 from r_2, specifically.

$$r_2 - r_1 = \left(\frac{c\Delta t}{2} - \frac{c(\Delta t - \tau)}{2}\right) = \frac{c\tau}{2} = \frac{l}{2} \qquad (5.12)$$

exactly one-half the pulse length, l. For example, the pulse duration of the WSR-88D in its short pulse mode is 1.57 μs. Multiplying the pulse duration by the speed of light produces a pulse volume whose pulse length $l = 500$ m. Therefore, the radial extent of the contributing volume for this mode of the WSR-88D is $l = 250$ m.

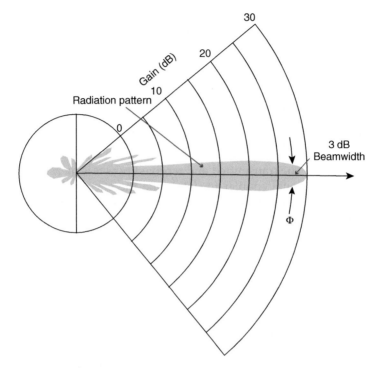

Figure 5.7   The radiation pattern of a directional antenna and the 3 dB beamwidth

We learned in Chapter 3 that a radar antenna focuses energy into a narrow beam, although some energy radiates off beam, primarily in sidelobes. By convention, the angular width of the main lobe, called the **beamwidth**, is defined as the angle between the two directions in a principal plane where the antenna gain function is one-half, or 3 dB less than, its maximum value (Figure 5.7). Beamwidths must be expressed in radians for numerical calculations; although in verbal or written communication, the values are normally expressed in degrees. The beamwidth of the WSR-88D, for example, is 0.95°. For beams that do not exhibit circular symmetry, the symbols Θ and Φ are used for the beamwidths in the horizontal (azimuth) and vertical (elevation) principal planes, respectively. Nearly all meteorological radar beams exhibit circular symmetry, so $\Theta = \Phi$. Henceforth, we will use the symbol Φ to represent beamwidth and assume circular symmetry.

Figure 5.8 shows the contributing volume within a conical beam. The contributing volume consists of a segment of a spherical shell bounded within the beamwidth of the conical beam. We can approximate the volume of this shell by assuming that it has the shape of a cylinder, rather than a segment of a tapered cone. The diameter of the cylinder is given by

$$D_{\mathrm{cyl}} = r\Phi \tag{5.13}$$

and the area of the base of the cylinder is

$$A_{\mathrm{cyl}} = \frac{\pi r^2 \Phi^2}{4} \tag{5.14}$$

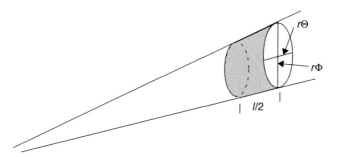

Figure 5.8　The contributing volume of a conical beam

The approximate volume of the contributing region is therefore

$$V_c = \frac{\pi l \Phi^2 r^2}{8} = \frac{\pi c \tau \Phi^2 r^2}{8} \tag{5.15}$$

How large is this volume? For a WSR-88D radar where $l = 250$ m and $\Phi = 0.95°$, at a range of 50 km, the contributing volume is

$$V_c = \frac{\pi(250\,\text{m})\left(0.95 \times \frac{2\pi}{360}\right)^2 (50,000\,\text{m})^2}{8} = 6.75 \times 10^7\,\text{m}^3 = 0.067\,\text{km}^3 \tag{5.16}$$

The concentration of raindrops in a typical rain shaft might be 1 raindrop per cubic meter. From (5.16), the contributing volume in that case would contain over 67 million raindrops!

5.3.2　The radar cross section of distributed targets

The radar cross section of a distributed target presents a more challenging issue than the case of a single target. The problem arises from the fact that electromagnetic waves scattered from objects within the contributing volume undergo constructive or destructive interference with one another. The resulting wave that arrives at the radar from the countless objects within the contributing volume is the **phasor sum** of all the individual backscattered waves (see Figure 1.8). Furthermore, interference between backscattered waves from each object in the contributing region depends on the relative position of the objects, and these positions change from pulse to pulse due to wind shear, turbulence, and the different rates at which the particles fall through the atmosphere. Even if the contributing volume contained the identically same particles, the backscattered power flux density will change from pulse to pulse because particles shuffle about between pulses and the interference effects change. In effect, the radar cross section of the particle ensemble changes from pulse to pulse. What is the average radar cross section, $\bar{\sigma}$, of an ensemble of objects? More importantly, is the radar cross section of a *distributed* target equal to the sum of the *individual* radar cross sections of the objects comprising the target?

Consider a radar that transmits an electromagnetic wave with an electric field given by

$$E(t) = E_0 e^{i\omega t}, \quad \text{for } 0 \le t \le \tau \tag{5.17}$$

so that a wave incident on the jth particle at range r_j is

$$E(r_j, t) = E_{inc}e^{i\omega\left(t-\frac{r_j}{c}\right)}, \quad \text{for } \frac{r_j}{c} \le t \le \frac{r_j}{c} + \tau \tag{5.18}$$

The backscattered electric field at the radar from the jth particle will be proportional to the amplitude of the incident wave and inversely proportional to the range.

$$E_j = \rho_j\frac{E_{inc}}{r_j}e^{i\omega\left(t-\frac{2r_j}{c}\right)}, \quad \text{for } \frac{2r_j}{c} \le t \le \frac{2r_j}{c} + \tau \tag{5.19}$$

where ρ_j, the constant of proportionality, will be determined later. The backscattered field from the object ensemble is therefore

$$E_r = \sum_j E_j = e^{i\omega t}\sum_j \rho_j\frac{E_{inc}}{r_j}e^{\left(\frac{i\omega 2r_j}{c}\right)} \tag{5.20}$$

Substituting the following relationship

$$\frac{\omega}{c} = \frac{2\pi}{\lambda} \tag{5.21}$$

in (5.20), we obtain.

$$E_r = \sum_j E_j = e^{i\omega t}\sum_j \rho_j\frac{E_{inc}}{r_j}e^{\left(\frac{i\pi 4r_j}{\lambda}\right)} \tag{5.22}$$

Noting that the power flux density is proportional to the square of the electric field, and using complex conjugate notation (*), we can then write

$$E_r E_r^* = \frac{E_{inc}E_{inc}^*}{r^2}\sum_j \rho_j e^{-i4\pi r_j/\lambda}\sum_k \rho_k e^{+i4\pi r_j/\lambda} \tag{5.23}$$

or

$$S_r = \frac{S_{inc}}{r^2}\sum_j\sum_k \rho_j\rho_k e^{-i4\pi(r_j-r_k)/\lambda} \tag{5.24}$$

Breaking (5.24) into two terms

$$S_r = \frac{S_{inc}}{r^2}\left(\sum_j \rho_j^2 + \sum_j\sum_{j\ne k} \rho_j\rho_k e^{-i4\pi(r_j-r_k)/\lambda}\right) \tag{5.25}$$

where in (5.25) the terms for which $j=k$ have been collected separately. We can determine the value of ρ_j by considering the special case where there is only one object. Then, (5.25) reduces to

$$S_r = \frac{S_{inc}\rho_1^2}{r^2} \tag{5.26}$$

Comparing to (5.3), we can deduce that

$$\rho_1^2 = \frac{r^2 S_r}{S_{inc}} = \frac{\sigma_1}{4\pi} \tag{5.27}$$

So (5.25) becomes

$$S_r = \frac{S_{inc}}{4\pi r^2} \left(\sum_j \sigma_j + \sum_{j\neq k} \sum \sqrt{\sigma_j}\sqrt{\sigma_k} e^{-i4\pi(r_j - r_k)/\lambda} \right) \tag{5.28}$$

If we only consider the first term, (5.28) can be written as

$$\overline{S_r} = \frac{S_{inc}}{4\pi r^2} \sum_j \sigma_j \tag{5.29}$$

from which we would conclude that

$$\overline{\sigma} = \sum_j \sigma_j \tag{5.30}$$

that is, the radar cross section of a distributed target is the sum of the radar cross sections of the individual objects composing the target. The problem is the second term, which quantifies the interference effects. This term is nonzero and leads to pulse-to-pulse fluctuations in the power flux density.

Fortunately, there is a way to minimize the effects of this term. If we consider a contributing volume full of raindrops, the spacing between each raindrop across the volume and each of its individual neighbors will be some fraction of a wavelength and will therefore lead to constructive or destructive interference with electromagnetic waves scattered by the neighboring raindrops. The second term in (5.28) might be positive (net constructive interference) or negative (net destructive interference) depending on the interference effects. If we leave enough time between pulses that raindrops in the volume can shuffle around to arrive at a new independent arrangement, and then take a second sample of the same volume with a new pulse, the interference patterns will be completely different, and again, the second term might be positive or negative. If we take enough independent samples, and then *average the measurements*, the negative and positive contributions of the second term in each sample cancel each other and the effect of the second term diminishes to zero. Effectively, by averaging a number of pulses, the radar cross section of a distributed target reduces to the sum of the radar cross sections of the individual objects composing the target, and (5.30) holds. The delay between pulses required for independent samples is called the **time to independence** or **decorrelation time**.

The decorrelation time depends on the wavelength of the radar, the particle size distribution, and the shear and turbulence in the sample volume. Particles need to shuffle a shorter distance to move an appreciable fraction of a wavelength for short-wavelength radars. The particle size distribution determines the spread in particle terminal velocities, and therefore the rate at which particles shift relative to one another. Shear and turbulence act to shuffle particles about relative to each other. Studies of decorrelation suggest that decorrelation occurs in a time between 2λ and 3λ, if λ is expressed in centimeters and time in milliseconds. For a K-band radar, the decorrelation time can be as short as 2 ms, whereas for an S-band radar, it might be as long as 30 ms. Practically, this puts limits on how fast an antenna should turn. For example, a typical turn rate of an S-band antenna might be $16°s^{-1}$. For a beamwidth of $1°$, the radar would sweep one pulse volume in 0.06 s. A pulse repetition frequency of $1000\,s^{-1}$ would yield 60 pulses during a $1°$ sweep. Of these, $60\,ms/30\,m = 2$ pulses would sample completely independent arrangements of particles. The other pulses would be partially correlated. This is not so bad, since (as we shall see in the next chapter) Doppler radars require correlated measurements to deduce the Doppler frequency.

5.3.3   The radar equation for a distributed target

The radar equation for a distributed target can be derived by substituting (5.30) into the radar equation for a solitary target (5.8) to give

$$\overline{P}_r = \frac{1}{64\pi^3}[P_tG^2\lambda^2]\left[\frac{\sum_j \sigma_j}{r^4}\right] \tag{5.31}$$

This equation is not quite accurate because the gain function varies across a distributed target. Generally, it is assumed that the gain function between the half power points of the main lobe conforms to a Gaussian distribution. Under this assumption, the correction to (5.30) involves inserting 2(ln 2) in the denominator of the leading constant:

$$\overline{P}_r = \frac{1}{64(2\ln(2))\pi^3}[P_tG^2\lambda^2]\left[\frac{\sum_j \sigma_j}{r^4}\right] \tag{5.32}$$

The summation of the radar cross sections of the particles spread through the contributing region is normally expressed using a quantity called the **radar reflectivity**. The radar reflectivity, η, is related to the radar cross section of a distributed target by

$$\overline{\sigma} = \sum_j \sigma_j = V_c\left(\sum_j \frac{\sigma_j}{V_c}\right) = V_c\eta \tag{5.33}$$

where V_c is the contributing volume. The radar reflectivity has units of m^{-1}. It is important here to emphasize that the radar reflectivity, η, is *not* the quantity shown on standard radar displays. That quantity is called the radar reflectivity factor. We will define that quantity in the next section. Substituting (5.33) and the expression for contributing volume in (5.15) into (5.32) we obtain.

Radar   Target

$$\overline{P}_r = \frac{c}{1024(\ln 2)\pi^2}[P_tG^2\lambda^2\tau\Phi^2]\left[\frac{\eta}{r^2}\right] \tag{5.34}$$

The above equation is the generic equation for a distributed target (regardless of the type of objects that compose the target). The equation indicates that the power received at the radar can be increased by increasing the pulse duration and/or the beamwidth, the effect of each being that more objects are included within the pulse volume. The disadvantage, of course, is that the pulse volume is larger and resolution is sacrificed. The power decreases with the square of range, rather than the fourth power of range in the equation for a solitary target, a difference again attributable to the pulse volume including more targets as it grows in volume with distance from the radar.

5.4   The weather radar equation

The **weather radar equation** is based on two critical assumptions about the nature of precipitation particles within the contributing volume. These are (1) the particles consist entirely of dielectric spheres and (2) the particles are sufficiently small relative to the wavelength of the radar that **Rayleigh scattering theory** can be used to describe their scattering behavior. We will discuss the problems that can arise with these two assumptions later in the section. For now, we will proceed forward assuming that the two assumptions hold true.

5.4.1   Radar cross section of a small dielectric sphere

If a spherical particle is small relative to the wavelength of the radar, the electric field of the incident electromagnetic wave can be considered to be uniform over it. A uniform electric field over a homogeneous dielectric sphere induces an electric **dipole**—two equal and opposite charges separated by a distance, with the electric field within the particle parallel to the incident electric field as shown in Figure 5.9. The radiation field of the oscillating dipole induced within the sphere by the incident wave is the scattered electric field. We are interested in the magnitude of the electric field in the backscattered direction. According to Rayleigh theory, the magnitude of the electric field in the plane perpendicular to the direction of polarization of the incident wave (this plane includes the backscattered direction) at range r from the sphere is given by

$$E_r = \frac{\pi^2 K D^3 E_{inc}}{2\lambda^2 r} \tag{5.35}$$

where D is the diameter of the particle and K is given by

$$K = \frac{\varepsilon_r - 1}{\varepsilon_r + 2} \tag{5.36}$$

Here, ε_r is the relative dielectric constant of the sphere, a dimensionless quantity that can be thought of as a measure of a substance's relative ability to store charge. The dielectric constant is a complex number, as the material composing the sphere can both scatter and absorb energy. The radar cross section can be obtained by dividing

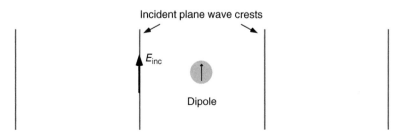

Figure 5.9   A dipole induced in a small spherical drop as a polarized plane wave propagates past the drop

both sides of Eq. (5.35) by E_{inc}, squaring the result to obtain the ratio of power flux densities and then multiplying both sides by $4\pi r^2$ to obtain

$$\sigma = 4\pi r^2 \frac{S_r}{S_{inc}} = \frac{\pi^5 |K|^2 D^6}{\lambda^4} \tag{5.37}$$

The absolute value signs for K are required to account for the possibility that K is complex. The value of $|K|^2$ is about 0.93 for water and 0.176 for ice. Note that the radar cross section of a small sphere is proportional to the sixth power of its diameter. This has enormous consequences for our meteorological interpretation of radar data. A 2 mm diameter raindrop, for example, will scatter back 64 times the amount of energy as a 1 mm raindrop. The radar cross section is inversely proportional to the fourth power of the radar wavelength. Decreasing the radar wavelength by a factor of 2 (from S to C band, e.g.) increases the radar cross section by a factor of 16. We will refer back to these two critical points about the radar cross section throughout the book.

5.4.2   The radar reflectivity factor

If a radar is sampling rain consisting of spherical droplets meeting the Rayleigh criteria, the radar cross section for the ensemble of spherical particles can be expressed as

$$\bar{\sigma} = \sum_j \sigma_j = \frac{\pi^5}{\lambda^4} |K|^2 \sum_j D_j^6 \tag{5.38}$$

where the subscript j denotes each of the drops within the contributing volume and the sum is over all the drops in the volume. In weather radar, it is essentially always assumed that the precipitation particles in the contributing region are water (a bad choice in winter), so the dielectric factor is taken outside the summation. If we divide by the contributing volume, we obtain the radar reflectivity

$$\eta = \frac{\sum_j \sigma_j}{V_c} = \left(\frac{\pi^5}{\lambda^4}\right) |K|^2 \frac{\sum_j D_j^6}{V_c} \tag{5.39}$$

The **radar reflectivity factor**, Z, is defined as

$$Z = \frac{\sum_j D_j^6}{V_c} \tag{5.40}$$

This quantity, the sum of the sixth power of the diameters of all of the raindrops within the contributing region divided by the volume of the contributing region, is the most important quantity in weather radar. It is the quantity, expressed in logarithmic units (dBZ), that you view when you look at radar displays (Figure 5.10). The radar reflectivity factor's basic units are $m^6/m^3 = m^3$, but it is conventionally expressed in units of $mm^6\ m^{-3}$, millimeters being the typical dimensions of raindrops and meters the dimensions of the contributing volume. The radar reflectivity factor is normally expressed in logarithmic units because it can vary by many orders of magnitude in nature. As a logarithm requires a unitless number, Z is normalized by $1\ mm^6\ m^{-3}$ so that.

$$dBZ = 10 \log\left(\frac{Z}{1\,mm^6\,m^{-3}}\right) \tag{5.41}$$

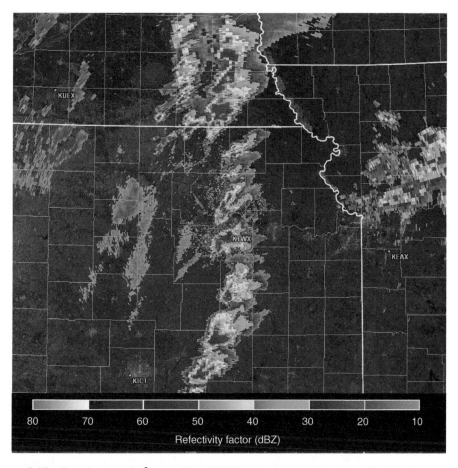

Figure 5.10   Data from the 0.5° scan of the KTWX radar showing the radar reflectivity factor during an outbreak of supercell thunderstorms

For example, when $Z = 10^{-2}$ mm^6 m^{-3}, the reflectivity factor is -20 dBZ, $Z = 1$ mm^6 m^{-3} is equivalent to 0 dBZ, and $Z = 10^4$ mm^6 m^{-3} is equivalent to 40 dBZ. The decibel notation (dB) is commonly used for the reflectivity factor. It is important to remember that when expressed in this manner, we mean "decibels with respect to a reference level of 1 mm^6 m^{-3}."

5.4.3   The weather radar equation

To convert the radar equation for distributed targets to the weather radar equation, we substitute (5.39) into (5.34) to obtain

$$\overline{P}_r = \frac{\pi^3 c}{1024(\ln 2)} \left[\frac{P_t G^2 \tau \Phi^2}{\lambda^2} \right] \left[\frac{|K|^2 Z}{r^2} \right] \tag{5.42}$$

and solve (5.42) for Z to give:

$$\text{Radar}\qquad\text{Target}$$

$$Z = \frac{1024(\ln 2)}{\pi^3 c}\left[\frac{\lambda^2}{P_t G^2 \tau \Phi^2}\right]\left[\frac{r^2 \overline{P_r}}{|K|^2}\right] \tag{5.43}$$

Equation (5.43) is called the weather radar equation. It relates the properties of the radar (wavelength, transmitted power, gain, pulse duration, and beam width) and properties of the target (composition, location, and the measured backscattered power) to the radar reflectivity factor. The radar reflectivity factor, in turn, provides a critical piece of information related to the raindrop drop size distribution because it is the sum of the sixth power of the diameters of the particles composing that distribution divided by the contributing volume.

Finally, it is important to account for the power losses that occur during transmission of the signal through the hardware including the waveguide, directional coupler, and antenna. These losses can be quantified in the radar equation by adding the term L in the numerator such that

$$\overline{P}_r = \frac{\pi^3 c}{1024(\ln 2)}\left[\frac{P_t G^2 \tau \Phi^2 L}{\lambda^2}\right]\left[\frac{|K|^2 Z}{r^2}\right] \tag{5.44}$$

Radar calibrations, as discussed earlier, are made by inserting a device to measure power at the directional coupler. The value of L can be determined for specific radar systems from these calibrations.

For a given radar, (5.44) can be simplified provided that pulse duration and transmitted power is maintained as constant. If the targets are assumed to be water, (5.43) can be written as

$$Z = C_r r^2 \overline{P}_r \tag{5.45}$$

where C_r is referred to as the **radar constant**, for that particular radar. It is worth noting here that the term *radar reflectivity* is commonly used by meteorologists, educators, the public, and even casually by members of the radar meteorology community to describe the radar reflectivity factor. The radar reflectivity, η, and the radar reflectivity factor, Z, are different quantities with different units. In meteorology, in virtually all cases, the radar reflectivity factor is the variable of interest—regardless of the terminology used.

5.4.4   The validity of the Rayleigh approximation

The weather radar equation was derived based on a number of assumptions about the nature of the targets and the characteristics of the radar radiating them. A key assumption was that Rayleigh scattering was applicable; that is, the scattering elements were homogeneous dielectric spheres with diameters small relative to the wavelength of the radar.

Table 5.1 lists various **hydrometeors** found in the atmosphere. Depending on the wavelength of the radar and the type of particle, only some strictly meet the

Table 5.1 Types of hydrometeors in cloud in relation to the Rayleigh criteria for size and shape

Hydrometeors / Diameter range	Cloud droplets 0.001–0.01 cm	Drizzle drops 0.01–0.05 cm	Raindrops 0.05–0.5 cm	Ice crystals 0.001–0.5 cm	Snowflakes 0.5–5 cm	Graupel 0.1–1 cm	Small hail 1–2 cm	Large hail 2–5 cm	Giant hail >5 cm
S-band (10 cm)	Yes	Yes	Yes	Yes	Mostly	Yes	Yes	Mostly	No
C-band (5 cm)	Yes	Yes	Yes	Yes	Sometimes	Mostly	Sometimes	No	No
X-band (3 cm)	Yes	Yes	Yes	Yes	Sometimes	Sometimes	No	No	No
K-band (1 cm)	Yes	Sometimes	Sometimes	Sometimes	No	Rarely	No	No	No
W-band (0.3 cm)	Yes	Sometimes	No	Sometimes	No	No	No	No	No
Shape	Spheres	Spheres	<1 mm = spheres; >1 mm = spheroids	Not spherical (exception: frozen drops)	Not Spherical	Spheres and cones	Quasi-spherical	Spherical and other shapes	Spherical and other shapes

Rayleigh criteria at different wavelengths. The Rayleigh approximation for particle size is valid at S-band for all but large hail. At progressively shorter wavelengths, fewer and fewer hydrometeors satisfy the Rayleigh approximation. At W-band, only cloud droplets, small drizzle drops, and small ice particles are sufficiently small to qualify as Rayleigh scatterers. For this reason—and the dependence of the radar cross section on λ^{-4}, which gives short-wavelength radars a distinct advantage observing cloud particles—W-band radars are often called "cloud radars." Short-wavelength radars also suffer from attenuation. For these reasons, short-wavelength radars are best to study cloud processes, and not very useful to map out the distribution of precipitation.

The criteria that the hydrometeors be spherical is only met by small raindrops, some frozen raindrops, and some graupel and hail. Larger raindrops become distorted into oblate spheroids (a hamburger shape). Ice particles have an enormous variety of shapes, and snowflakes tend to be flattened in the horizontal dimension. Graupel can sometimes be conical, and hail often has lobes of ice growing on one side or the other.

Complicating the situation more is that many particles in the atmosphere are mixed phase, particularly near the melting level. Mixed phase particles also occur in hail shafts, particularly where hail is growing rapidly. Hail falling below the melting level can also develop a water coating.

Finally, in practice, many targets observed with meteorological radars are non-meteorological. In summer, the lower atmosphere is full of insects that contribute to the returned power. Point targets such as aircraft and birds are routinely sampled by meteorological radars. Ground targets, such as trees, buildings, and wind turbines, complicate the radar signal.

In practice, none of these complexities can be taken into account. Operational radars (and nearly all research radars) apply the weather radar Eq. (5.43) assuming Rayleigh scattering, implicitly recognizing that the criteria for Rayleigh scattering is unlikely to be met in many cases. When it is obvious that the Rayleigh criteria will not be met, radar meteorologists often use the term **equivalent radar reflectivity factor**, Z_e, to describe the quantity derived from power measurements and (5.43).

5.5   Summary

Despite the many assumptions required to derive the weather radar equation, and the caveats about Rayleigh scattering, the radar reflectivity factor has been demonstrated to have tremendous application in meteorology. It is critical to estimating rainfall rates and total rainfall, identifying where flash floods are possible, pinpointing locations where hail and tornadoes are likely, and mapping out the location, structure, intensity, and movement of both small and large storm systems. The radar reflectivity factor has become an essential tool in the arsenal of meteorologists—and remains today one of the most important quantities used in mesoscale meteorological research and applications.

Important terms

attenuation

beam axis

beamwidth

contributing volume

decorrelation time

dipole

distributed target

effective area

equivalent radar reflectivity factor

gain

gain function

hydrometeors

isotropic antenna

lossless

phasor sum

power flux density

radar constant

radar cross section

radar equation

radar reflectivity

radar reflectivity factor

Rayleigh scattering theory

target

time to independence

weather radar equation

Review questions

1. Go through the chapter and make a list of every assumption made in deriving the weather radar equation. Under what conditions, if any, do you consider each assumption invalid?

2. Antenna gain is normally quoted in decibels. If an antenna's maximum gain is 35 dB, by what factor is the power increased at the center of the beam compared to an isotropic antenna?

3. Suppose a radar has a beamwidth of one degree and operates with a pulse length of 500 m. Compare the size of the contributing volume at distances of 20, 50, and 100 km from the radar.

4. Why should a radar use a number of pulses, rather than a single pulse, to measure the radar reflectivity of a distributed target?

5. List the advantages and disadvantages of using a short-wavelength radar to measure the radar reflectivity factor.

6. Suppose a research aircraft determines that raindrops in a particular cloud have an average spacing of 0.5 m in the direction of a radar beam from a ground-based radar. The mean wind measured at cloud level was $10\,\text{m s}^{-1}$, and turbulent fluctuations in the wind on the scale of the droplet spacing were $0.1\,\text{m s}^{-1}$. The updraft in the cloud is just the right speed to keep the drops suspended at the flight level so that they are neither falling nor rising. If the radar observing these drops operated at S-band, about how long would it take for the pulse-to-pulse signal from the drops to decorrelate (i.e., how much time would the drops require to shuffle about into an independent arrangement). Hint: Consider how long it takes a single drop to move a half wavelength.

7. What are the units commonly used for the radar reflectivity factor? If one were to express the radar reflectivity factor in standard SI units, what factor would the common units have to be multiplied by?

(continued)

8. What is the Rayleigh approximation? For what type of hydrometeors is the Rayleigh approximation violated at S-band?

9. Assume that two radars, one operating at X-band and the other at S-band, have identically aligned beams with identical gain functions. The hydrometeors both radars are observing are sufficiently small and spherical that they satisfy the Rayleigh criteria at both wavelengths. The radars both transmit a sufficient number of pulses that Eq. (5.29) holds true. Should the radar reflectivity factor measured by each radar be the same, or should it differ significantly from the other radar?

10. Examine today's composite radar display from the National Weather Service or another source. What is the range of the radar reflectivity factor in the various weather systems occurring today?

Challenge problems

11. An aircraft flies through a cloud with a device that measures the raindrop size distribution. The table below shows the measurements. To obtain this drop size distribution, the aircraft sampled 6.15 m³ of air.

    a. If this sample is representative of the contributing volume simultaneously sampled by a radar in the same location, what will the value of the radar reflectivity factor measured by the radar be (show work, and report your answer in dBZ).

    b. What conclusions can you draw about relationship between a raindrop distribution and the radar reflectivity factor?

Diameter range (mm)	Mid-point of range (mm)	Droplet concentration (m^{-3})
0.3–0.5	0.4	455
0.5–0.7	0.6	347
0.7–0.9	0.8	299
0.9–1.1	1.0	253
1.1–1.3	1.2	238
1.3–1.5	1.4	187
1.5–1.7	1.6	54
1.7–1.9	1.8	27
1.9–2.1	2.0	16
2.1–2.3	2.2	5
2.3–2.5	2.4	3
2.5–2.7	2.6	2
2.7–2.9	2.8	3
2.9–3.1	3.0	1
3.1–3.3	3.2	2

12. Using parameters for the WSR-88D radar provided in this chapter and your answer to part (a) of the previous problem, calculate the average power returned to the radar from a series of pulses if the sample in part (a) were collected at 60 km range from the radar. (Hint: Be very careful with units.)

13. Consider a monodisperse rain drop size distribution with $D = 500 \ \mu m$.

    a. Calculate Z for a concentration of $1 \ L^{-1}$.

    b. Assume that the drops suddenly freeze into perfect ice spheres.

        i. By how much will the returned power change?

        ii. If the radar operator continues to use the value of $|K|^2$ for water in the radar equation after the drops freeze, by how much will the radar reflectivity factor change when the drops freeze (give answer in dBZ).

14. A weather radar is transmitting 100 kW of power:

    a. If the radar has an isotropic antenna, what is the power flux density when a pulse passes 120 km from the radar?

    b. If the radar has a gain of 30 dB, what is the power flux density at the center of the beam at a distance of 120 km.

15. A mosquito flying across the center of a radar beam will have its internal temperature raised 4 °C if it is irradiated by $1000 \ W \ m^{-2}$, causing it to have heart failure. Assuming that the radar emits a peak power of 1 MW, and the maximum gain of the radar is 40 dB, within what radius from the radar will the poor mosquito be cooked?

6

Radial Velocity—The Doppler Effect

Objectives

By the end of this chapter, you should understand the following:

- The Doppler effect, and how a Doppler frequency shift occurs.
- What is meant by radial velocity, and the relationship between radial velocity and the Doppler frequency shift.
- Why Doppler radars use pulse-to-pulse phase changes to measure radial velocity.
- Why Doppler radars can only measure radial velocity within a finite range.
- The Doppler dilemma and techniques to circumvent it.
- What is meant by a Doppler spectrum.
- The relationship between a Doppler spectrum and its moments.
- The pulse-pair processor algorithm to obtain Doppler spectral moments.

6.1 Introduction

Have you ever heard a train blow its whistle as it approaches and then moves away from you—although the train's sound is the same, you hear a high pitch on approach and a lower pitch on departure. On approach, sound waves are compacted because they and the train are moving in the same direction. Successive waves travel shorter distances, compressing their arrival times and increasing the frequency with which they pass a stationary observer. As the train moves away, sound waves still move toward the observer but are stretched because the train and the waves are moving in opposite directions. In this case, the frequency is decreased, lowering the pitch. The faster the train is moving, the greater the frequency shift a stationary observer hears. The frequency change is called the **Doppler frequency shift** after the Austrian physicist Hans Christian Doppler, who first explained the phenomenon.

Radar Meteorology: A First Course, First Edition. Robert M. Rauber and Stephen W. Nesbitt.
© 2018 John Wiley & Sons Ltd. Published 2018 by John Wiley & Sons Ltd.
Companion website: www.wiley.com/go/Rauber/RadarMeteorology

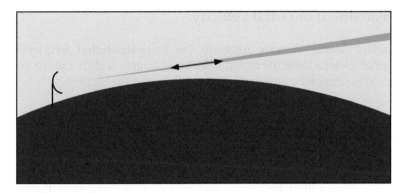

Figure 6.1   A Doppler radar can only measure the component of the target's motion along the direction of the beam

Microwaves emitted by radar also undergo a very slight frequency shift when they are scattered by raindrops or other targets moving toward or away from the radar. As with sound, the shift is toward higher frequencies when the targets are moving toward the radar and lower frequencies when they are moving away. The faster the targets move, the greater the frequency shift will be. A **Doppler radar** determines the Doppler frequency shift by measuring the phase of electromagnetic waves returned from a series of microwave pulses. From the pulse-to-pulse phase change, the radar determines the **radial velocity**, defined as the reflectivity-weighted velocity *along the direction of the beam* of all the targets within a pulse volume. As Figure 6.1 shows, the radial velocity represents a single component of motion of the targets, specifically the radial component; Doppler radars cannot detect the tangential components of target motion perpendicular to the radar beam. There are in fact four components to target motion, assuming that the targets are meteorological and move with the wind. These are the east–west wind component (u), the north–south wind component (v), the up–down wind component (w), and the target's terminal velocity in still air (w_t).

Radars detect other targets besides precipitation. Energy transmitted by radars is scattered by flying insects, birds, bats, and any other airborne objects. The dominant source of clear air signals in summertime is from insects. We can think of an insect's motion in two parts: the motion of an insect due to its own energy and the motion of an insect due to the wind. As radars view many insects simultaneously, and insects fly in random directions, their motion due to their own energy tends to average out to zero velocity. The radar measures their common motion, which is the wind blowing the insects along. Radars also detect energy scattered by small variations in the refractive index associated with changes in atmospheric density and moisture within turbulent regions of the atmosphere, the phenomena called Bragg scattering described in Chapter 1. These turbulent blobs of air also move with the wind. Energy scattered from insects and from turbulent regions of the atmosphere constitutes a **clear air signal** that, together with the **precipitation signal**, is used by Doppler radars to determine information about the wind. Radars can also detect surface targets that contribute to the radial velocity. Although most surface targets have zero velocity, some, such as ocean waves, trucks, and wind turbines, do move and can lead to ambiguities in interpreting the meteorological component of the signal.

6.2   Measurement of radial velocity

Doppler radars cannot directly measure the Doppler-shifted frequency of the returned signal to determine the radial velocity of targets within a pulse volume. To understand why, consider a simple continuous wave of the form

$$E_t(t) = E_0 \cos[\omega t + \phi_0] \tag{6.1}$$

where E_0 is the amplitude of the wave and ω is the angular frequency given by

$$\omega = 2\pi f_t \tag{6.2}$$

where f_t is the transmitted frequency and $\omega t + \phi_0$ is the phase of the wave. The received signal will have a form

$$E_r(t) = E_1 \cos[\omega(t - \Delta t) + \phi_1] \tag{6.3}$$

where E_1 is the amplitude and $\omega(t - \Delta t)\phi_1$ is the phase of the returned signal and

$$\Delta t = \frac{2r}{c} \tag{6.4}$$

is the round-trip time it takes for the wave to travel at the speed of light, c, to a target located at radial distance, r. The received frequency, f_r, is proportional to the time derivative of the term in square brackets in (6.3), that is

$$f_r = \frac{1}{2\pi} \frac{d}{dt} \left[\omega \left(t - \frac{2r}{c} \right) + \phi_1 \right] = f_t - \frac{2f_t}{c} \frac{dr}{dt} = f_t - \frac{2f_t}{c} v_r = f_t - f_d \tag{6.5}$$

where f_d is the Doppler frequency shift and v_r is the target radial velocity. The Doppler shift leads to lower received frequencies for receding targets. By convention, most Doppler radar systems, including the WSR-88Ds, assign positive values to receding velocities. The sign convention in (6.3) was chosen so that the frequency, f_r, in (6.5) is lower and v_r is positive when targets are receding from the radar.

The shift in frequency of the returned wave due to target motion is quite small. For example, an S-band radar operating at exactly 3,000,000,000 Hz (3.0 GHz) observing targets in a tornado that are receding at $100 \, \mathrm{m \, s^{-1}}$ from the radar would experience a Doppler shift of only

$$\frac{2f_t}{c} v_r = \frac{2 \times 3 \times 10^9 \, \mathrm{s^{-1}}}{3 \times 10^8 \, \mathrm{m \, s^{-1}}} (100 \, \mathrm{m \, s^{-1}}) = 2000 \, \mathrm{s^{-1}} = 2000 \, \mathrm{Hz} \tag{6.6}$$

so that the returned signal would be 3,000,002,000 Hz. For a more ordinary $10 \, \mathrm{m \, s^{-1}}$ receding wind, the returned frequency would be 3,000,000,200 Hz. A 200 Hz frequency corresponds to a wavelength of

$$\lambda_d = \frac{c}{f_d} = \frac{3 \times 10^8 \, \mathrm{m \, s^{-1}}}{200 \, \mathrm{s^{-1}}} = 1.5 \times 10^6 \, \mathrm{m} = 1500 \, \mathrm{km} \tag{6.7}$$

A radar pulse has a typical radial dimension of meters, so it contains only a tiny fraction of the Doppler-shifted component of the wave as the returned signal arrives back at the radar. For this reason, it is impossible to determine the radial velocity by direct measurement of the Doppler-shifted frequency.

6.2.1   *Phase measurements and radial velocity retrieval*

To estimate the radial velocity of targets within a pulse volume, Doppler radars instead measure the **phase change** that occurs *between* electromagnetic waves received from successive pulses. Consider first a single target. During one pulse period, T_r, a target will move a distance of

$$d = T_r v_r \tag{6.8}$$

that is some fraction of the radar wavelength, λ. As the electromagnetic wave must travel out and back to the target, the phase shift, $\Delta\phi$, during one pulse period will be twice that distance, or

$$\frac{\Delta\phi}{2\pi} = \frac{2T_r v_r}{\lambda} \tag{6.9}$$

This equation can be solved to obtain the radial velocity

$$v_r = \frac{\lambda}{2T_r}\left(\frac{\Delta\phi}{2\pi}\right) \tag{6.10}$$

One can visualize how a series of phase measurements at intervals of the pulse period, T_r, can allow one to reconstruct the Doppler frequency by considering Figure 6.2. Equation (6.10) is the foundation of a signal-processing algorithm called the *Pulse Pair Processor* that is used in nearly all radars to calculate the radial velocity. We will explore how this algorithm works in practice later in the chapter.

Unfortunately, (6.10) does not provide a unique solution for the radial velocity. Consider, for example, a stationary target where $\Delta\phi = 0$ and $v_r = 0$ and a target moving at a radial speed such that $\Delta\phi = 2\pi$. The second returned wave with a phase change of 2π would appear identical to the first wave with a phase change of zero. The same is true of waves that are integer multiples of 2π (e.g., -4π, -2π, 0, 2π, and 4π). In Figure 6.3, we see that more than one sinusoid can fit the phase measurements. Indeed, there are an infinite number of sinusoids that will fit the

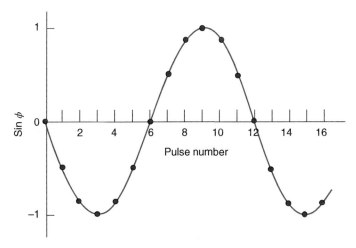

Figure 6.2   Illustration of recovery of the Doppler frequency from a sequence of phase measurements. The blue dots denote the measurements. The red line is a fit to the measurements

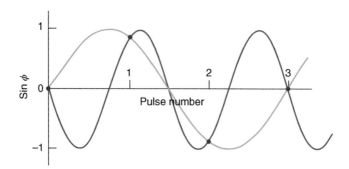

Figure 6.3 Illustration of ambiguity in recovery of the Doppler frequency from phase measurements. The blue dots denote the measurements. The red and green lines are equally valid fits to the measurements

phase measurements, each representing a different radial velocity for the target. This ambiguity characterizes all Doppler radar measurements.

6.2.2   Velocity ambiguities and their resolution

An ambiguity in retrieved radial velocity comes about when a target moves a distance that corresponds to a phase change of π or greater. A target moving in the opposite direction with the same radial velocity would produce a phase change of $-\pi$ or less. Two targets, one producing exactly a phase change of π and the other $-\pi$, are exactly 2π out of phase and therefore produce identical signals.

The only way that the ambiguity in interpretation of the Doppler velocity can be avoided would be to choose radar parameters, specifically the pulse repetition frequency (PRF) and wavelength, so that all actual radial velocities expected to occur in the region to be sampled fall within a range so that no phase increments greater than π occur. Referring to (6.10), this implies that

$$|\Delta\phi| = \left|\frac{4\pi v_r T_r}{\lambda}\right| < \pi \tag{6.11}$$

which leads, after substitution for T_r, to

$$|v_r| < \frac{\lambda}{4T_r} = \frac{\lambda(\mathrm{PRF})}{4} = v_{\max} \tag{6.12}$$

where $v_{\max}$, the maximum unambiguous velocity measurable by the radar, is called the **Nyquist velocity**. The range $-v_{\max}$ to $+v_{\max}$ is called the **Nyquist interval** and represents the range of velocities that can be measured unambiguously by a Doppler radar. For example, if the Nyquist velocity was $10\,\mathrm{m\,s^{-1}}$, then a true radial velocity of $+9.9\,\mathrm{m\,s^{-1}}$ would be reported as $+9.9\,\mathrm{m\,s^{-1}}$, but a true velocity of $+10.1\,\mathrm{m\,s^{-1}}$ would be measured as $-9.9\,\mathrm{m\,s^{-1}}$. Figure 6.4 extends this example to a wider range of true radial velocities. In this manner, true radial velocities that are beyond the maximum unambiguous velocity are **folded** back into the observable range.

Equation (6.12) relates the radar wavelength and PRF to the maximum radial velocity that can be determined unambiguously with a radar. The equation demonstrates

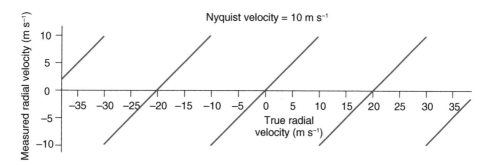

Figure 6.4 The true versus measured radial velocity of a target for a radar with a Nyquist velocity of 10 m s⁻¹. Note that all true radial velocities beyond the range −10 to 10 m s⁻¹ are folded back into that range

a clear advantage for long-wavelength radars. The equation also presents a dilemma for radar operations. Recall from Chapter 3 that the maximum unambiguous range of a radar is related to the PRF by

$$r_{max} = \frac{c}{2(\text{PRF})} \tag{6.13}$$

As the PRF is in the numerator of (6.12) and denominator of (6.13), a good choice of PRF for observing echoes at a distance range will be a poor choice for determining radial velocity. Multiplying (6.12) by (6.13), we obtain

$$r_{max}v_{max} = \frac{c\lambda}{8} \tag{6.14}$$

which formally states the **Doppler dilemma**: for a given wavelength radar, a poor choice for r_{max} will be a good choice for v_{max} and vice versa. We can understand the implications of this equation better by examining Figure 6.5, which shows the Doppler dilemma for four common wavelengths. This figure illustrates that a short-wavelength radar such as K_a-band (0.86 cm) has severe restrictions on resolving both range and velocity. Operating with a PRF corresponding to a 60 km maximum range, a K_a-band radar will only be able to resolve radial velocities less than 6 m s⁻¹ without folding. In most weather conditions, winds easily exceed 6 m s⁻¹, and in severe storms can exceed 60 m s⁻¹, so a K_a-band radar would suffer from severe velocity ambiguities if used in most weather conditions. Mobile radars used to study tornadoes often operate at X-band. From Figure 6.5, one can see that to unambiguously resolve velocities less than 36 m s⁻¹ near a tornado, the radar would have to be parked within 22 km of the storm. S-band radars have the least restriction, but still suffer from folding, particularly when used at long range where beams can intersect the jet stream aloft, for example.

Velocity folding makes it difficult to interpret radial velocity patterns on radar radial velocity imagery. For example, Figure 6.6 shows reflectivity and radial velocity data from the WSR-88D in Norman, OK, during the May 3, 1999 tornado outbreak. Note the folded region of velocities where the values abruptly shift to the opposite end of the scale. This can be particularly problematic when trying to identify rotation. For example, the radial velocities near the hook echo of the supercell just northwest of the radar in Figure 6.7 are likely to be folded. There are also folded velocities present near the center of the storm to the southwest of the radar.

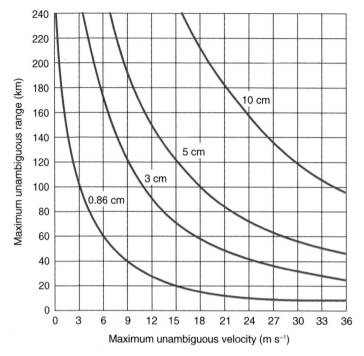

Figure 6.5 The relationship between maximum unambiguous range and velocity for common radar wavelengths

To avoid ambiguity, folded velocities must be **unfolded** to retrieve the true velocity. Unfolding can, in some cases, be accomplished with computer algorithms that may take advantage of additional information, such as ancillary wind measurements. In some cases, folding is sufficiently complicated that it must be done manually using graphics software for each scene. In some cases, it is still impossible to unfold veloci-ties. Consider, for example, Figure 6.8, an range-height indicator (RHI) scan of a severe thunderstorm near Newcastle, TX, observed by an X-band radar during the VORTEX experiment in 1994. The Nyquist velocity in this case was 12 m s^{-1}, implying that the maximum unambiguous range of the radar was 90 km. Folds are obvious in the radial velocity data in locations where red pixels are adjacent to blue pixels. Velocities in the upper part of the storm are extremely complicated. The storm contains hail, and at the high elevation angles, the radar senses a large component of the particle fall velocities. The turbulent winds and complicated particle size distribution aloft lead to velocities so complicated that unfolding velocity data accurately in this part of the storm is not possible. Figure 6.9 shows the same data with the recoverable velocities unfolded. Note that the maximum radial velocities exceeded 30 m s^{-1} in the storm.

The Doppler dilemma creates a second problem. To widen the Nyquist interval and avoid velocity folding, Doppler radars are typically operated at a high PRF. This leads to a small unambiguous range and a high likelihood of range-folded (second-trip) echoes. The range-folded velocities, which have Doppler radial veloci-ties characteristic of the distant storm, superimpose on the radial velocities of storms within the radar's unambiguous range, confounding the Doppler velocity retrieval and making the data unusable. WSR-88D radars, particularly before 2007 when

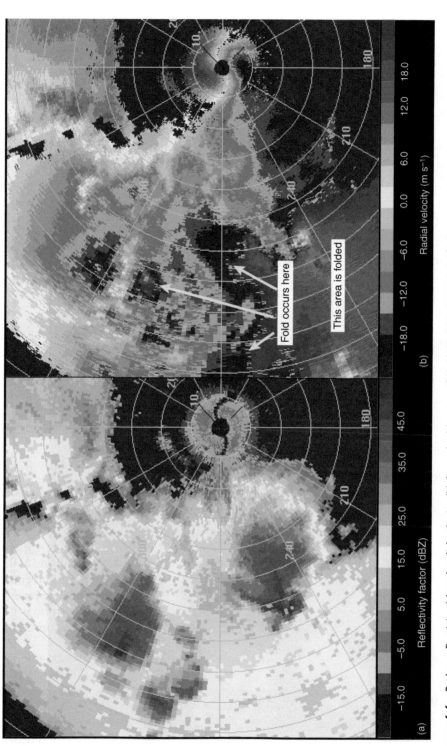

Figure 6.6 Radar reflectivity (a) and radial velocity (b) from the Norman, OK, WSR-88D (KTLX) 5.2° elevation scan at 23 : 18 : 55 UTC on May 3, 1999. Velocity folds are indicated on the radial velocity image (Radar image courtesy Nolan Atkins)

Radial velocity (m s⁻¹)

18.0   12.0   6.0   0.0   −6.0   −12.0   −18.0

(b)

Fold occurs here

This area is folded

Reflectivity factor (dBZ)

45.0   35.0   25.0   15.0   5.0   −5.0   −15.0

(a)

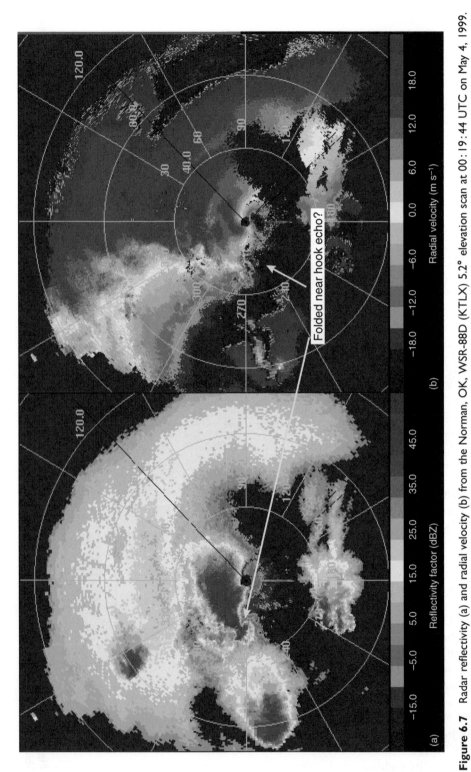

Figure 6.7 Radar reflectivity (a) and radial velocity (b) from the Norman, OK, WSR-88D (KTLX) 5.2° elevation scan at 00:19:44 UTC on May 4, 1999. Note the complicated structures in the radial velocity field near the hook echo of the supercell near the radar (Radar image courtesy Nolan Atkins)

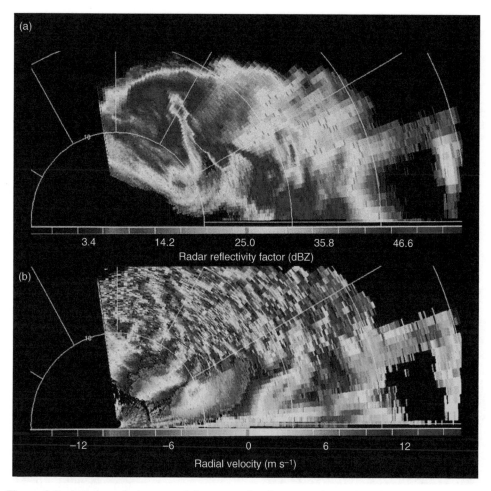

Figure 6.8 RHI through the core of the severe thunderstorm observed at Newcastle showing the reflectivity factor (a) and measured radial velocity (b). The radial velocities are folded in many locations. Turbulence in the top half of the storm is severe enough that the velocity pattern lacks coherence (Radar image courtesy: Nolan Atkins)

new algorithms were introduced, coded these unrecoverable velocities beyond the unambiguous range as purple, which meteorologists colloquially called **purple haze**.

A number of techniques have been developed to circumvent the Doppler dilemma, recover Doppler velocities over a wider range, and remove range-folded echoes. The first of these involves carrying out two complete plan position indicator (PPI) scans at the same elevation, the first with a low PRF to determine the reflectivity field without range folding and the second with a high PRF to recover the radial velocity field. The true radial velocity for the scan with the first PRF is given by

$$v_{r,true} = v_{1,meas} \pm 2n_1 v_{1,max} \tag{6.15}$$

where $v_{r,true}$ is the true radial velocity, $v_{1,meas}$ is the measured radial velocity with the first PRF, n_1 is the number of folds, and $v_{1,max}$ is the Nyquist velocity for the first PRF. The second PRF yields a different v_{max}, number of folds, and measured radial velocity

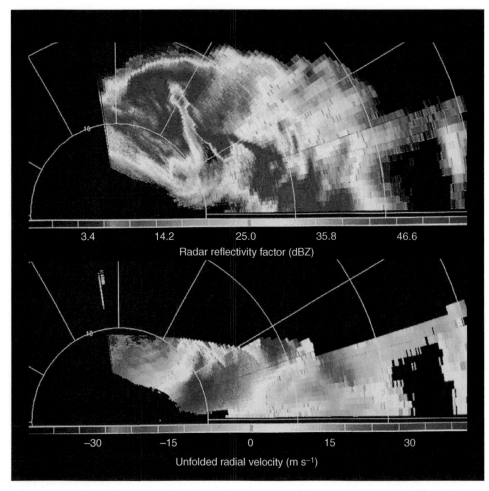

Figure 6.9   Same RHI through the core of a severe thunderstorm updraft as Figure 6.8, except that the radial velocities have been unfolded. The velocities in the turbulent region were too complicated to recover and were removed (Radar image courtesy: Nolan Atkins)

such that

$$v_{r,true} = v_{2,meas} \pm 2n_2 v_{2,max} \tag{6.16}$$

The two equations involve three unknowns ($v_{r,true}$, n_1, n_2), but if the number of folds is not too large, the equations can be easily solved and a look-up table can be used to obtain the true radial velocity and the number of folds for each PRF. The procedure extends the range of velocities for which an unambiguous inversion can be obtained by the factor

$$N = \frac{PRF_1}{PRF_2 - PRF_1} \tag{6.17}$$

Another method used to deal with range folding and velocity folding is to transmit trains of pulses first with a low PRF and then with a high PRF. The procedure to recover the radial velocities is similar to above, but only one complete scan is required instead of two. As an example, the NCAR airborne radar system, ELDORA, which used to fly on the NCAR Electra and Navy WP-3D aircraft, had a wavelength of 3.2 cm and used two PRFs with a ratio of $4:5$. If only one PRF was used at $1000\,s^{-1}$, v_{max} would be 8 m s^{-1}. With two PRFs of 1000 and $1200\,s^{-1}$, $v_{max} = 8 \times 1000/(1200{-}1000) = 40$ m s^{-1}. Using PRFs of 3200 and $4000\,s^{-1}$, v_{max} can be extended to 102 m s^{-1}, which made the airborne radar suitable for studying tornadoes at distances less than r_{max}, which for $4000\,s^{-1}$ is 37.5 km. Other PRF pairs could also be used to optimize v_{max} and r_{max} for the particular meteorological conditions. The tail radar systems on the NOAA P-3 fleet of aircraft that fly into hurricanes also have this capability.

Recently, a technique has been developed where pulses are phase-coded, that is, transmitted with a known phase that is different for each PRF. Advanced signal processing of these data allow separation of the velocity contributions from first- and second-trip echoes based on the phase of the returned signal. This technique, called the Sachidananda–Zrnic algorithm after its developers, is now employed on low-level scans with the WSR-88Ds and permits both unfolding and recovery of radial velocities in regions contaminated by second-trip echoes. The algorithm has greatly reduced the purple haze problem common on radial velocity displays of the WSR-88D before 2007 and extended the range of velocities that can be determined unambiguously with the WSR-88D.

6.3   Doppler spectra

The contributing volume representing a single radar measurement of a WSR-88D radar is 6.5×10^7 m^3 at 50 km range and can contain tens of millions of raindrops, snowflakes, hailstones, and/or ice particles. The particle sizes and terminal velocities will vary within the volume, and the projection of these terminal velocities in the beam direction will lead to a spread in radial speeds. In addition, wind shear, especially in the vertical, will cause some particles to move at faster radial speeds relative to others. Turbulent eddies, as well as sustained updrafts and downdrafts, will also contribute to the spread in particle radial speeds. Each particle scatters electromagnetic energy back toward the radar, and each wavelet produced by a scattering event has a frequency that is Doppler-shifted according to the radial speed of the particle. These wavelets interfere with one another to produce the returned wave sampled at the radar receiver.

Between pulses, the particles within a contributing volume move about relative to one another. If they shuffle too far, the next pulse will sample a completely independent arrangement, such that the returned wave, again a product of interference of wavelets from each of the particles, will have a random phase from the previous pulse. As a Doppler radar determines the Doppler frequency and radial velocity from successive phase measurements, a random phase would render the measurements useless to determine radial velocity. For this reason, Doppler radars must transmit pulses with a PRF high enough that the measurements are dependent; that is, the particles shuffle only a small fraction of a wavelength in the time between the arrival of successive pulses.

Even when this condition is met, we should not expect that the radial velocity determined from the phase change between two successive pulses will be the same as that from the adjacent pulse-pair in the train of pulses because the interference that occurs between the wavelets will change with each successive pulse as a result of the particle rearrangement.

6.3.1   Doppler spectra of weather and other targets

Consider a contributing volume of a radar containing millions of particles each with a distinct radial velocity. The **Doppler power spectrum** of this ensemble of particles consists of the reflectivity-weighted distribution of particle radial speeds. The power returned at each Doppler frequency is related to the size, concentration, and phase of each of the particles moving at various radial speeds. We represent Doppler spectra mathematically as a power spectrum, $S(v_r)$. Figure 6.10 shows an example of a Doppler spectrum measured by a vertically pointing radar. The contributing volume is centered 1477 m above a location where a light steady rain was occurring. The Doppler power spectrum in this sample represents the distribution of the reflectivity-weighted fall speeds (terminal velocities + vertical air motion) of the particles in the contributing volume. Because of the unusually wide radar beamwidth of the particular radar that collected these measurements (9.5°), there is also some contribution to the power spectrum from particle motion due to the horizontal wind. The Nyquist velocity of the radar is 18 m s^{-1}, and the spectrum extends across the Nyquist interval. The flat parts of the spectrum represent noise, whereas the meteorological signal is represented by the raised part of the distribution. Note that the spectrum is quantized in bins. The number of bins is determined by the number of pulses used to construct the spectrum.

The Doppler spectrum of a contributing volume is determined by processing a time series of M discrete samples of the echo at intervals of the pulse repetition period, T_r. If we represent each of these samples by its electric field, $E_r(t)$, we can use a Fourier transform to relate the frequency and time domain. Specifically, we can

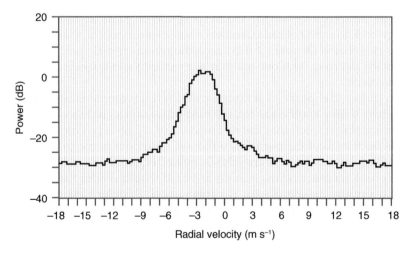

Figure 6.10   A Doppler spectrum measured by a vertically pointing radar in a light rain shower. Note the logarithmic scale for power

write the **discrete Fourier transform** as

$$E(mT_r) = \frac{1}{M} \sum_{m=0}^{M-1} F(kf_0) \cos[2\pi kf_0 mT_r] \tag{6.18}$$

where

$$F(kf_0) = \sum_{m}^{M-1} E_r(mT_r) \cos[2\pi kf_0 mT_r] \tag{6.19}$$

In these equations, F is the amplitude of the individual frequency components contributing the total electric field. There are M frequency components in the spectrum separated by intervals of $f_0 = 1/MT_r$. In practice, radars that record the Doppler spectrum use the **Fast Fourier Transform** algorithm to measure the power at each of the component frequencies.

Doppler spectra are almost never recorded by operational radars but are now being recorded more routinely by some research radars. The primary use of Doppler spectra has been in cloud physics applications. For example, Figure 6.11 shows four Doppler spectra recorded at different altitudes in the same rain shower as in Figure 6.10. The spectrum of velocities recorded can be directly related to the particle size distributions as fall velocity is a function of particle size. The degree to which these deviate from expected fall velocities of raindrops and snowflakes can be used as a rough measure of storm updrafts and downdrafts. The change in the power spectra with altitude can be related to the growth of the particles as they fall to the surface and can pinpoint the location of the melting layer, where particle fall velocities increase by a factor of 4–6. Note that ground clutter, marked by the second mode at zero velocity, is evident in the lowest altitude.

6.3.2 Moments of the Doppler spectrum

Doppler spectra are not required in radar operations and in most research. What data are required are the returned power, from which we derive the radar reflectivity factor, the mean radial velocity, and the spectral width. These quantities, called the **spectral moments**, are derived from the Doppler spectra and defined mathematically as

$$\overline{P}_r = \int_{-v_{max}}^{+v_{max}} S(v_r)\,dv \tag{6.20}$$

$$\overline{v}_r = \frac{\int_{-v_{max}}^{+v_{max}} (vS(v)\,dv)}{\int_{-v_{max}}^{+v_{max}} (S(v)\,dv)} = \frac{\int_{-v_{max}}^{+v_{max}} (vS(v)\,dv)}{\overline{P}_r} \tag{6.21}$$

$$\sigma_v^2 = \frac{\int_{-v_{max}}^{+v_{max}} ((v-\overline{v})^2 S(v)\,dv)}{\int_{-v_{max}}^{+v_{max}} (S(v)\,dv)} = \frac{\int_{-v_{max}}^{+v_{max}} ((v-\overline{v})^2 S(v)\,dv)}{\overline{P}_r} \tag{6.22}$$

Physically, we can understand the relationship between these quantities and Doppler spectra by referring to Figure 6.12. The **zeroth moment** (6.20), the area under the curve of $S(v_r)$ in Figure 6.12, is the **mean echo power**, $\overline{P}_r$, from which we can calculate the radar reflectivity factor. Although the area under the curve *above* the noise floor constitutes the echo from the target, the total returned power

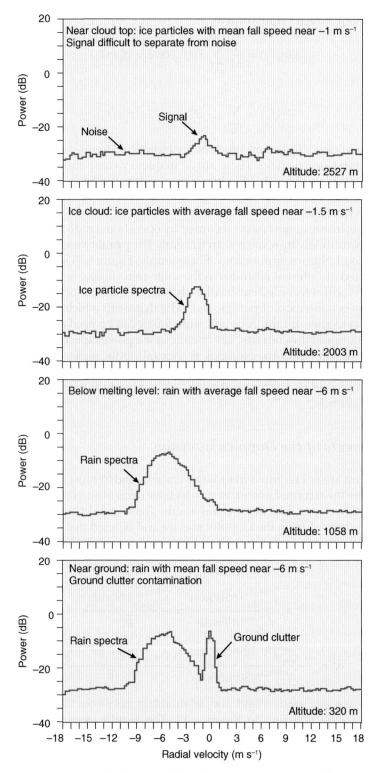

Figure 6.11   Doppler spectra measured in a vertical column as ice particles form near cloud top, grow as they fall, melt to become rain, and approach the ground

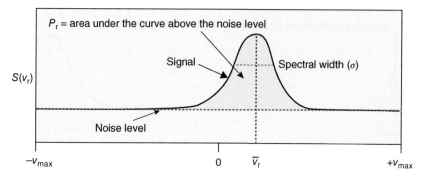

Figure 6.12   An idealized Doppler spectrum showing the relationship of the spectrum to the spectral moments

measured by the radar includes noise power. This is not an issue except when trying to distinguish the signal from very weak echoes (note that the vertical scale is logarithmic in Figure 6.10). The **first moment**, the **mean radial velocity** (6.21), is the reflectivity-weighted mean radial speed. For a beam pointed parallel to the earth's surface, this would be a measure of the mean radial wind component. For a beam pointed normal to the earth's surface, it would be a measure of the reflectivity-weighted mean fall speed of the particles, which includes their terminal velocities and any vertical air motion. For all other elevations, it is a combination of both vertical and horizontal components. The **second moment** is a measure of the variance of the spectrum. The square root of (6.22), σ_v, is called the **spectrum width**. The spectrum width is a measure of the spread of radial motions of particles within the pulse volume. Wind shear, turbulence, and variations in fall velocity are the primary contributors to the spread.

   We can understand how errors can arise in the estimation of the moments by considering example spectra for problematic situations. Figure 6.13 shows three idealized spectra that each illustrates a different problem. Figure 6.13a shows an example where the radial velocity of the particles in the contributing region bridges the Nyquist velocity. In this situation, the mean radial velocity would be biased toward zero, even though none of the particles in the contributing volume have the calculated mean radial velocity. The second panel shows a strong ground clutter return in addition to a meteorological spectrum. The ground clutter again will bias the velocity toward zero and also increase the mean echo power. The third example results from a very weak signal, as might be obtained in non-precipitating clouds or near cloud top of a deep stratiform ice cloud. As the signal is barely above the noise, the radial velocity will have a strong contribution from the noise in the spectrum, biasing calculations of mean Doppler velocity toward zero (because $\bar{v}_r$ for a uniform spectrum would be zero). The effect on spectrum width would be to increase its value toward that of a uniform noise spectrum.

6.4   Measurement of the Doppler moments

Calculations of spectral moments can be accomplished with the complex version of the discrete Fourier transform and complex algebra. These calculations are

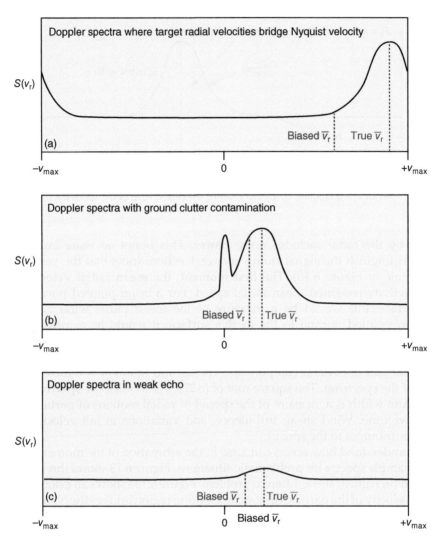

Figure 6.13 Examples of Doppler spectra that produce biased estimates of the mean radial velocity

computationally demanding and time consuming, requiring recording of a time series at each range gate, a Fourier analysis, calculating the moments, and then discarding the spectral data. An alternative, computationally efficient method uses the **autocorrelation function**, which permits discarding data from a pulse continuously after one calculation. The autocorrelation function is the foundation of a technique called the **pulse-pair processor**.

Before examining the pulse-pair concept, recall the structure of the phase detector of a Doppler radar (see Figure 2.15). In complex notation, the signal from I and Q channels for a single pulse can be expressed as

$$R_1 = \frac{A_0 A_1}{2} \exp[i(\omega_d t + \phi_1)] \tag{6.23}$$

where $\omega_d = 2\pi f_d$, the real part of R_1 represents the in-phase component, and the imaginary part the quadrature component. For a train of n pulses

$$R_n = \frac{A_0 A_n}{2} \exp[i(\omega_d t + \phi_n)] \qquad (6.24)$$

The autocorrelation function is given by

$$C = \frac{1}{M} \sum_{n=1}^{M} R_n R_{n+1}^* \qquad (6.25)$$

where for an sample in the sum

$$R_{n-1} R_n^* = \frac{A_0^2 A_{n-1} A_n}{4} \exp[i(\phi_n - \phi_{n-1})] \qquad (6.26)$$

and the star denotes the complex conjugate. The argument of the complex autocorrelation (at lag T_r) is an angle essentially equivalent to the mean phase difference. Therefore, the mean radial velocity can be obtained from the argument of the complex autocorrelation function.

To understand conceptually how the autocorrelation function works, consider the phase diagram in Figure 6.14, where the x axis is the real axis and the y axis is the imaginary axis. The red arrow represents the amplitude A of an oscillating wave, the phase is ϕ, and the Doppler angular frequency is ω_d. The autocorrelation function can be visualized on such phase diagram. Each component of the autocorrelation function determined from a pulse-pair is one estimate of the phase. By putting the vectors tail to head (Figure 6.15), the mean phase angle (shown in green), which is related to the mean Doppler velocity, can be determined.

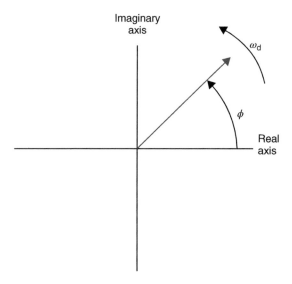

Figure 6.14 Simple phase diagram showing the amplitude vector rotating at an angular frequency ω_d. The phase at a particular time is ϕ

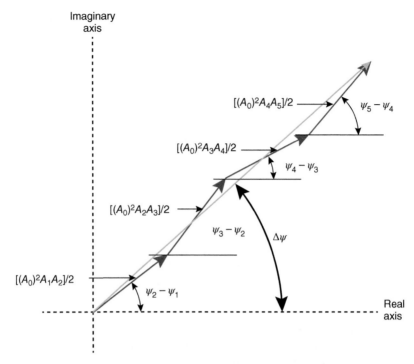

Figure 6.15   Determination of the mean phase change (green) from the components of the autocorrelation function (red)

Processing using autocorrelation provides the other moments as well. The zero-lag autocorrelation is related to the echo power and the reflectivity factor. The ratio of the autocorrelation at lag T_r to the autocorrelation at lag zero is directly related to the echo time to independence and inversely to the width of the Doppler spectrum.

6.5   Summary

Doppler radars provide measurements of the radial speed of targets, which gives information about the horizontal wind at low elevation angles and the fall speed of particles at very high elevation angles. They also provide measurements of spectrum width, which can be related to turbulence and wind shear. Radial velocity data can be used directly to interpret the wind field in the vicinity of the radar. In coming chapters, we will examine how radial velocity measurements can be used to determine a vertical profile of the wind over the radar, to identify rotation and divergence in a flow field that may be associated with tornadoes or downbursts and to find fronts and other mesoscale boundaries. We will also examine how two or more Doppler radars, simultaneously viewing a storm from different angles, can map out the complete four-dimensional structure of the wind field at high resolution within an evolving storm.

Important terms

autocorrelation function
clear air signal
discrete Fourier transform
Doppler dilemma
Doppler frequency shift
Doppler power spectrum
Doppler radar
Fast Fourier Transform
first moment
folded
mean echo power
mean radial velocity

Nyquist interval
Nyquist velocity
phase change
pulse-pair processor
precipitation signal
purple haze
radial velocity
second moment
spectral moments
spectrum width
unfolded
zeroth moment

Review questions

1. Explain why Doppler radars cannot directly measure the frequency shift associated with the Doppler effect and instead must rely on pulse-to-pulse measurements of phase to reconstruct the Doppler frequency.

2. Is the Doppler frequency shift toward higher frequencies or lower frequencies for objects approaching the radar?

3. One of the PRFs used by the WSR-88D radar is 1014 Hz. Assume that the radar has a wavelength of 10 cm. For this PRF, what would be the mean pulse-to-pulse phase change that occurs for a target moving at $10\,\mathrm{m\ s^{-1}}$ away from the radar?

4. Which radar wavelength (K, X, C, S) is most suitable to make radial velocity measurements in strong wind conditions? Why?

5. What was the Nyquist velocity of the WSR-88D radar in Norman, OK, on May 3, 1999 based on the radial velocity image in Figure 6.6?

6. What is velocity folding, and how would a fold appear on a display of radial velocity?

7. What is the Doppler dilemma? Describe two methods that radar meteorologists have developed to circumvent limitations imposed on measurement of radial velocity by the Doppler dilemma.

8. List reasons why targets within the contributing volume of a radar might have a range of radial velocities.

9. Physically, what is meant by the term "moments" of a Doppler power spectrum?

10. The Fast Fourier Transform and the Autocorrelation Function both can be used to determine the moments of the Doppler spectrum. What is the advantage of the autocorrelation approach?

Challenge problems

11. Examine the Doppler spectra in Figure 6.11. The spectrum in Figure 6.10 was obtained in the same vertical column at the same time. What altitude range was this spectrum from? What physical process might be happening at that altitude to produce the spectrum in Figure 6.10.

12. The fastest wind speed ever recorded in a microburst was $67\,m\,s^{-1}$ at Andrews Air Force Base on August 1, 1983. (a) How close to the airport would a 5 cm wavelength Doppler radar have been located to observe this wind event without folding? (b) What other complications might have arisen that might have corrupted the radial velocity data, based on your answer to (a).

13. Suppose a WSR-88D radar located near the coast of the Gulf of Mexico uses two Pulse Repetition Frequencies, 326 and $652\,s^{-1}$. The radar has a wavelength of 10 cm. The lower PRF is used for reflectivity measurements and the higher PRF is used for velocity measurements.

    a. What is the maximum unambiguous range and velocity for the lower PRF?

    b. What is the maximum unambiguous range and velocity for the higher PRF?

    c. What is the transmitted frequency of the radar?

    d. What is the received frequency for the lower PRF for a target moving at the maximum unambiguous velocity?

    e. What is the received frequency for the higher PRF for a target moving at the maximum unambiguous velocity?

    f. Suppose a category 2 hurricane (43–49 m s^{-1} max velocity) is approaching the radar. The strongest winds on the right side of the eye are blowing directly toward the radar. The radar is using the lower PRF. If the measured radial velocity at the location of the true maximum radial wind in the eyewall is $+5.9\,m\,s^{-1}$, what is the true radial wind speed, and how many times is the data folded?

    g. The PRF is switched to the higher PRF. What is the measured radial velocity now at the location of the maximum radial wind in the hurricane and how many times is the data folded?

14. Your local police department has just purchased a brand new "pulsed Doppler K-band (1 cm wavelength) traffic radar with an amazing PRF of $5000\,s^{-1}$." The police were told that the radar was "sure to catch speeders zooming through school zones." The reading on the radar is calibrated by the manufacturer so that velocities of cars moving toward the radar (at speeds below the Nyquist velocity) have positive values. One morning, you decide to break the law and speed through a school zone where the speed limit is 20 mph. The police radar your car.

   a. The policeman looks confused. You seem to be going backward even though he can see clearly that your car is approaching him. At what range of normal driving speeds will you appear to be going backward based on the policeman's radar reading?

   b. You decide to blast through anyway above the speed limit, but stay below 100 mph so that you do not spend the rest of your life in jail. What is the absolute fastest speed you can drive so that the policeman cannot use the radar data in court to convict you for speeding?

15. Suppose that a 3 cm Doppler radar operating at a PRF of $1000 \, s^{-1}$ measures the following phase changes between 10 consecutive pulse pairs from a particular pulse volume: $8.1°, 9.3°, 7.9°, 7.4°, 8.3°, 9.0°, 8.7°, 7.8°, 8.1°, 9.1°, 8.9°$. The power returned in each pulse in this case was identical to all the other pulses.

   a. What is the mean radial velocity of the targets in the pulse volume?

   b. What would the average phase shift have to exceed for the radial velocity to be folded?

7

Dual-Polarization Radar

Objectives

By the end of this chapter, you should understand the following:

- The difference between dual-polarization radars that are polarization agile versus polarization diverse.
- How linear polarization of electromagnetic waves can be applied to the problem of radar remote sensing of meteorological and non-meteorological targets.
- How the size, shape, number distribution, and composition of precipitation particles lead to their intrinsic and propagation dual-polarization properties.
- The physical processes in precipitation leading to variations in differential reflectivity.
- Why the differential phase of returned energy can change due to precipitation.
- The challenges in the estimation of specific differential phase in precipitation.
- The factors that influence the value of the co-polar correlation coefficient in precipitation.
- Why intrinsic polarization variables change as a function of radar wavelength.
- How to visually identify hydrometeor types in dual-polarization radar variables.
- How to automatically classify hydrometeor types.

7.1 Introduction

By taking advantage of the polarizability of light, active remote sensors (including radars) can use the characteristics of microwave energy in the **polarization plane** to retrieve information about targets of interest. In meteorological radar, **polarization radars** use our knowledge of how hydrometeors and other objects interact with the polarization state of the incident beam in order to estimate the size, shape, and type of hydrometeors; the mixing of hydrometeor types; and the separation of hydrometeors and non-meteorological echoes. The power of these tools for operational meteorology, progress in the research, and demonstration of the use of dual-polarization radars has justified the deployment of dual-polarization technology to operational and research radars around the world at weather radar wavelengths (e.g., S, C, and X-band). Recently, dual-polarization millimeter-wave radars (e.g., W, K_a, and

Radar Meteorology: A First Course, First Edition. Robert M. Rauber and Stephen W. Nesbitt.
© 2018 John Wiley & Sons Ltd. Published 2018 by John Wiley & Sons Ltd.
Companion website: www.wiley.com/go/Rauber/RadarMeteorology

K$_u$-band) are also being deployed to study clouds and precipitation processes; however, as we will see, there are additional challenges and opportunities in interpreting these measurements. In order to understand what information dual-polarization radars can provide, and their use in operational and research meteorology, we will first examine the parameters available from a dual-polarization radar, then demonstrate their use for applications, and give guidelines for their interpretation. In this chapter, we begin with the **intrinsic polarization variables** that are derived from power measurements that a dual-polarization radar provides and then investigate **propagation polarization variables**. Note that in this text, we will only describe a linear polarization basis (using orthogonal horizontal and vertical polarization bases) for polarization radars, although it is possible to construct arbitrary polarizations for remote sensing measurements (circular polarization being a primary alternative to linear polarization). Almost all operational and research meteorological radars use a linear basis for polarization.[1]

Several variables are available to a dual-polarization radar beyond reflectivity, Doppler velocity, and spectrum width, which have been discussed in previous chapters. Most conventional radars measure these parameters at horizontal polarization (H), whereas polarization radar measures the same parameters at both horizontal and vertical polarization (V). Dual-linear polarization radars receive signals at horizontal and vertical polarization (we define horizontal as the direction parallel with the horizon and vertical as perpendicular to the horizon) and have a set of variables that are defined to allow for physical interpretation of the returned signals from each polarization state. When displayed, these variables allow users to visually interpret additional information from the returned signals, or use the variables quantitatively in **dual-polarization retrievals**. The hardware characteristics (polarization diverse vs polarization agile) and mode of operation of a dual-polarization radar allow for a range of information available from the characteristics of the return signal, including

1. the returned average powers at each polarization,
2. variation in average phase and the statistical consistency of the pulse-to-pulse variations in phase at each polarization within a range gate,
3. the depolarization of averaged returned power relative to a transmit polarization (only available on polarization-agile radars).

In this chapter, we will examine how this information is commonly used in **polarimetric variables** related to the amplitudes of the measured powers and phases from a polarization radar.

7.2   The physical bases for radar polarimetry

The interpretation of polarization variables, such as reflectivity and Doppler measurements, relies on a physical understanding of how meteorological and

---

[1]For a more engineering-oriented text on dual-polarization radar, the reader is referred to Bringi, V. and Chandrasekar, V. (2001) *Polarimetric Doppler Radar: Principles and Applications.* Cambridge University Press, 636 pp.

non-meteorological targets interact with electromagnetic energy at the radar wavelength, but considering the polarization of the energy. As stated in Chapter 1, electromagnetic waves are defined by their **amplitude, wavelength** or **frequency,** and **phase**. In interpreting dual-polarization measurements, as before, we are concerned with the properties of the electromagnetic energy as

1. it is transmitted and received by the radar,
3. it interacts with the matter within the range resolution volume,
4. it changes as the waves propagate to and from the radar range resolution volume.

Changes in these properties individually, and how they change relative to each other as a function of polarization state, are used to infer, quantify or **retrieve** the properties of the matter using polarization radar measurements, as we know the characteristics of the transmitted and retrieved radar pulses. Thus, the ability of dual-polarization measurements to provide information on the properties of meteorological and non-meteorological targets relies on both the radar signal characteristics and the propagation and interaction of the waves along the radar beam. When referring to the returned power, we use the term **co-polar power** to refer to power returned in the same polarization state as the transmitted power and the term **cross-polar power** to refer to power returned in the state orthogonal to the transmitted power.

In Eq. (1.15), we mathematically defined a plane electromagnetic wave. Here, we expand that definition to

$$\vec{E}_I(t) = Re(\vec{E}_0 e^{i\omega t}) = |A_{I,H}| \cos(\omega t + \phi_H)\vec{i} + |A_{I,V}| \cos(\omega t + \phi_V)\vec{j} \qquad (7.1)$$

where we now consider the time-varying horizontal and vertical components of a linearly polarized wave. The components of the wave, noted by subscripts H and V, respectively, are illustrated in Figure 7.1, with unit vectors $\vec{i}$ and $\vec{j}$ defined in the H and V polarization plane; recall that $\omega = \frac{2\pi}{f}$ is the angular frequency. A radar system will transmit a wave that can be approximated by Eq. (7.1), which can be considered as the incident wave on a particle or collection of particles (thus the I subscript on the electric field and amplitude terms).

We can define the interaction of the incident wave with a single particle and the scattered wave $(\vec{E}_S)$ through the definition of the backscattering matrix **S**, where

$$\vec{E}_S(t) = \mathbf{S}\vec{E}_I(t) \qquad (7.2)$$

$\vec{E}_S$ will have a form like Eq. (7.1), except the frequency will be modulated by the Doppler shift due to the component of particle motion of the particles toward or away from the radar as well as interactions with matter both to and from the radar resolution volume. Breaking the scattered field into its individual components, and representing the wave equation in a complex notation, we can write

$$\begin{bmatrix} E_S^H \\ E_S^V \end{bmatrix} = \mathbf{S} \begin{bmatrix} E_I^H \\ E_I^V \end{bmatrix} \qquad (7.3)$$

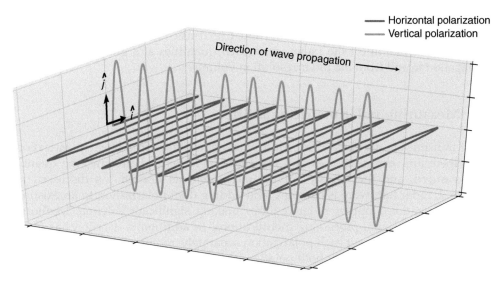

Figure 7.1 Three-dimensional visualization of a horizontally (blue) and vertically (green) linearly polarized electromagnetic wave

From matrix algebra, the scattering matrix **S** must be a 2×2 matrix. It takes the form

$$\mathbf{S} = \begin{bmatrix} S^{HH} & S^{HV} \\ S^{VH} & S^{VV} \end{bmatrix} \tag{7.4}$$

where the superscripts indicate the incident and scattered polarization amplitudes (note that the elements of **S** are typically complex numbers). Note that precipitation is termed a **reciprocal media**, such that $S^{HV} = S^{VH}$. Therefore, three parameters can characterize the interaction of the incident wave with targets along the beam. **S** is dependent on the size, shape, composition (e.g., ice vs water, ice density, internal crystal structure, degree of melting, or riming), and orientation of the particle under consideration, as well as radar wavelength, and therefore causes the observed differences in **intrinsic polarization variables** in various types of targets.

To account for a real radar configuration, where the radar beam propagates through non-vacuum and illuminates an ensemble of particles within the radar resolution volume, we can write the complete transformation amplitude matrix. In Eq. (7.3), we replace the subscripts incident and scattered on a single particle with the energy transmitted by the radar (T) and received (R), with the scattering matrix **S** now representing the electromagnetic interactions with an ensemble of particles in the radar resolution volume, and the transmission matrix **T** (with components T^H and T^V) representing the interactions in both amplitude and phase with the matter between the radar and the radar resolution volume:

$$\begin{bmatrix} E_R^H \\ E_R^V \end{bmatrix} = \begin{bmatrix} T^H & 0 \\ 0 & T^V \end{bmatrix} \mathbf{S} \begin{bmatrix} T^H & 0 \\ 0 & T^V \end{bmatrix} \begin{bmatrix} E_T^H \\ E_T^V \end{bmatrix} \tag{7.5}$$

T can represent one-way path-integrated attenuation, due to atmospheric gases (dry air and water vapor) and precipitation (both of which can change the amplitude and phase of the propagating electromagnetic wave). Thus, to interpret polarimetric variables received at the radar, it is important to understand how **S** (intrinsic to the particles in the radar resolution volume) and **T** (extrinsic to the particles in the radar resolution volume) affect the radar's measured quantities at a given range during propagation to and from the radar resolution volume.

7.3 Measuring polarimetric quantities

When transmitting and measuring electromagnetic energy in a linear polarization basis, there are several choices and trade-offs in terms of hardware, the quantities measured, and the "purity" of the measurements desired. As discussed in Chapter 2, radar hardware is complex, and adding dual-polarization to an existing single-polarization Doppler radar requires making significant antenna, feedhorn, waveguide, and signal-processing modifications. In addition, a decision has to be made whether to transmit each polarization separately, receiving alternate pulses at each polarization, or transmit both polarizations simultaneously, allowing reception of each polarization simultaneously. Figure 7.2 shows three configurations for making dual-polarization measurements.

If **alternate transmission and reception** of horizontal and vertical polarizations, which give the full complement of polarization measurements, are desired, there are two options. The most expensive option is to have two identical transmitters that are configured to transmit at orthogonal polarizations (the waveguides are oriented orthogonally at the feedhorn to accomplish this). The CSU-CHILL radar, with two klystron transmitters, is configured as such but requires good calibration between the two independent transmitters. An alternative is to use only one transmitter, but use a fast switch that splits the transmit power into the horizontal and vertical polarizations. Radars such as the NCAR S-PolKa (which uses a single klystron transmitter) and the NASA NPOL (which uses a single magnetron transmitter) can operate in this configuration. The advantage here is that only one transmitter is needed, which saves significant costs. The disadvantages are that the switch must be maintained (as switches are operating at the pulse repetition frequency (PRF) of the radar, they can be subject to failure), as well as some loss of cross-polarization isolation within the switch. Radars applying either of these techniques are referred to as **polarization-agile** radars.

The third option, which is employed in most operational and research radars (WSR-88Ds uses this mode and is an available mode in research radars such as S-PolKa and NPOL), is to operate in a mode called **simultaneous transmit and receive** or STAR mode. Such radars are not polarization agile because they do not transmit independent polarizations but are termed **polarization diverse** because they receive H and V channels. STAR mode polarization radars transmit essentially at 45° polarization because they employ a power splitter from their single transmitter that splits the power equally among polarization channels so that power emanates from both polarization channels simultaneously. The advantage for this hardware configuration is that there is no need for a second transmitter, or a fast switch that may require significant maintenance and cost over time. This is a much more economical solution to build a dual-polarization radar compared with other configurations. One other advantage of STAR mode operation is that half as many pulses are required to estimate the polarimetric moments compared with an alternate transmission mode,

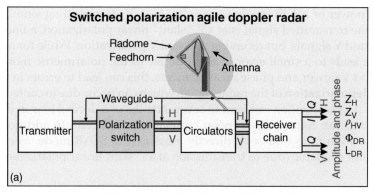

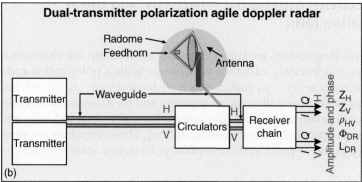

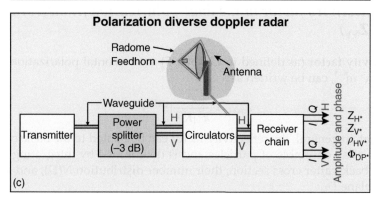

Figure 7.2 Block diagram of a switched polarization-agile Doppler radar (a), dual-transmitter polarization-agile Doppler radar (b), polarization-diverse Doppler radar (c). Polarimetric variables measured by each system are shown at the right. Asterisks denote variables measured when transmission is made at 45° slant polarization

which means the antenna can spin faster for the same azimuth resolution. This is a big advantage in operational situations where understanding the nature of rapidly evolving storms is of interest.

However, there are also disadvantages to STAR mode polarization. One disadvantage is that by transmitting both polarizations simultaneously, the transmitter's power is equally split between the H and V polarization channels, which equates to a

50% loss of power or a 3 dB loss in sensitivity. Also, by transmitting simultaneously, essentially the transmitted signal is at 45° "slant" linear polarization, a linear combination of H and V signals, but receiving at H and V polarization. While for most applications, this leads to a small error in the estimate of the polarimetric moments that require H and V power and phase measurements, this can lead to errors in the case of significant **depolarization** of the radar signal, which can occur due to canted, oriented ice particles. Depolarization can lead to several tenths of a decibel bias in differential reflectivity and can reduce the co-polar correlation coefficient. In addition, the linear depolarization ratio cannot be directly measured in STAR mode, as cross-polar powers are not available (due to transmission at 45° slant linear polarization).

7.4   Reflectivity, differential reflectivity, and linear depolarization ratio

From the scattering matrix, we can write relations between the elements of the scattering matrix and intrinsic variables we observe with a polarization radar. We will start with the radar reflectivity factor, which we have already discussed, but not in the framework of dual-polarization measurements, and then discuss differential reflectivity and linear depolarization ratio (the latter is typically observed only with research radars having a specific hardware configuration). These variables are simply related to reflectivity in the co-polar and cross-polar polarization states, respectively.

7.4.1   Reflectivity factor in the dual-polarization framework (Z_{HH} and Z_{VV})

The **reflectivity factor** (as defined in Chapter 5) at horizontal polarization (Z_{HH}), in units of mm^6 m^{-3}, can be written as

$$Z_{HH} = \frac{\lambda^4}{\pi^5 |K|^2} |S^{HH}|^2 \tag{7.6}$$

Recall that, in general, the reflectivity factor can be related to the backscatter cross section (σ_b) of an ensemble of particles and is determined by integrating the product of their backscatter cross section, their number distribution $N(D)$, and size in the horizontal plane D_H:

$$Z_{HH} = \frac{\lambda^4}{\pi^5 |K|^2} \int_{D_{H,min}}^{D_{H,max}} \sigma_b(D_H) N(D_H) dD_H \tag{7.7}$$

In the Rayleigh scattering limit, the reflectivity is proportional to the sixth moment of the particle size distribution

$$Z_{HH} = \int_{D_{H,min}}^{D_{H,max}} N(D_H) D_H^6 dD_H \tag{7.8}$$

Most non-polarization radars, including WSR-88Ds before their dual-polarization upgrade, measure Z_{HH}, as this was how the feedhorn is oriented. We can write equivalent equations for the reflectivity factor at vertical polarization Z_{VV} in Eqs. (7.7) and (7.8) by replacing all H with V.

7.4.2   Differential reflectivity (Z_{DR})

From Eqs. (7.7) and (7.8) (and their counterparts in the vertical polarization plane), it is clear that if one assumes that the same contributing volume is illuminated by the radar beam equally at each polarization state, and pulse-to-pulse fluctuations in the return signal have been averaged with a sufficient number of samples, polarization-dependent changes in the radar reflectivity factor must result from changes in the radar cross section at the polarization being considered. For a radar that is linearly polarized, we can define the **differential reflectivity factor** Z_{DR} as

$$Z_{DR} = 10 \log_{10} \left(\frac{|S^{HH}|^2}{|S^{VV}|^2} \right) = 10\log_{10} \left(\frac{Z_{HH}}{Z_{VV}} \right) \tag{7.9}$$

which is the ratio of the polarization-dependent reflectivity factors. Z_{DR} has logarithmic units of decibels (dB). If the radar reflectivity factors are equivalent at each polarization, we see that $Z_{DR} = 0\,dB$, whereas if $Z_{HH} > Z_{VV}$, then $Z_{DR} > 0\,dB$, and if $Z_{VV} > Z_{HH}$, then $Z_{DR} < 0\,dB$. In the Rayleigh limit, if the properties of the hydrometeors in the contributing volume are assumed to be identical among polarizations, and propagation effects such as attenuation are ignored, the only factor that can change Z_{DR} is a change in the mean **reflectivity-weighted axis ratio** of the targets in the radar resolution volume (proportional to D^6 computed along the horizontal and vertical polarization planes). Thus, Z_{DR} is an intrinsic polarization variable because it depends on the shape of the particles within the radar resolution volume. Note that Z_{DR} is independent of number concentration and the volume of the particles in the radar resolution volume; it is simply related to the ratio of the backscattering cross sections at each polarization. Hereafter, we will refer to Z_{HH} as Z_H interchangeably.

In the non-Rayleigh scattering regime, measurements of Z_{DR} are more complicated due to the break down of the D^6 assumption, attenuation, and differential attenuation, all of which complicate the quantitative use of Z_{DR} at millimeter wavelengths. However, in some situations, Z_{DR} can be of use for the study of clouds (including ice clouds) and light precipitation at short wavelengths.

Accurate Z_{DR} measurements in precipitation require that the radar beams at the different polarizations are geometrically matched; sidelobes, beam blockage, anomalous propagation, and clutter contamination are equal at each polarization; and minimal, **differential attenuation** (defined as the attenuation difference between the H and V polarization plane caused by non-spherical absorbing particles, such as melting hail or large raindrops) is minimal or corrected for; and most importantly outside of these data quality issues, the absolute calibration of the H and V channels must be within 0.1–0.2 dB! You may ask, how can we ever hope to calibrate these two signals to within 0.1 dB if we are lucky if we can calibrate Z_{HH} to within 1 dBZ?[2]

In which situations would we expect the radar cross sections, or in the Rayleigh limit, the hydrometeor diameters in each polarization plane to differ? The natural variations of the size and composition of scatterers in the atmosphere, whether they are meteorological (e.g., raindrops, snowflakes, hailstones, graupel, and pristine ice

---

[2]Read more about how reflectivity and differential reflectivity can be calibrated: Ryzhkov, A.V., Giangrande, S.E., Melnikov, V.M., and Schuur, T.J. (2005) Calibration issues of dual-polarization radar measurements. *J. Atmos. Oceanic Technol.*, **22**, 1138–1155.

particles) or non-meteorological (e.g., birds, insects, aircraft, and clutter), make it possible to differentiate the backscatter cross sections at a given polarization and use that information to deduce properties of the scatterers.

7.4.3   Raindrop shapes and sizes

It turns out that for raindrops, the shape of a drop (which, despite popular belief, is not a teardrop shape!) in the polarization plane of the radar depends primarily on the size of the drop, but also to other factors as described below. How do we measure the size and shape of raindrops? A **disdrometer** is an instrument designed for this task. Early disdrometers were very simple: exposing a pan covered in flour, where the drops created small flour balls, or a piece of filter paper covered in water-soluble ink for a given amount of time that could be later measured to estimate the droplet concentration and sizes. As you could imagine, the process of analyzing many drops was quite tedious and contained sources of error. Fortunately, automatic techniques have been developed, including use of imaging technology that can be applied to a laboratory or field setting. High-speed cameras on the ground, or on aircraft probes, combined with sophisticated image-processing software, can be used to observe the behavior of precipitation particles as they fall, thus allowing us to automatically characterize particle sizes, shapes, and terminal fall speeds (see Chapter 13).

One example instrument that is used to measure drop size and shape at the surface is called the **two-dimensional video disdrometer** (or 2DVD). 2DVD instruments have been used in many locations around the world to characterize precipitation particles in different meteorological regimes. The 2DVD uses two high-speed line-scan cameras orthogonal to each other to capture images of each particle falling through an aperture of 10 cm on a side. The disdrometer provides information on the sizes and shapes, estimates the *particle size distribution* (a histogram of drop size normalized to illustrate the concentration of particles per unit volume of air per diameter size range; MKS units m^{-4}), and tracks each particle through the captured images to estimate the terminal fall velocities of precipitation particles. Figure 7.3 shows a picture of a 2DVD instrument in the field and illustrates the instrument measurement principle.

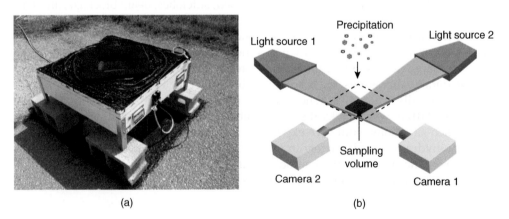

(a)   (b)

Figure 7.3   (a) Photograph of a two-dimensional video distrometer. (Courtesy NASA.) (b) Illustration of the measurement principle of the two-dimensional video disdrometer. For scale, the sampling area is 10 cm × 10 cm

Laboratory and field studies have confirmed that the average shape of a raindrop as it grows tends to develop into a "hamburger bun" shape as the particle diameter increases. Figure 7.4a shows measurements of the characteristic shape and size of raindrops of various diameters. As you can see, the droplets that reach even 3–4 mm in diameter start to depart significantly from a spherical droplet shape. We quantify the deviation of the droplet from a sphere via the quantity called the *axis ratio*, which we define as the ratio of the shorter axis to the longer axis of the particle:

$$A = \frac{D_{\text{short}}}{D_{\text{long}}} \tag{7.10}$$

Raindrops in light wind conditions typically fall with their long axes oriented more or less horizontally due to aerodynamic forces. As the drop becomes **oblate** (oblate is defined as an object with an axis larger in the horizontal direction compared with the vertical direction), A decreases from a spherical value of 1. A **prolate** drop, which would be unusual in the atmosphere, would have its long axis oriented in the vertical direction. Figure 7.4b shows the results of measurements of raindrop axis ratios from two laboratory studies, revealing that axis ratios are near unity for small drizzle-size droplets, whereas the axis ratios fall below 0.8 for droplets greater than 4 mm in diameter. Although ratios keep falling with increasing diameter, indicating a more oblate shape for larger droplets, two studies have shown considerable disagreement as to the axis ratios of drops >4 mm in diameter.

Oblate raindrops do not always fall with their long axis oriented with the horizontal direction. The angular deviation from the long axis of the particle from the horizontal plane is quantified by the **canting angle** of the particle. This angle has been quantified using theoretical studies and disdrometer observations and has been found to vary due to atmospheric turbulence acting on the droplet. Figure 7.5 shows

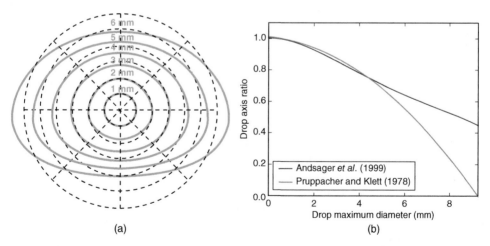

(a)   (b)

Figure 7.4   (a) Axis ratio as a function of drop diameter (mm) for a raindrop falling in still air. (b) Axis ratio versus maximum diameter from two laboratory studies (Panel (a). Adapted from Beard, K.V. and Chuang, C. (1987) A new model for the equilibrium shape of raindrops. *J. Atmos. Sci.*, **44**, 1509–1524. © the American Meteorological Society, used with permission. Lines in (b) from Andsager, K., Beard, K.V. and Laird, N.F. (1999) Laboratory measurements of axis ratios for large raindrops. *J. Atmos. Sci.*, **56**, 2673–2683, and Pruppacher, H. R. and Klett, J. D. (1978): *Microphysics of clouds and precipitation*. Reidel, Boston, 714 pp.)

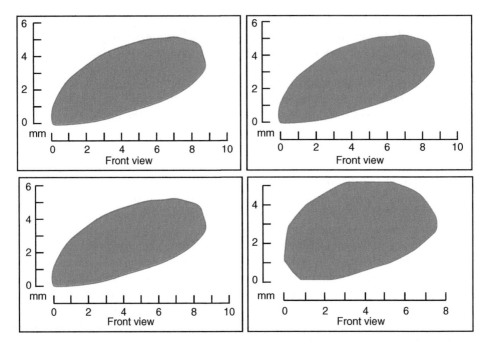

Figure 7.5   Illustration of canting angle of four raindrops observed with a two-dimensional video disdrometer (Adapted from Huang, G., Bringi, V.N., and Thurai, M. (2008) Orientation angle distributions of drops after an 80-m fall using a 2D video disdrometer. *J. Atmos. Oceanic Technol.*, **25**, 1717–1723. © the American Meteorological Society, used with permission)

example images of canted drops collected with a 2DVD. Each of the two cameras in the disdrometer independently measures the canting angle, and these measurements are presented in histogram form in Figure 7.6. In this study, the distribution of canting angles conforms to a normal distribution with mean canting angle near 0°, but with a standard deviation of nearly 7°.

Complicating matters further, **drop oscillations** alter drop shapes as they fall, such that droplets can have shape variations about their mean axis ratios. Figure 7.7 shows high-speed camera images of drops undergoing oscillations. According to high-speed camera measurements carried out in a vertical wind tunnel, oscillations in a drop of diameter 4.6 mm can cause variations in the axis ratio between approximately 0.67 and 0.8, modify the canting angle by as much as 6°, and change the equivalent diameter (defined as the equivalent spherical diameter as measured in a 2-D plane) by nearly 0.1 mm. Axis ratio, droplet canting, and oscillation are important considerations when comparing observed versus simulated radar measurements. Physical characteristics of the raindrops within a radar resolution volume must be assumed in order to interpret radar characteristics based on simulations of raindrop characteristics.

Other complicating factors cause raindrops to fall in modes other than discussed above. Drops can break up due to collision with another particle or suffer hydrodynamic failure as the drops become too large. In the latter case, the deformation to an oblate shape reaches the point that the drop becomes too large for surface tension to hold the drop together, and the drop breaks up into two or more droplets. In the absence of an ice core to maintain the droplet's shape (which can happen when hail or

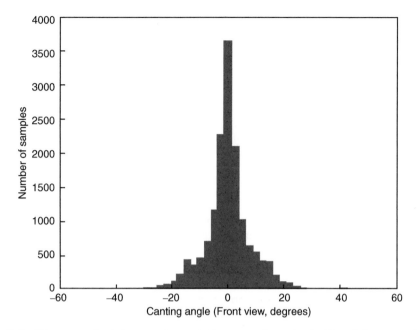

Figure 7.6 Histogram of raindrop canting angle from two-dimensional video disdrometer observations (Adapted from Huang, G., Bringi, V.N., and Thurai, M. (2008) Orientation angle distributions of drops after an 80-m fall using a 2D video disdrometer. *J. Atmos. Oceanic Technol.*, **25**, 1717–1723. © the American Meteorological Society, used with permission)

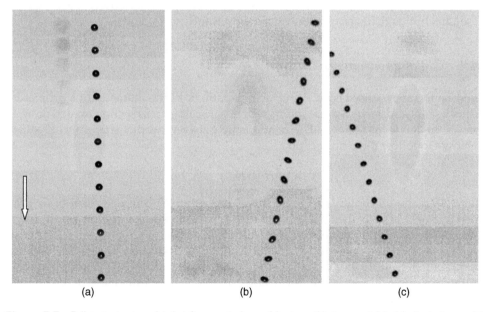

Figure 7.7 Fall trajectories of 1.4–1.9 mm raindrops (a) at equilibrium and (b), (c) displaying multimode oscillations. These images are obtained by digitally combining the instantaneous raindrop images taken 1 ms apart. White arrow indicates the direction of gravity vector and the horizontal size of an image frame (43 mm) gives the scale (Adapted from Testik, F.Y., Barros, A.P., and Bliven, L.F. (2006) Field observations of multimode raindrop oscillations by high-speed imaging. *J. Atmos. Sci.*, **63**, 2663–2668. © the American Meteorological Society, used with permission)

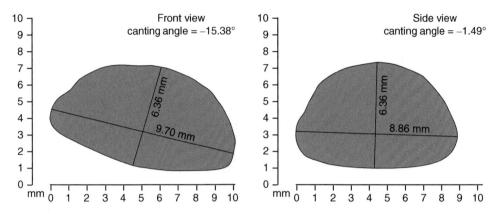

Figure 7.8   Two-dimensional video disdrometer images of a large droplet observed on April 23, 2011 in north-central Oklahoma. The equivalent diameter of the droplet was observed to be 8.7 mm (Patrick Gatlin, NASA MSFC)

graupel is melting), drop breakup is observed to occur when the drop diameter typically exceeds 6–8 mm, which defines an effective **maximum drop diameter**, although fully melted droplets have been observed by a 2DVD exceeding 9 mm during a field project in Oklahoma! Figure 7.8 shows an image of this large raindrop as imaged by a 2DVD.[3] Partially melted particles can exceed this size, but typically do not have the classical "hamburger bun" droplet shape that fully melted droplets have due to the ice core of the drop stabilizing the internal flows within the droplet. The assumption of maximum droplet diameter is important in simulating radar observables because these rare, but large drops can contribute significantly to Z_H and Z_{DR}.

7.4.4 Z_{DR} measurements in rain

Based on the above discussion, raindrop axis ratio links the diameter of a droplet in the horizontal and vertical axes, and its radar backscatter cross section in each dimension (in the Rayleigh regime). The relationship between droplet shape and Z_{DR} can be shown to follow the relationship

$$10^{\left(-\frac{Z_{DR}}{10}\right)} = A_Z^{\frac{7}{3}} \tag{7.11}$$

where A_Z is the reflectivity-weighted mean axis ratio. With the physical basis provided by Eq. (7.11), it is then plausible to relate characteristics of the size of raindrops in the radar resolution volume to Z_{DR}. With scattering simulation codes, such as the T-matrix method, we can simulate the terms of Eq. (7.5) to produce simulated radar observables. Results from a T-Matrix simulation carried out at S-, C-, and X-band are shown for a single droplet, assuming droplet shape characteristics listed in the legend, for Z_H and Z_{DR} (Figure 7.9).

---

[3]Learn more about large raindrops in this study: Gatlin, P.N., Thurai, M., Bringi, V.N., Petersen, W., Wolff, D., Tokay, A., Carey, L., and Wingo, M. (2015) Searching for large raindrops: a global summary of two-dimensional video disdrometer observations. *J. Appl. Meteor. Climatol.*, **54**, 1069–1089.

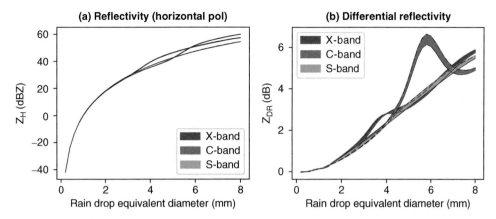

Figure 7.9 T-matrix simulations of a raindrop (a) reflectivity at horizontal polarization and (b) differential reflectivity versus drop equivalent diameter at S-band (green), C-band (blue), and X-band (red). The dashed and solid lines denote different axis ratio parameterizations. Droplet canting was modeled with a Gaussian distribution with a 7° standard deviation

For a distribution of raindrops, a relation between the median volume diameter (D_0, the diameter at which equal parts of the volume distribution are divided) and Z_{DR}, again using a raindrop diameter-shaped model derived from disdrometer measurements in Florida thunderstorms with an assumed canting angle distribution, and valid at S-band, is:

$$D_0 = 1.81 Z_{DR}^{0.486} \tag{7.12}$$

High values of Z_{DR} at S-band can indicate the presence of "big drops" with large median diameters. In precipitation, observations of Z_{DR} can give insights into the microphysical and kinematic processes occurring within the storms.

At radar wavelengths shorter than S-band, non-Rayleigh effects can modify the relationship between drop size characteristics and Z_{DR} compared with S-band. For large drops, the diameters are different in the horizontal and vertical plane relative to the radar wavelength (note that a particle diameter about 0.2 times the radar wavelength is the cutoff between the Rayleigh and the non-Rayleigh regime). Departures from Rayleigh scattering at shorter wavelengths lead to differing backscatter cross sections that fluctuate with drop diameter depending on the wavelength being considered. At C-band (5 cm wavelength), 5–8 mm drops in convective storms can cause significant fluctuations in Z_{DR}, typically leading to larger values where large droplets exist.[4] Figure 7.10 illustrates a convective storm in Oklahoma observed with an S-band and a C-band radar at nearly the same time. Note regions of high Z_{HH} where C-band Z_{DR} is observed to be several decibels higher than S-band Z_{DR} in the same regions of the storm.

In addition, in high concentrations of large drops (or oblate melting hail), a phenomenon called **differential attenuation** can affect measurements of Z_{DR} (see Chapter 9). Differential attenuation occurs in dual-polarization radar measurements

---

[4]More about this effect can be found here: May, P.T., Keenan, T.D., Zrnić, D.S., Carey, L.D., and Rutledge, S.A. (1999) Polarimetric radar measurements of tropical rain at a 5-cm wavelength. *J. Appl. Meteor.*, **38**, 750–765.

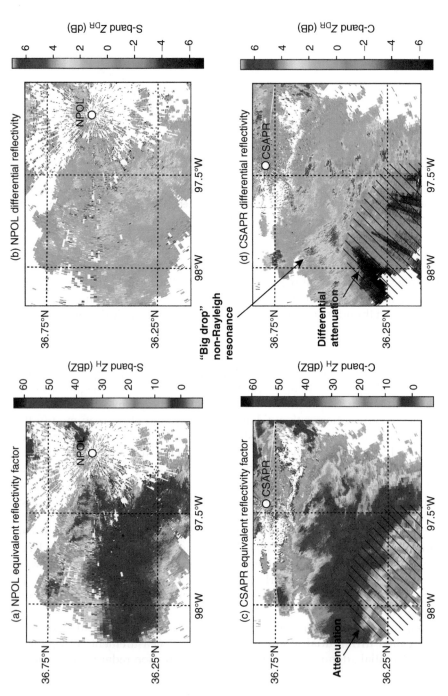

Figure 7.10 Comparison of nearly time-matched 0.7° PPI scans from the NASA NPOL S-band radar, sited near Marland, Oklahoma (panels a and b), and the Department of Energy Atmospheric Radiation Measurement Climate Research Facility CSAPR C-band radar near Nardin, Oklahoma (panels c and d). Panels a and c are reflectivity at horizontal polarization, whereas panels b and d are differential reflectivity. The NPOL sweep began at May 23, 2011 at 22:21 UTC, whereas the CSAPR sweep began at 22:16 UTC. Regions of attenuation and differential attenuation are noted with hatching in the CSAPR images

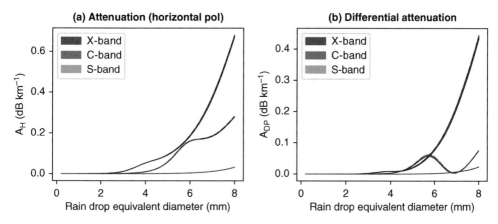

Figure 7.11 As in Figure 7.9, but for (a) specific attenuation at horizontal polarization and (b) specific differential attenuation

because, due to the orientation of the particles in the radar resolution volume, more power is absorbed (attenuated) in one polarization plane than in another plane. Large droplets and melting hail have their long axis typically oriented in the horizontal plane (consistent with $Z_{DR} > 0$). This will lead to more attenuation in the horizontal versus the vertical plane, effectively causing the observed Z_{DR} to decrease with increasing attenuation along the beam. The simulations shown in Figure 7.11 show that attenuation and differential attenuation is typically small at S-band but becomes significant at C-band and shorter wavelengths. Note the region in Figure 7.10 where differential attenuation leads to a negative bias in Z_{DR} that increases with range in the C-band radar measurements, clearly indicating differential attenuation. Although attenuation is an obstacle for using Z_H for estimating rainfall,[5] differential attenuation and non-Rayleigh effects can be an obstacle in using measurements of Z_{DR} for estimating raindrop size characteristics, classifying hydrometeor types such as hail, and estimating rainfall. However, as we will see, shorter wavelength radars have some advantages in terms of the behavior of their propagation polarization variables that can mitigate some of the shortcomings of having to correct Z_{DR} measurement issues.

7.4.5 Z_{DR} measurements in ice and mixed-phase precipitation

Hail and graupel tend to have a more spherical shape than rain, although the shape of hail particles can be quite complicated depending on growth mechanisms. Figure 7.12 illustrates the typical observed characteristics of high-density and melting graupel and hail particles. Even if hail particles are not perfectly spherical, hail tends to tumble as it falls, which yields a very broad distribution

---

[5]Read more about using polarimetric self-consistency methods for correcting for attenuation: Testud, J., Le Bouar, E., Obligis, E., and Ali-Mehenni, M. (2000) The rain profiling algorithm applied to polarimetric weather radar. *J. Atmos. Oceanic Technol.*, **17**, 332–356.

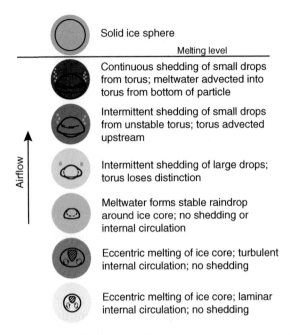

Solid ice sphere

Melting level

Continuous shedding of small drops from torus; meltwater advected into torus from bottom of particle

Intermittent shedding of small drops from unstable torus; torus advected upstream

Intermittent shedding of large drops; torus loses distinction

Meltwater forms stable raindrop around ice core; no shedding or internal circulation

Eccentric melting of ice core; turbulent internal circulation; no shedding

Eccentric melting of ice core; laminar internal circulation; no shedding

Airflow

Figure 7.12   Vertical wind tunnel observations of the time evolution of a melting hail particle (Adapted from Rasmussen, R.M., Levizzani, V., and Pruppacher, H.R. (1984) A wind tunnel and theoretical study on the melting behavior of atmospheric ice particles: III. Experiment and theory for spherical ice particles of radius > 500 μm. *J. Atmos. Sci.*, **41**, 381–388. © the American Meteorological Society, used with permission)

of canting angles. Therefore, hail *that is not coated with significant amounts of water* will tend to appear geometrically spherical to the radar. This physical fact leads to the observation that if hail is not coated with significant amounts of accreted or melted water, it will tend to have Z_{DR} near 0 dB. At S-band, this is a key distinction for reflectivities at which hail tends to be present ($Z_H > 50–55$ dBZ): regions with high reflectivity and large Z_{DR} are physically linked with rain, whereas regions with high reflectivity and low, near-zero, or even negative Z_{DR} can be classified as hail. Negative or even fluctuating Z_{DR} can be caused by non-Rayleigh scattering in very large hail >4 cm at S-band and can complicate interpretations of Z_{DR} for potential hail damage estimation. Figure 7.13 shows observations of a hail core in a supercell thunderstorm observed from the Front Range, Colorado WSR-88D radar (east of Denver), where within high-reflectivity regions in the storm, the regions of rain and hail are clearly discriminated using Z_{DR}. In the plot of a ray at 262.5° azimuth through the core of this same storm (Figure 7.14), the high-reflectivity region (>55 dBZ) between 27 and 42 km range is accompanied by near-zero Z_{DR}. This polarization characteristic of hail is widely used operationally for the detection of hail at S-band.

Hail tends to be mixed with rain in convective storms, and coexists with supercooled cloud water droplets or raindrops in convective updrafts. Thus, the coexistence of water-coated hail, graupel, and liquid water is common in convective updrafts. This can lead to some ambiguity in the behavior of the polarimetric

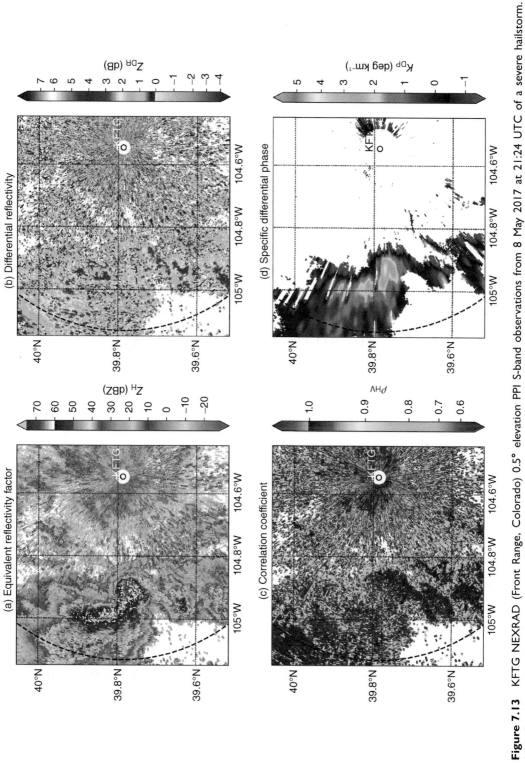

Figure 7.13 KFTG NEXRAD (Front Range, Colorado) 0.5° elevation PPI S-band observations from 8 May 2017 at 21:24 UTC of a severe hailstorm. (a) Reflectivity at horizontal polarization, (b) differential reflectivity, (c) co-polar correlation coefficient, and (d) specific differential phase

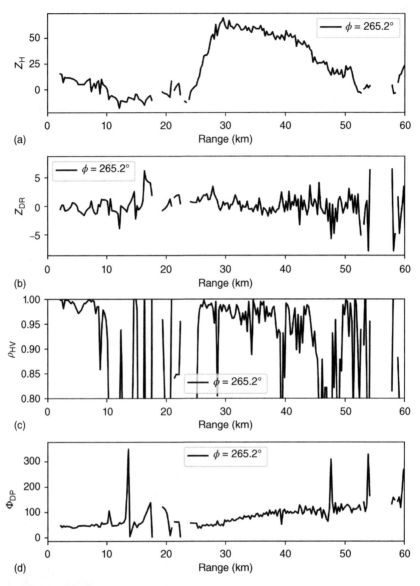

Figure 7.14 KFTG NEXRAD (Front Range, Colorado) 0.5° elevation ray plot at 265.2° azimuth from the severe hailstorm shown in Figure 7.13. (a) Reflectivity at horizontal polarization (dBZ), (b) differential reflectivity (dB), (c) co-polar correlation coefficient, and (d) differential phase (deg)

variables in these regions, as the physical characteristics of each of these particle types are quite different. In convective mixed-phase regions, few *in situ* measurements are available to validate radar retrievals. However, one common observation in deep convective updrafts containing mixed-phase regions is the presence of a Z_{DR} column. These are regions that extend from the surface through the depth of the mixed-phase region where liquid and frozen raindrops as well as melting hail with oblate shapes exist and lead to positive Z_{DR} columns that extend above the level of the 0 °C isotherm.[6]

At wavelengths shorter than S-band, the physical characteristics of these particles, non-Rayleigh effects, and differential attenuation lead to difficulties in determining regions of melting hail using Z_{DR}. For example, as shown in Figure 7.10, water-coated melting hail with a diameter greater than 5 mm, which may have a **taurus** of water, likely contributes to significant non-Rayleigh effects and differential attenuation at C-band that would not be observed at S-band. These effects lead to nonzero values of Z_{DR} that complicate interpretations of Z_{DR} at C- and X-band for detecting hail. Thus, at wavelengths shorter than S-band, non-Rayleigh effects can complicate the interpretation of Z_{DR} for hail detection.

In pristine ice crystals and snow aggregates, Z_{DR} measurements can be used to detect information about the orientation and shape of the particles in the radar resolution volume. Pristine ice particles, such as dendrites and plates, are observed to typically fall with their long axis parallel to the horizon. The complex shapes of dendrites in particular leads to positive Z_{DR} in these regions. This produces relatively large values of Z_{DR} in these regions near the dendritic growth layer between −13 and −18 °C as shown in the quasi-vertical profile (QVP) diagram in Figure 7.15. A QVP diagram is a time–height cross section from a ground-based radar that displays azimuthally averaged radar variables as a function of height. Typically, high elevation plan position indicator (PPI) scans are used (>5 °C) in generating a QVP.

However, dendrites and other pristine ice particles typically aggregate to form larger snowflake aggregates. These particles can have a very complex non-spherical shape. When aggregation begins with the first few crystals joining together, it is thought that the forces on the particles and the typical shape of these particles lead to the formation of a chain with a small axis ratio, oriented horizontally. Once aggregation continues, however, the aggregates tend to become more spherically shaped and tend to cant/flutter, leading to commonly observed values of Z_{DR} near 0 dB in snow aggregates. In snow, Z_{DR} values near 0 dB in high values of Z_{HH} (>30 dBZ) usually mean that moderate to heavy snow is occurring. In snow and ice precipitation, attenuation is small, and thus attenuation and differential attenuation are typically negligible.

7.4.6 Linear depolarization ratio (L_{DR})

Linear depolarization ratio (L_{DR}, units of dB) is another intrinsic polarization variable that can be used to infer the properties of hydrometeors. L_{DR} is defined by

$$L_{DR} = 10 \log_{10} \left(\frac{|S^{VH}|^2}{|S^{VV}|^2} \right) = 10 \log_{10} \left(\frac{Z_{VH}}{Z_{VV}} \right) \tag{7.13}$$

where Z_{VH} is the reflectivity factor measured in a cross-polar manner, that is, when transmitting in one polarization (i.e., vertical polarization) and receiving in the opposite polarization (i.e., horizontal polarization). As discussed, only dual-polarization

---

[6]Learn more about the physical processes leading to differential reflectivity columns: Kumjian, M.R., Khain, A.P., Benmoshe, N., Ilotoviz, E., Ryzhkov, A.V., and Phillips, V.T. (2014) The anatomy and physics of ZDR columns: investigating a polarimetric radar signature with a spectral bin microphysical model. *J. Appl. Meteor. Climatol.*, **53**, 1820–1843.

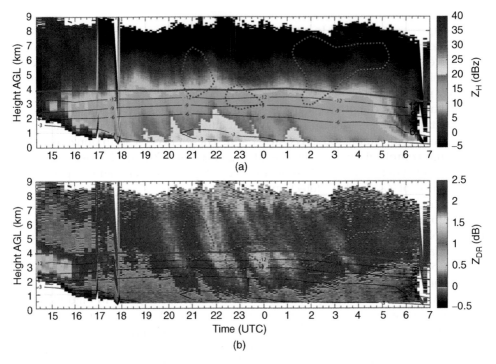

Figure 7.15 Time–height quasi-vertical profiles (QVPs) of (a) reflectivity at horizontal polarization (dBZ) and (b) differential reflectivity (dB) observed from February 15, 2014 to February 16, 2014, taken by the KBOX NEXRAD (Boston, Massachusetts) at 10° elevation. Contours of RAP model-analyzed temperature starting at −15 °C (magenta) in 3 °C increments (black) and pressure vertical velocity starting at −1 Pa s⁻¹ in −1 Pa s⁻¹ increments (heavy dotted) are overlaid (Adapted from Kumjian, M.R. and Lombardo, K.A. (2017) Insights into the evolving microphysical and kinematic structure of North-eastern U.S. Winter Storms from dual-polarization Doppler radar. *Mon. Weather Rev.*, **145**, 1033–1061. © the American Meteorological Society, used with permission)

radars that transmit each polarization separately, or alternatively, can measure L_{DR} directly. Presently, this capability is limited mostly to research radars. Radars that transmit both polarizations simultaneously and receive polarizations separately, such as the WSR-88Ds and most other operational radars, cannot measure L_{DR} because, as shown in Eq. (7.8), it is required that the transmitted wave and the received wave be orthogonal to each other, not polarized at 45° as is done in simultaneous transmission.

L_{DR} measures the degree to which the transmitted signal is *depolarized*; that is, to what degree the intrinsic scattering properties of the hydrometeors within the radar resolution volume cause the transmitted wave to lose its "pure" transmitted polarization to the orthogonal polarization. At weather radar frequencies, the amount of depolarization is very small (about 15–40 dB less than co-polar power), and so measurements of this quantity require excellent cross-polar isolation of the two polarizations (i.e., little leakage between channels). This "leakage" can arise from the design of the antenna, feedhorn, and radome. In practice, about −35 dB is the smallest value of L_{DR} that can be measured.

In rain and snow, L_{DR} will not typically exceed small values (i.e., −26 dB). In regions of tumbling quasi-spherical hail, L_{DR} will increase to as high as about −15 dB. Figure 7.16 shows a vertical cross section from the CSU-CHILL radar system, which

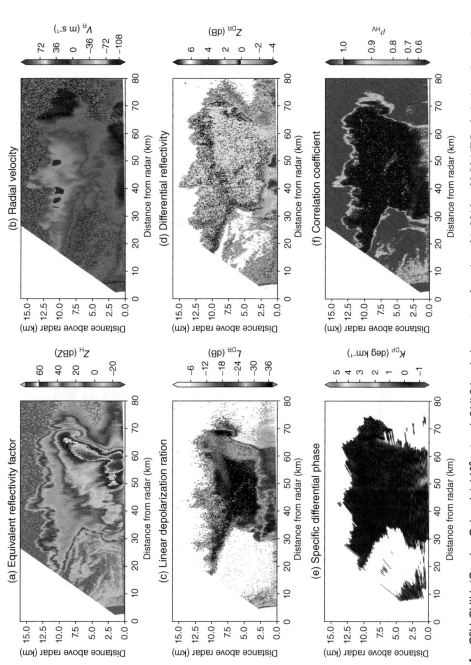

Figure 7.16 CSU-CHILL (Greeley, Colorado) 160° azimuth RHI S-band observations from June 21, 2010 at 22:20 UTC of a hail-producing deep convective system. (a) Reflectivity at horizontal polarization, (b) Doppler radial velocity, (c) linear depolarization ratio, (d) differential reflectivity, (e) specific differential phase, and (f) correlation coefficient

is a dual-transmitter polarization-agile system capable of measuring L_{DR}. What is seen in the cross section is a deep convective system, showing a convective region on the right side of the figure, with an anvil/stratiform precipitation on the left side. The Doppler velocity panel shows descending outflow moving away from the radar at low levels and a classic divergence signature above 8 km, consistent with an organized convective system. This system produced significant hail at the surface, and the polarimetric signatures are consistent with those reports. Figure 7.17 isolates

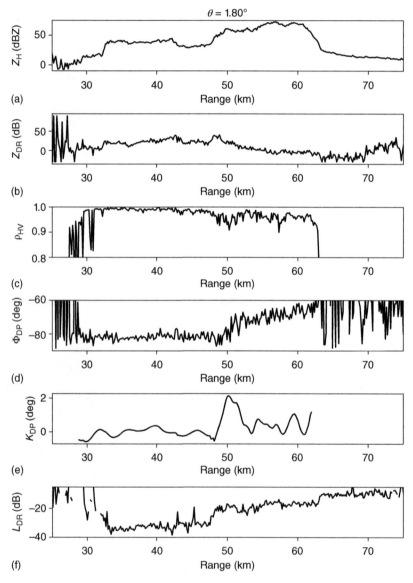

Figure 7.17   CSU-CHILL (Greeley, Colorado) 1.25° elevation 160° azimuth ray plot from the convective system shown in Figure 7.16. (a) Reflectivity at horizontal polarization, (b) differential reflectivity, (c) correlation coefficient, (d) differential phase, (e) specific differential phase, and (f) linear depolarization ratio

the observed radar variables at a ray at 1.08° elevation along this range-height indicator (RHI). The signatures indicate that there are a variety of precipitation types falling along this ray. Between 30 and 45 km range, moderate rain is evident. This is due to a signature of moderate Z_H, moderate to high Z_{DR}, and low L_{DR}. The precipitation transitions to rain mixed with hail near 50 km range near the surface (as evidenced by high Z_H, high Z_{DR}, and elevated L_{DR}—and reduced ρ_{HV}, to be discussed later), whereas by 55–60 km range near the surface, the high Z_H, low Z_{DR}, and elevated L_{DR} display a classic hail signature.

Polarimetric fields displayed on an RHI allow one to probe some of the microphysical processes occurring in a storm. In Figure 7.16, aloft, a convective updraft has created high Z_H values extending into the upper troposphere by lofting large hydrometeors to high altitudes. The precipitation process in deep thunderstorm updrafts is mixed phase, including wet and dry hail growth, as well as the lofting of supercooled cloud droplets and raindrops important for rapid particle growth. In these regions, evident in the profile in Figure 7.16 near 58–65 km range in the RHI, several polarimetric signatures are evident. A column of elevated L_{DR} indicates the presence of tumbling hail and mixed-phase processes, which contribute to depolarization of the radar beam's linear polarization. In addition, the presence of lofted raindrops above the melting level near 65 km range and 5–7 km above the radar level is revealed with elevated Z_{DR} (and K_{DP}) values. A region of hail mixed with rain produces this signature, coincident with the top of a Z_{DR} (and K_{DP}) column or sometimes called an **L_{DR} cap**. This signature indicates a mixed-phase region, consistent with T-matrix simulations of tumbling quasi-spherical hail and graupel where depolarization of the incident scattered wave is occurring. A region of hydrometeor melting is evidenced in the quasi-horizontal layer near 2.5 km altitude above the radar between 30 and 50 km range, with high L_{DR}, and a transition from low to high Z_H and Z_{DR} with decreasing height, consistent with an increased dielectric factor of snow versus rain (although a classic "bright band" is not present), and reduced ρ_{HV} (to be discussed later, although here as in the convective region, lower ρ_{HV} usually indicates a mixture of hydrometeor types in meteorological echo).

L_{DR} has been related to the detection and sizing of hail at the surface; however, its utility in detecting hail characteristics over techniques using Z_{DR} has shown limited promise to date. This, combined with the added expense of adding a second transmit channel to the radar system, and requiring excellent cross-polar isolation, makes the added costs prohibitive for most operational radars. It is clear in the example above that much of the information contained in L_{DR} is apparent in combinations of the other polarimetric variables, which are available on a polarization radar that is not agile.

7.5 Polarization and phase

So far, we have discussed intrinsic polarization variables that are dependent on the co-polar or cross-polar returned power. The received wave at each polarization has a phase in each polarization (ϕ_{HH} and ϕ_{VV}), which differs from the transmitted wave due to interaction with the particles within the resolution volume, as well as during the propagation along the radar beam's path. In terms of intrinsic polarization variables, there are two variables related to the statistical properties of the phase measured across radar pulses in a given resolution volume and one related to the phase shift between polarizations that occurs upon scattering. We will discuss the latter when we discuss differential propagation phase shift below.

As discussed previously, the radar transmits and receives a number of pulses in each radar resolution volume in order to attempt to remove statistical fluctuations in each pulse. In a Doppler radar system, the phase shift relative to the transmitted wave of a number of samples is used to calculate the Doppler spectrum, from which moments such as Doppler velocity and spectral width are calculated. However, a dual-polarization radar also measures the phase shift of each pulse at each polarization. Assuming that the component of the Doppler shift of each pulse due to particle motion should be equal at each polarization, the remaining part of the phase shift is due to polarization-dependent interactions with matter along the radar resolution volume. Characteristics of the phase properties of the received pulses or changes in the received phase along the ray path can tell us about the properties of the media within each radar resolution volume.

There are two ways that this differential phase shift between polarizations can occur in meteorological echo. One is due to the phase shift associated with absorption, which is a propagation effect, and the other is due to non-Rayleigh scattering, which produces a local modification of the differential phase. Here we describe each of these, termed the **propagation differential phase shift** and **backscatter differential phase shift.**

7.5.1   Propagation differential phase shift (ϕ_{DP})

The illustration in Figure 7.18 illustrates the physical process of how matter can interact with the phase and amplitude of electromagnetic energy. Recall that the complex index of refraction, m, of matter can be given by $m = n + ik_c$. In panel a of Figure 7.18, an electromagnetic wave is assumed to propagate in a vacuum, where n and k_c are 1 and 0, respectively. The index of refraction of a medium is related to the speed of the electromagnetic energy in that medium. More specifically, the velocity of light in a medium (v_c) is related to the real part of the complex index of refraction:

$$n = \frac{c}{v_c} \qquad (7.14)$$

In a vacuum, n is 1 and so the velocity of light is simply equal to the speed of light in a vacuum. In air, even at sea level where it is most dense and the speed of light is the slowest, the real part of the refractive index of air at radar frequencies differs from a vacuum by only approximately 0.03%. However, in precipitation, n increases to a value of about 8 (unitless) at S-band (note that it is a function of temperature and wavelength). This indicates that the speed of light is quite a bit slower in water than air. In precipitation, the amount of water relative to air is only about 0.1%, so the slowdown is not large. However, this slowdown, as illustrated in Figure 7.18b, will cause a phase shift ($\Delta\phi$) between waves traversing matter with one effective real part of the complex refractive index (panel a) and matter with a different refractive index (panel b).

With a non-polarimetric radar, this **phase shift** at a given polarization is small, and it is swamped in magnitude by the phase shift due to motion of the objects in the radar resolution volume. However, with a dual-polarization radar, we can measure

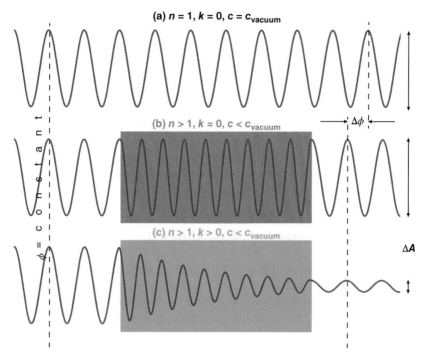

(a) $n = 1, k = 0, c = c_{vacuum}$

(b) $n > 1, k \approx 0, c < c_{vacuum}$

(c) $n > 1, k > 0, c < c_{vacuum}$

$\Delta\phi$

ΔA

$\phi = constant$

Figure 7.18 Illustration of the differential phase shift ($\Delta\phi$) and attenuation (ΔA) effects of passing through matter of different complex indices of refraction

the Doppler shift at each polarization from each pulse, which we call the **differential phase**. Essentially, we can cancel out the motion of the objects in the radar resolution volume over a number of pulses (because we measure phase at each polarization) and then look at the deviation from the phase value associated with the passage of the microwaves through liquid water at each polarization, this causing the observed differential phase shift. Why do we care about this phase shift? If there is a phase shift of one polarization relative to another, then based on the principle above (ignoring the effects of non-Rayleigh scattering that will be discussed below), more matter must be slowing down the electromagnetic wave at one polarization relative to another. We have shown above, for example, that raindrops are oblate, which effectively puts more matter with a slower electromagnetic "speed limit" at one polarization relative to another. This causes the "traffic jam" depicted in Figure 7.18, leading to a phase shift illustrated in the figure. This process is called a **differential phase shift** or ϕ_{DP}—as the beam propagates through a collection of oblate raindrops, a larger slowdown of the wave happens at one polarization relative to the other, so the phase shift increases with range.

In addition, note that the complex part k of the complex index of refraction is proportional to the absorption of an electromagnetic wave at that frequency. Thus, for a value of n that is greater than 1 and a value of k that is greater than 0, there will be a differential phase shift and absorption (which we call attenuation in radar meteorology). If we know the complex index of refraction value for matter, such as for rain, and as both propagation differential phase and attenuation given by that index of refraction are proportional to the mass of matter in the radar resolution, it is possible to use differential phase shift to estimate, and correct, the amount of attenuation in Z_H, and potentially differential attenuation in Z_{DR}. In rain, it turns out that k_c has a value of 0.9 at S-band (at 10 °C), whereas at X-band, k_c increases to 2.44 at the same temperature. These data physically illustrate that attenuation and differential phase increase together with decreasing wavelength at weather radar wavelengths. Although attenuation is relatively small at S-band, so is the differential propagation phase shift, so rainfall estimates using differential phase have a relatively small signal compared with shorter wavelengths. By contrast, at shorter radar wavelengths, the differential phase shift is larger and can be used to estimate rain rates at lower rain rates, but there is more attenuation in reflectivity and differential reflectivity, making those estimates more uncertain and in need of more correction. This conundrum is yet another of the trade-offs when designing a radar system.

7.5.2   *Backscatter differential phase shift (δ)*

The second type of differential phase shift that can occur is due to non-Rayleigh scattering. Recall that one of the characteristics of non-Rayleigh scattering is that energy is no longer scattered more or less equally in all directions as in Rayleigh scattering. During the fluctuations of the backscattering cross section, changes in the amount of backscattered energy are also accompanied by changes in the backscattered phase. This is caused by constructive and destructive interference as the wavelength of the energy approaches the size of the particle. Essentially, this causes the backscattered phase to be modified relative to the initial phase. This quantity is called **backscatter differential phase** (δ). This effect differs from ϕ_{DP} in that it is a local effect rather than a propagation effect. This means that when the scatterers in the radar resolution volume produce δ (which is always a positive differential phase shift), that it only occurs where those hydrometeors exist and is not observable further in range like the propagation effects that lead to increasing ϕ_{DP} with range. Thus, measured values of total differential phase will show a temporary spike where there is a typically positive contribution from δ and then return to the local value of ϕ_{DP} caused by the propagation along the beam.

Although there is information in δ in terms of detecting where non-Rayleigh scattering exists in the beam, δ, when it is present, typically turns out to be a headache in terms of estimating ϕ_{DP} to estimate precipitation mass content and rate. As the non-Rayleigh effects are strongly dependent on hydrometeor diameter, relating values of δ to a quantitative physical process is difficult because the drop size distribution is typically unknown, and δ is small in magnitude, and thus may be difficult to separate from noise in differential phase measurements. Therefore, attempts are typically made to filter out δ when retrieving rainfall using differential phase measurements.

7.5.3   *Specific differential phase (K_{DP})*

As stated above, the useful property of propagation differential phase (ϕ_{DP}) values is that they are nearly directly proportional to the amount of mass in the radar resolution volume for a given observed particle oblateness–diameter model. What this means is that these measurements allow for an estimate of the precipitation water content, provided that the phase shift is sufficiently large enough to measure (i.e., there is sufficient mass to produce a measurable phase shift).

To estimate the phase shift at each range gate along the beam, we define **specific differential phase** (K_{DP}, units are $deg\,km^{-1}$) as

$$K_{DP} = \frac{1}{2}\frac{d(\phi_{DP})}{dr} \tag{7.15}$$

The term $d(\phi_{DP})/dr$ is the range derivative of ϕ_{DP}. The factor of $1/2$ arises in Eq. (7.15) because a phase shift occurs on the way to the radar resolution volume and back; propagation differential phase shift is a two-way effect, like attenuation. Figure 7.19a shows T-matrix calculations of K_{DP} for raindrops as a function of diameter at S-, C-, and X-band. Note the strong frequency dependence of K_{DP} and the non-Rayleigh effects at C-band for large drops around an equivalent diameter of 6 mm. Note that these results apply over a drop size distribution, so such large fluctuations are typically canceled out such that the effective C-band K_{DP} falls in between S- and X-band. These results emphasize the fact that K_{DP} is roughly inversely proportional to wavelength for typical drop size and drop shape distributions.

Figure 7.19b displays T-matrix simulations of backscatter differential phase (δ). The effects of backscatter differential phase δ are negligible at S-band in rain (note that this may not be the case with large, melting hail or melting snow aggregates >1–2 cm in diameter, where the effective scattering properties of the particles may approach that of water). At C-band, δ becomes apparent for drops exceeding 6 mm in equivalent diameter, then fluctuates at equivalent diameters above 7 mm. At X-band, $\hat{\delta}$ is more proportional to equivalent diameter, which, in some sense, makes δ easier to predict at X-band compared with C-band given an assumed drop size distribution.

When do we expect K_{DP} to be significant in real hydrometeors? Physically, K_{DP} can be calculated from the following equation:

$$K_{DP} = \frac{180\lambda}{\pi}\int_{D_{min}}^{D_{max}} Re[f^H - f^V]N(D)dD \tag{7.16}$$

In Eq. (7.16), the notation Re[ ] denotes the real part of the difference in scattering matrix values, which is multiplied by the number concentration (in units of inverse

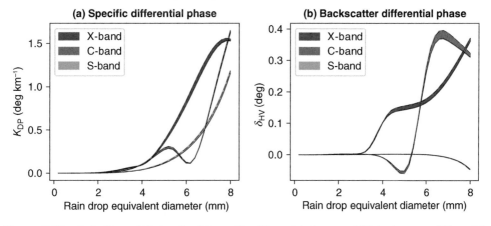

Figure 7.19   As in Figure 7.9, but for (a) specific differential phase and (b) backscatter differential phase

length times volume, i.e., m^{-4}) in each bin, integrated over the extreme values of the drop diameter distribution ($D_{\min}$, $D_{\max}$). Here, f^H and f^N are the *forward* scattering amplitudes (i.e., how much of the electromagnetic wave is scattered in the forward direction) of the particles, which can be estimated from scattering calculations for a particular hydrometeor type (as in Figure 7.19). It turns out that if one assumes a drop shape diameter relationship, K_{DP} can be shown to be directly proportional to mass-weighted droplet axis ratio (A_M):

$$K_{DP} = \frac{180}{\lambda} \cdot 10^{-3} \cdot C \cdot LWC \, (1 - A_M) \tag{7.17}$$

In Eq. (7.17), LWC is the liquid water content (g m^{-3}), and the value of C is approximately 3.75, which is independent of wavelength. Mass-weighted droplet axis ratio (A_M) can be calculated using

$$A_M = \frac{\int_{D_{\max}}^{D_{\min}} M(D) \, A(D) \, N(D) dD}{\int_{D_{\max}}^{D_{\min}} M(D) \, N(D) \, dD} \tag{7.18}$$

where $M(D)$ is the mass of a spherical droplet of a given diameter and $A(D)$ is the axis ratio value in a given particle diameter size bin. From Eqs. (7.17) and (7.18), it is clear from this theoretical perspective that K_{DP} is inversely proportional to wavelength and directly proportional to the mass-weighted "oblateness" of the droplets $(1 - A_M)$ times the liquid water content. Of course, according to Eq. (7.16), if prolate particles are present, then a negative K_{DP} is possible (however, this only happens in rare situations, like when strong electric fields in thunderstorm anvils orient non-spherical ice crystals with their long axis vertically). This means that K_{DP} should be positive for oblate drops and is controlled by both the number concentration of drops and their mass-weighted axis ratios. With a model of the relationship between mass-weighted axis ratio and droplet diameter, it is clear that K_{DP} enables retrievals of characteristics of the droplet particle size distribution in rain, which enables accurate estimates of liquid water content and precipitation.

From observations, situations that lead to positive values of K_{DP} at S-band (ignoring non-Rayleigh δ effects) and values near zero are shown in Figure 7.20 on the left and right sides, respectively. According to Eq. (7.16), positive K_{DP} is the product of number concentration and a factor related to the oblateness of the drops. As discussed above, this condition is commonly satisfied when droplets become larger than, say, 4 mm in convective rain situations. Oblate drops can exist below or above the melting level, as convective updrafts can loft large raindrops to altitudes above the melting level. Situations where hail is water coated and forms a taurus of water in an oblate manner (as shown in Figure 7.12) also produce positive K_{DP}. Third, high concentrations of horizontally oriented aggregates of ice crystals can also produce positive K_{DP}. This has been observed to happen in regions of active initial particle aggregation, where large number concentrations of crystals first join together. Aerodynamically, these particles tend to fall horizontally, giving an oblate presentation to the radar. As they grow, larger aggregates, with diameters exceeding 10 mm, tend to fall more randomly and their number concentration decreases (because ice particles are joining together), leading to near 0 deg km^{-1} K_{DP}. Note that when non-Rayleigh effects are present, for example, in hail at C-band, or at wavelengths less than K$_u$-band, these

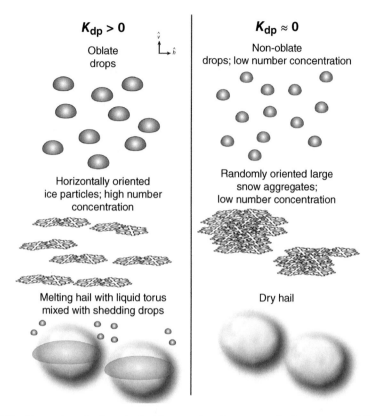

Figure 7.20 Illustration of microphysical factors that, in practice, lead to positive differential phase (left panel) and small to zero differential phase (right panel)

rules of thumb are not necessarily applicable due to the fluctuating forward scattering power amplitudes in such conditions.

7.5.4   Retrieval of K_{DP}

Although desirable properties of K_{DP} are related to physical properties we want to measure in the atmosphere (liquid water content, number concentration, mass-weighted mean axis ratio), unfortunately, we cannot measure K_{DP} directly with a polarimetric radar. In practice, we measure the total differential phase shift Φ_{DP} (note the Φ is used here to denote measured differential phase shift vs the lowercase phi, which denotes the differential phase shift due to absorption discussed above) and must take a range derivative to calculate K_{DP}. It turns out that measurements of Φ_{DP} by a real radar system are somewhat noisy. For a typical radar system, sampling with 64 pulses yields a standard deviation of Φ_{DP} of 2–4° in rain when averaging 32 samples, which is the WSR-88D sampling strategy. One can average more samples, but then the antenna scan rate must be reduced by a factor of 2 to achieve the same effective spatial resolution, which may be undesirable when scanning rapidly evolving storms. Thus, extracting the signal to estimate ϕ_{DP} and δ to ultimately estimate K_{DP} is challenging.

We can write down this retrieval problem as

$$\Phi_{DP} = \phi_{DP} + \delta + \varepsilon \qquad (7.19)$$

where Φ_{DP} is measured by the radar and ϕ_{DP}, δ, and ε (which is the error due to noise in the total differential phase measurement) must be extracted from the measurement. From Eq. (7.15), we are actually interested in calculating the range derivative of ϕ_{DP} for estimating K_{DP}, and δ, and ε are fluctuating, reducing the ability to estimate the true value of the derivative. For small values of K_{DP}, the noise may overwhelm the signal, making K_{DP} difficult or impossible to retrieve. This is particularly challenging at S-band, where K_{DP} is small at rain rates below approximately $10\,\mathrm{mm\,h^{-1}}$. Fortunately at S-band, δ is relatively small (except in the radar bright band at high elevation angles, or in large melting hail), so that makes the retrieval more tractable. Even though ε is the only noise contributor in light precipitation, it can cause the range derivative of ϕ_{DP} to be <0. At shorter wavelengths, δ is not negligible in rain with large drops or hail and complicates the retrieval of K_{DP}. However, as discussed above, K_{DP} scales with wavelength, so the contribution of the signal to the range derivative of ϕ_{DP} is larger at shorter weather radar wavelengths.

Figure 7.21 illustrates measurements of S-band Φ_{DP}, its 11 gate-averaged standard deviation σ_{DP}, and a retrieval of K_{DP} using a technique that uses a noise filter called a

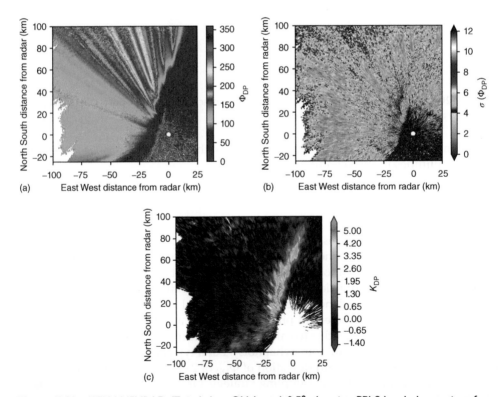

Figure 7.21   KTLX NEXRAD (Twin Lakes, Oklahoma) 0.5° elevation PPI S-band observations from April 21, 2017 at 14:31 UTC of a leading line-trailing stratiform convective system. (a) Differential phase (deg), (b) 11-gate moving window standard deviation of differential phase (deg), and (c) specific differential phase (deg km^{-1})

Finite Impulse Response filter on Φ_{DP} to smooth out ε and estimate K_{DP}. This technique[7] works over a specified number of gates (here 11) to remove the noise from the signal and then estimate the range derivative of Φ_{DP}. Other approaches are available in the literature to perform this filtering; each has strengths and weaknesses. Here, as is common, we ignore contributions of δ at S-band, although δ could be parameterized based on some other radar parameter at each gate and included in the retrieval. In addition, Φ_{DP} can become folded: if there is more than 360° of phase shift between the two measured polarizations (i.e., in long paths of heavy rain, the phase shift can wrap around); clearly, this is a problem when estimating Eq. (7.15). Therefore, quality control of Φ_{DP} is sometimes required.

In Figure 7.21, a leading line-trailing stratiform convective system in central Oklahoma is shown, which contained high values of K_{DP} exceeding 5 deg km^{-1} within the convective line. As K_{DP} is highly related to liquid water content, it can be used as an accurate precipitation estimator in such situations (see Chapter 13). However, in lighter precipitation behind the leading convective line, notice that the retrieval of K_{DP} is noisy. This is due to the inability of the filter to overcome the noise in the measurement of Φ_{DP}. Thus, K_{DP} does not provide a reliable estimate of precipitation-related quantities in such regions at S-band. Shorter wavelengths have a higher signal-to-noise ratio and have improved K_{DP} performance in such regions. These ideas are summarized in Figure 7.22, which is a ray plot at S-band through a convective storm, as evidenced by high values of Z_H exceeding 50 dBZ in panel (a). In panel (b), the measured Φ_{DP} is shown. Outside of precipitation Φ_{DP} is more or less random, which enables it to be used as a quality control variable (a range gate-averaged value of σ_{DP} greater than a threshold, shown in the third row, can be used to identify non-meteorological echo in other variables; here, 12 deg km^{-1} is used and applied to the K_{DP} retrieval). Finally, the 11-gate Finite Impulse Response retrieval of K_{DP} for this ray is shown in the fourth column. Note that the retrieval produces a negative value of K_{DP} on the edge of the cell; this is not uncommon and is not due to prolate precipitation particles, it is simply noise entering into the filter.

Another example of K_{DP} is shown in Figure 7.23, for a convective system near the Tulsa, Oklahoma WSR-88D. Here, you can see the association of high Z_H, high Z_{DR}, and high K_{DP} in convective cores. This is a classic signature of heavy rain—a large sum of the drop diameters to the sixth power, large oblate drops, and large number concentrations of drops. Going back to Figure 7.13, do you find the same association for this storm? Your answer is probably yes and no. Note that there are some regions with high Z_H, low Z_{DR}, and varying values of K_{DP}. This is a storm with large hail. This is an example where K_{DP} can help discriminate regions of hail versus hail mixed with rain at S-band. Regions of high Z_H with low Z_{DR} and low K_{DP} means that there is a low concentration of oblate liquid drops, more spherical liquid drops, or no liquid drops. Conversely, regions of low Z_{DR} and elevated K_{DP} mean that there are oblate drops or melting hail in the region. In the high-reflectivity core of this storm (exceeding

---

[7]Read more about retrieving specific differential phase using the Finite Impulse Response filter: Hubbert, J. and Bringi, V.N. (1995) An iterative filtering technique for the analysis of copolar differential phase and dual-frequency radar measurements. *J. Atmos. Oceanic Technol.*, **12**, 643–648.

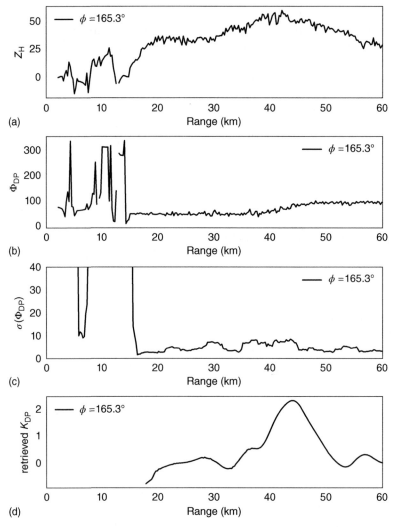

Figure 7.22   KVNX NEXRAD (Vance Air Force Base, Oklahoma) 0.5° elevation 165.3° azimuth ray plot from May 20, 2011 at 09:44 UTC through a leading line-trailing stratiform convective system. (a) Reflectivity at horizontal polarization (dBZ), (b) differential reflectivity (dB), (c) 11-gate moving window standard deviation of differential phase (deg), and (d) retrieved specific differential phase (deg km^{-1})

55 dBZ), you can see variations in Z_{DR} and K_{DP} consistent with these physical process inferences.

At C-band, the increased sensitivity of K_{DP} allows for a lower rain rate threshold for the application of K_{DP}-based rain rate estimators, but the influence of δ may lead to errors in the estimation of K_{DP}. Figure 7.24 shows, in the same format as in Figure 7.23, data from the CSAPR2 radar through the region of heavy rain and differential attenuation at an azimuth of 165° (see Figure 7.10). Here, note the possibility of the influence of δ near 52 km range—the temporary increase then subsequent leveling-off of Φ_{DP} in a high-reflectivity region could be due to δ. Also note that at C-band, there is likely attenuation in Z_H through this convective core, as evident in

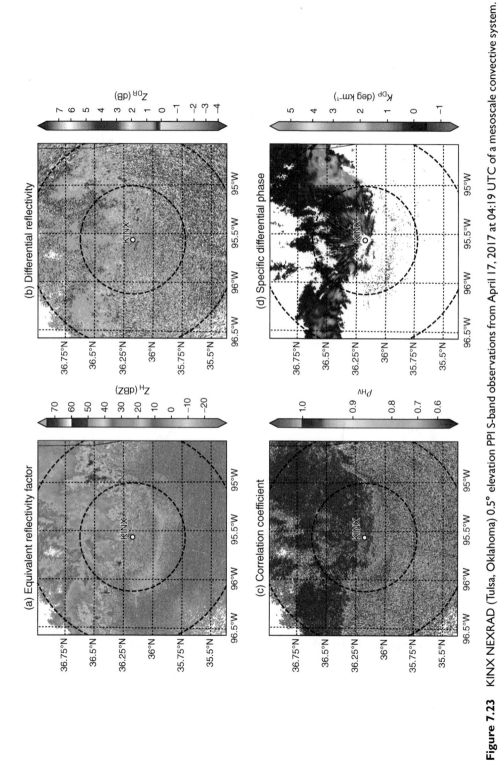

Figure 7.23 KINX NEXRAD (Tulsa, Oklahoma) 0.5° elevation PPI S-band observations from April 17, 2017 at 04:19 UTC of a mesoscale convective system. (a) Reflectivity at horizontal polarization, (b) differential reflectivity, (c) co-polar correlation coefficient, and (d) specific differential phase

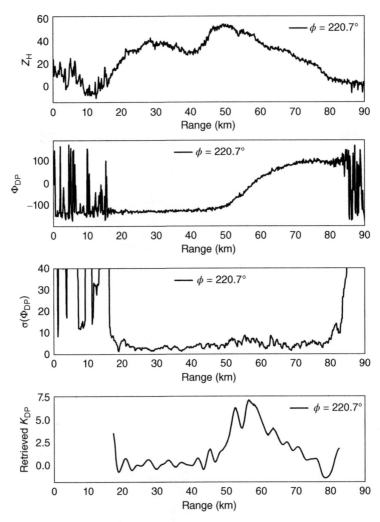

Figure 7.24 Department of Energy Atmospheric Radiation Measurement Climate Research Facility CSAPR C-band radar 0.7° elevation 220.7° azimuth ray plot of the convective system shown in Figure 7.10, panels b and d. (a) Reflectivity at horizontal polarization, (b) differential reflectivity, (c) co-polar correlation coefficient, and (d) specific differential phase

Figure 7.10. Note that the highest values of K_{DP} occur behind the highest Z_H values (compare 50 and 56 km range). Also, higher values of K_{DP} occur at farther ranges, with lower values of Z_H than before the convective core (compare 30 and 65 km range). This example reinforces the fact that K_{DP} is insensitive to attenuation and, based on the physical reasoning above, can be used to, in some cases, correct attenuation and differential attenuation, provided that the effects of noise and δ can be overcome, particularly in rain where the relationships between Z_H, Z_{DR}, and K_{DP} can be well constrained for a given drop size and drop shape parameterization. Furthermore, K_{DP}, like Doppler radial velocity measurements, do not need to be calibrated. For quantitative rainfall estimation in heavy rain, and especially at C- and X-band, K_{DP} is an important variable.

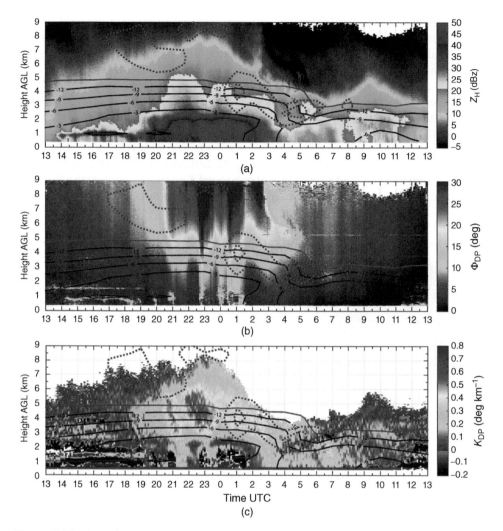

Figure 7.25 As in Figure 7.15, except showing data from the KOKX NEXRAD (Upton, New York) from February 8, 2013 to February 9, 2013. Plotted are (a) reflectivity at horizontal polarization, (b) differential phase, and (c) specific differential phase (Adapted from Kumjian, M.R. and Lombardo, K.A. (2017) Insights into the evolving microphysical and kinematic structure of Northeastern U.S. winter storms from dual-polarization Doppler radar. *Mon. Weather Rev.*, **145**, 1033–1061. © the American Meteorological Society, used with permission)

Another application of K_{DP} observations is in winter storms.[8] Figure 7.25 shows results from a QVP analysis of a winter storm. Here, Z_H, Φ_{DP}, and retrieved K_{DP} are shown. Between 13 and 19 UTC, light stratiform rain was occurring at the surface, and a radar bright band was located around 1 km above ground level (AGL). Note that there is a significant increase in Φ_{DP} in the bright band. At high elevation, it turns

---

[8]A summary of S-band radar observations of winter storms can be found here: Kennedy, P.C. and Rutledge, S.A. (2011) S-band dual-polarization radar observations of winter storms. *J. Appl. Meteor. Climatol.*, **50**, 844–858.

out that melting snowflakes (diameter ~10 mm or more) produce significant values of δ at S-band. In this retrieval of K_{DP}, δ is ignored (as it is typically done at S-band), and high values of K_{DP} are shown (but this is actually δ). Later in this storm, a region of deep stratiform snow developed between 20 and 03 UTC. In this case, large concentrations of snow aggregate particles formed in the region with temperatures near $-15\,°C$, where reflectivity is increasing downward indicating particle growth; this is the region at which dendritic growth of ice particles is most favored. This enhances K_{DP} in this region to values near 0.8 deg km^{-1} and is a fingerprint of dendritic particle growth and aggregation in ice clouds.

In convective storms, as mentioned above, the lofting of raindrops and the existence of melting hail can lead to columns of enhanced K_{DP} and Z_{DR} above the freezing level. Figure 7.26 shows an example of **K_{DP} columns** (as indicated by the vertical

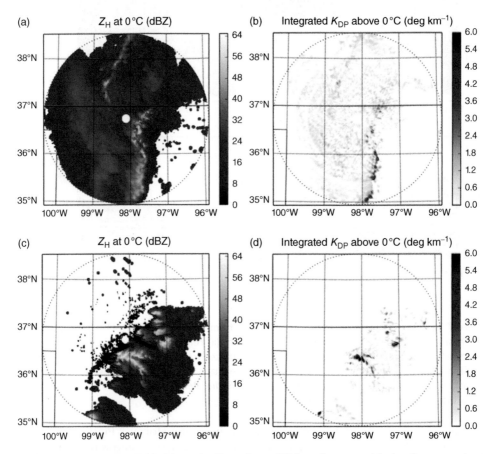

Figure 7.26   KVNX NEXRAD (Vance Air Force Base, Oklahoma) images of (a,c) reflectivity at horizontal polarization at approximately the melting level, (b,d) specific differential phase integrated in a slab above the melting level. The panels a and b are from May 20, 2011 at 09:48 UTC, whereas the panels (c) and (d) are from May 23, 2011 at 21:53 UTC (Adapted from van Lier-Walqui, M., Fridlind, A.M., Ackerman, A.S., Collis, S., Helmus, J., MacGorman, D.R., North, K., Kollias, P., and Posselt, D.J. (2016) On polarimetric radar signatures of deep convection for model evaluation: columns of specific differential phase observed during MC3E. *Mon. Weather Rev.*, **144**, 737–758. © the American Meteorological Society, used with permission)

integral of K_{DP} above the freezing level) in two convective storms observed by the Vance Air Force Base WSR-88D in Oklahoma. Note that in both of these cases, the high-reflectivity cores are relatively broad (associated with falling heavy precipitation), but the K_{DP} columns above the freezing level help localize where the convective updrafts loft drops and produce melting hail. When analyzed together, K_{DP} and Z_{DR} columns (Figure 7.27) are highly coincident in time and are both clear indicators of mixed-phase microphysical processes occurring above the freezing level. At C- and X-band, observing K_{DP} columns to identify convective mixed-phase regions is easier because of differential attenuation effects on Z_{DR} at those wavelengths as discussed above.

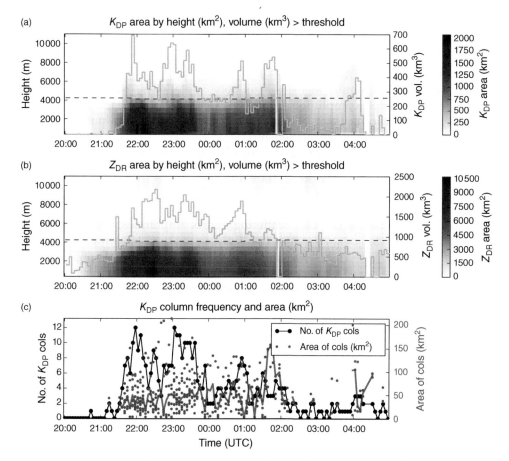

Figure 7.27 Analysis of polarimetric observations from the S-band Vance WSR-88D on May 23, 2011. (a) Area with specific differential phase >0.75 deg km^{-1} at each level (filled colors), volume with specific differential phase >0.75 deg km^{-1} above the melting level (gray line), and the melting level (dotted line). (b) As in (panel a), but for differential reflectivity >1 dB. (c) Number of specific differential phase columns detected (black) and area of each column (red) (Adapted from van Lier-Walqui, M., Fridlind, A.M., Ackerman, A.S., Collis, S., Helmus, J., MacGorman, D.R., North, K., Kollias, P., and Posselt, D.J. (2016) On polarimetric radar signatures of deep convection for model evaluation: columns of specific differential phase observed during MC3E. *Mon. Weather Rev.*, **144**, 737–758. © the American Meteorological Society, used with permission)

7.5.5   *Co-polar correlation coefficient (ρ_{HV})*

In addition to using the differential phase measurements available on a dual-polarization radar, another use of phase information is to examine the statistical consistency of the phase measurements at a given radar resolution volume. In statistics, the correlation coefficient (sometimes called the *Pearson correlation coefficient*) is used to quantitatively measure the degree to which a relationship is linear (which by definition, has a correlation coefficient of 1) or random (which has a correlation coefficient of 0). This can be illustrated by the general formula for Pearson's correlation coefficient (r) for a sample of data:

$$r = \frac{\sum_{i=1}^{n}(x_i - \bar{x})(y_i - \bar{y})}{\sqrt{\sum_{i=1}^{n}(x_i - \bar{x})^2 \sum_{i=1}^{n}(y_i - \bar{y})^2}} \tag{7.20}$$

where n is the number of samples, x_i and y_i are each of the i data samples, and $\bar{x}$ and $\bar{y}$ are the sample means (e.g., $\bar{x} = \frac{1}{n}\sum_{i=1}^{n} x_i$).

Here, we perform a similar analysis as is done in statistics, except we consider and compare the complex amplitudes of the return radar signal at each polarization. Essentially, we need to measure to what degree the phases of each pulse within a resolution volume form a linear relationship.

Mathematically, we define the **co-polar correlation coefficient** (unitless) as

$$\rho_{HV} = \frac{\langle |S^{VV} S^{HH^*}| \rangle}{\sqrt{\langle |S^{HH}|^2 \rangle \langle |S^{VV}|^2 \rangle}} \tag{7.21}$$

Here, note that we use the co-polar scattering matrix amplitudes, and the asterisk denotes the complex conjugate operator (multiplying the complex component of the scattering amplitude by -1), and the angle brackets $\langle \rangle$ denote the average over all pulses in the average (i.e., 32, 64). In practice, this can be calculated in several ways; however, similar to the Pearson's correlation coefficient in Eq. (7.20), ρ_{HV} can be calculated using

$$\rho_{HV} = \frac{\sum_{i=1}^{n}(S_i^{VV} - \overline{S^{VV}})(S_i^{HH^*} - \overline{S^{HH^*}})}{\sqrt{\sum_{i=1}^{n}(S_i^{VV} - \overline{S^{VV}})^2 \sum_{i=1}^{n}(S_i^{HH^*} - \overline{S^{HH^*}})^2}} \tag{7.22}$$

where i here is the number of pulses to be considered in the average. Note that the National Weather Service refers to the co-polar correlation coefficient as "CC" in practice rather than ρ_{HV}.

Figure 7.28 illustrates measurements of the dual-polarization complex amplitudes of the radar signal from the X-band Doppler on Wheels (DOW), which is a simultaneous transmit and receive radar. Here, the measurements were taken at the University of Illinois at Urbana-Champaign campus in Urbana, Illinois, at vertical incidence where light rain was observed at the surface, but the melting layer was only 500 m above the surface. Above the melting layer, dendrite aggregates were falling. Within the melting layer (shown in Figure 7.28a), the value of ρ_{HV} calculated with Eq. (7.22) is about 0.9, with large variations in the individual pulse I and Q data (as measurements of $|S^{VV} S^{HH^*}|$) at each polarization. In contrast, 1 km above the melting layer in snow (Figure 7.28b), the values of I and Q data are highly correlated, with a value of $\rho_{HV} = 0.999$. The information that is contained in ρ_{HV} tells us about the consistency of

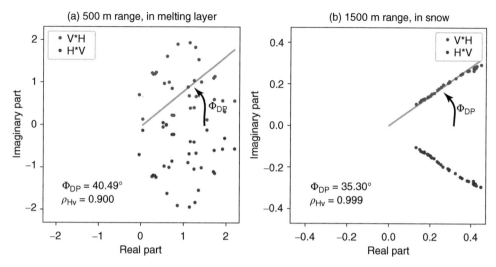

Figure 7.28 Doppler on Wheels (DOW) complex pulse pair time series measurements observed at the University of Illinois at Urbana-Champaign campus from March 3, 2016 at 16:37 UTC. Data from (a) range = 500 m within the bright band and (b) range = 1500 m within snow. The red and blue points show the H*V and V*H points. The angle depicted is an illustration of the estimate differential phase. Thirty-two pulses are shown from each sample

the I and Q data from pulse to pulse, which is related to the consistency of the phase returned at each pulse. Mixtures in sizes and shapes of the particles or inhomogeneous δ within the radar resolution volume will lower ρ_{HV} values. In addition, the angle relative to the x-axis of the best fit line shown in the figure is equal to the value of Φ_{DP} for a simultaneous transmit and receive radar (for an alternate transmit and receive radar, the change in amplitudes caused by the Doppler shift between pulses must be removed first).

To illustrate how mixtures of particles can lower ρ_{HV}, Figure 7.29 shows a simulation of data from parent populations of data with the parameters given in the legend. In each panel, 32 samples are drawn at random from a distribution to produce the data in each plot, and then the ρ_{HV} and Φ_{DP} values are calculated using the sampled data. For the rain case in Figure 7.29a, the ρ_{HV} decreased only slightly, whereas a small deviation in Φ_{DP} was noted. However, in Figure 7.29b, the data includes 16 samples drawn from rain and 16 from melting hail. It was assumed that each of these particle populations had the same intrinsic ρ_{HV} values but had values of intrinsic phase differing by 5° (perhaps, there was higher liquid water content in the melting hail). By sampling these populations together, we can see that the ρ_{HV} value decreased to 0.973, and the estimated phase was 45.6°! You can see how in this exercise, a mixture of particles lowered ρ_{HV} and caused apparent noise in Φ_{DP}.

Many of the examples in the figures presented in this chapter include ρ_{HV} as a panel, so you can examine how the covariations in the polarimetric parameters manifest themselves in real storms. For example, in the hail storm presented in Figure 7.13, it is evident that the region of low Z_{DR} values has depressed ρ_{HV} values near 0.90, in comparison with higher values in the rain-only regions surrounding the storm. This provides further evidence that in much of the core of this storm, we can infer that hail is mixed with rain. In Figure 7.16, the RHI through the convective

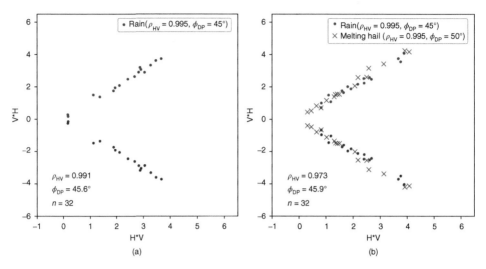

Figure 7.29 Simulation of complex pulse pair measurements in (a) rain only with 32 samples and an inherent co-polar correlation coefficient of 0.995 and (b) in rain mixed with melting hail, each with a co-polar correlation coefficient of 0.995, but with an intrinsic differential phase shift of 45° and 50°, respectively

storm shows depressed ρ_{HV} in regions of particle melting and large hail (where non-Rayleigh effects may be occurring, which also lowers ρ_{HV}—additional evidence of this is provided by the three-body scatter spike in Z_H to the right side of the panel (see Chapter 14 for more discussion of this feature of hailstorms)—notice that ρ_{HV} is low in this region and can be used as a filter to remove the spike. In most cases, ρ_{HV} can be an excellent tool for quality control as shown in this figure—most precipitation regions have $\rho_{HV} > 0.7$–0.8, and so it is an easy way to remove non-meteorological echo if desired.

Figure 7.30 shows another example of how ρ_{HV} and its capability to identify hydrometeor mixtures is useful—identifying transitions between rain and snow in winter storms. In this PPI from the Boston, Massachusetts WSR-88D, the region of low ρ_{HV} extending southwest–northeast across Cape Cod is a clear indication of a transition between rain, mixed-phase precipitation (sleet), and snow. This is reinforced by the radar bright band being collocated in Z_H, as well as higher values of Z_{DR} present in the rain region to the southeast. K_{DP} values at S-band in winter storms show small signals, but do show some small values (<1 deg km^{-1} in some of the heavier snowbands and in the bright band). The lower ρ_{HV} values are also useful in RHIs in identifying melting as well.

Probably, the most widely recognized example of mixtures of non-meteorological objects lowering ρ_{HV} is in the tornado debris ball. The same physical principle applies: various objects in a tornado's circulation will have a wide variety of phases sampled in the radar resolution volume, which will greatly decrease the value of ρ_{HV}. Examples will be discussed in Chapter 14. The same principle allows ρ_{HV} to discriminate ground clutter (which also has poorly correlated phase measurements) from precipitation using a ρ_{HV} threshold.

Keep in mind when examining such signatures in determining precipitation type, especially in winter storms, that the radar beam distance above the surface increases

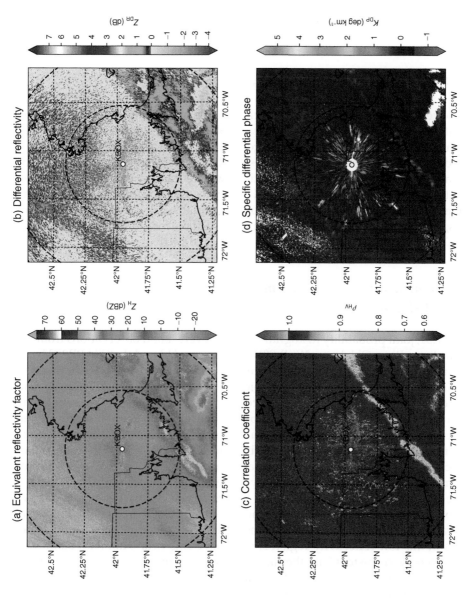

Figure 7.30 KBOX NEXRAD (Boston, Massachusetts) 0.5° elevation PPI S-band observations from February 5, 2016 at 13:15 UTC of an east coast cyclone with a clear rain–snow transition extending southwest–northeast across Cape Cod. (a) Reflectivity at horizontal polarization, (b) differential reflectivity, (c) co-polar correlation coefficient, and (d) specific differential phase

with range. Just as evaporation or storm overshoot can lead to discrepancies in radar-estimated rainfall and surface-measured rainfall, the same geometric effect can lead to issues in interpreting dual-polarization radar information. Thus, melting (and/or refreezing in the case of shallow temperature inversions) could be occurring to make the signatures apparent on a PPI not very reliable when trying to glean information about surface precipitation type. It is always important to compare the inferences you make from dual-polarization radar with surface reports of precipitation!

7.5.6   *Using polarimetric variables together*

In the above discussion of each polarimetric variable, it becomes clear that there are interrelationships among the polarimetric variables. These interrelationships yield complimentary information about the targets within the radar resolution volume. If we can design physical models that represent the relationships between variations of the variables and meteorological quantities we are interested in retrieving, then quantitative retrievals can be designed and implemented. However, as we have seen, while many of the theoretical descriptions of each variable provide information about the sizes, shapes, and types of hydrometeors, many of the measurements also contain artifacts, and some require calibration (Z_H and Z_{DR} in particular).

7.5.7   *Covariation of the polarimetric variables: an example at S- and C-band*

In Figure 7.31, joint probability histograms of low-level S-band observations from the Vance Air Force Base, Oklahoma WSR-88D (panels a and b) and the CSAPR2 C-band radar (panels c and d) are shown. Darker colors in the shaded plots mean more observations in the regions of each of the variables considered. In the left column, Z_H and Z_{DR} are considered, whereas in the right column, Z_H and retrieved K_{DP} are shown. For each panel in the plot, the histograms on the border of the figure indicate the relative occurrence of each variable individually. It should be noted that these plots are not exactly matched in time and in space, but the data reveal some interesting characteristics of the covariability of these measurements at each radar wavelength for similar meteorological conditions in a convective storm.

   In the panels of Z_H versus Z_{DR}, at low values of reflectivity (below 20 dBZ), we would expect a Z_{DR} of nearly 0 dB due to near-spherical small drops if the radars are indeed calibrated for Z_{DR}. In fact, this is one technique for calibrating Z_{DR}; the mean of points in this low-reflectivity region should have Z_{DR} near 0 dB. At S-band, Z_{DR} begins to increase consistent with the simulations shown in Figure 7.9. At S-band, the increase is more or less linear with drop diameter. Keep in mind the simulations in Figure 7.9 are for a single drop, and they must be integrated over a size distribution to obtain a value consistent with observations. However, at high values of Z_H, near point (2) in the figure, we can see that values of Z_{DR} of 3–4 dB are observed. Compare the S-band observations with the C-band observations in the corresponding panel in the second row. You can see near point (2) that the values of Z_{DR} at C-band are similar to those at S-band, and the joint histogram values are similar in this region.

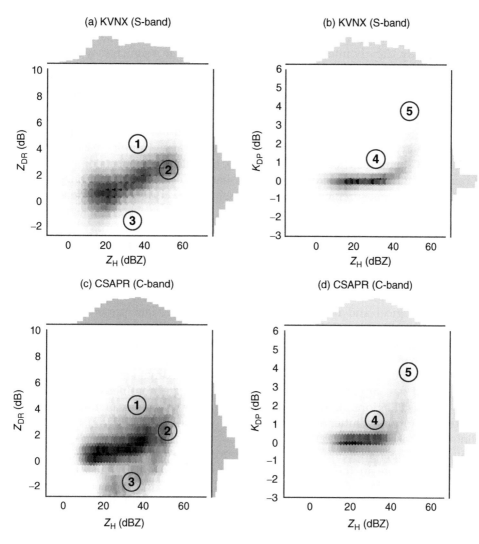

Figure 7.31   Relationships between S-band and C-band polarimetric variables. (a and c) Reflectivity at horizontal polarization versus differential reflectivity and (b and d) reflectivity at horizontal polarization and specific differential phase. From May 23, 2011, S-band observations (panels a and c) are from a 0.9° elevation PPI observed by the KVNX NEXRAD (Vance Air Force Base, Oklahoma) at 22:15 UTC, whereas C-band observations are from a 0.7° elevation PPI from the Department of Energy Atmospheric Radiation Measurement Climate Research Facility CSAPR C-band radar near Nardin, Oklahoma at 22:16 UTC. The shaded panels in each figure give the joint probability distribution of the data, whereas the traces on the edges of each figure display the probability distribution on one axis only

However, there are differences in the statistical distributions. Extreme high values of Z_H are higher in frequency of occurrence at S-band compared with C-band, indicating, at C-band, attenuation and the onset of non-Rayleigh effects at high Z_H. The population of data points at locations (1) and (3) are also different between C-band and S-band. Point (1) is indicative of non-Rayleigh resonance from large raindrops at C-band, where the distribution of Z_{DR} values at C-band extends well above the values observed at S-band. Near point (3), low to negative values of Z_{DR} and moderate values of Z_H indicate a signature of attenuation and differential attenuation in Z_H and Z_{DR}, respectively. Clearly, interpreting C-band radar data in a similar way as at S-band leads to issues in relating processes to the observations. These discrepancies do not mean that C-band data are not usable for understanding convective storms; it just takes more care and careful considerations of the measurements.

In the plots of Z_H versus K_{DP} on Figure 7.31b,d, qualitatively good consistency is seen between S-band and C-band. Both plots show non-linear increases in Z_H with K_{DP}. The primary differences between S- and C-band arise from increased sensitivity in K_{DP} at C-band (as shown in Figure 7.19), as well as the effects of attenuation and non-Rayleigh effects at reducing Z_H at C-band. However, as mentioned above, differential phase measurements have the possibility to correct for attenuation and differential attenuation in cases such as this.

7.5.8 Using dual-polarization variables to discern meteorological versus non-meteorological echo and non-uniform beam filling

Figure 7.32 shows an example where polarimetric radar measurements help separate meteorological and non-meteorological echo. The data presented are from the Melbourne, Florida WSR-88D during Hurricane Matthew in 2015, which grazed the east coast of Florida. The Hurricane's eyewall is clearly visible, with high Z_H, high K_{DP}, and moderate Z_{DR} (which is a typical signature of tropical rainfall—large number concentrations of small- to medium-sized drops). In this case, Z_{DR} and ρ_{HV} are showing signatures that indicate non-meteorological echo in two regions. Regions of low ρ_{HV} (i.e., <0.7) near the radar in this image are indicative of ground clutter contamination. If you look closely, you can see the outline of the barrier islands offshore of Melbourne. The WSR-88D signal-processing algorithm attempts to remove these areas from Z_H and Z_{DR} fields. The low ρ_{HV} region offshore in the hurricane eye is not filtered out in the Z_H and Z_{DR} fields. In fact, this low Z_H region (around 10 dBZ) with low ρ_{HV} (<0.6) contains high values of Z_{DR}, exceeding 4–8 dB! This is not consistent with rain or any meteorological scatterers for that matter (refer to Figure 7.31). It turns out that these scatterers are large birds! It has been commonly observed that birds take refuge in the calm winds of the hurricane eye. These birds have a horizontal backscatter cross section much greater than their vertical cross section, have relatively low concentration, and have widely varying phase characteristics. Thus, dual-polarization radar can not only identify cloud microphysical properties but also properties of non-meteorological scatterers (including biological scatterers as well).

One additional phenomenon should be mentioned: **non-uniform beam filling**. Figure 7.33 shows an intense squall line approaching the radar from the west. Microphysical variations across each radar beam within convection (likely large melting hail mixed with rain) create propagation effects (the spikes leading away from the radar) as radar beams pass through these regions, reducing Z_{DR} (appearing like differential attenuation) and ρ_{HV} (caused by depolarization of some of the pulses) that continues beyond the non-uniform region. The values of Z_{DR} and ρ_{HV} are not reliable in these regions, and are not representative of polarimetric variable values in stratiform rain.

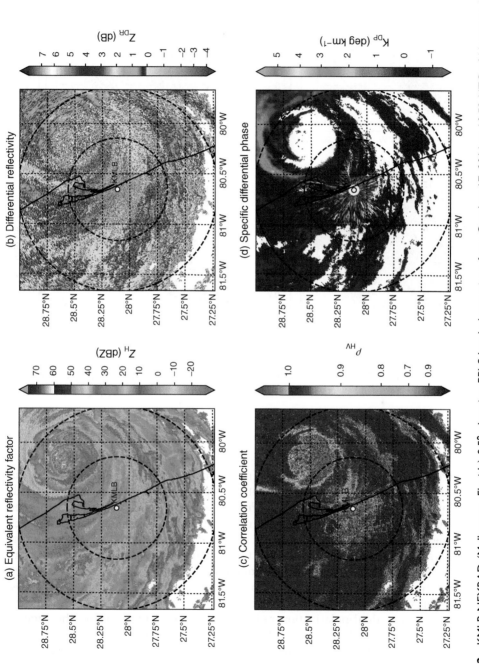

Figure 7.32 KMLB NEXRAD (Melbourne, Florida) 0.5° elevation PPI S-band observations from October 17, 2016 at 10:52 UTC of Hurricane Matthew (2016). (a) Reflectivity at horizontal polarization, (b) differential reflectivity, (c) co-polar correlation coefficient, and (d) specific differential phase

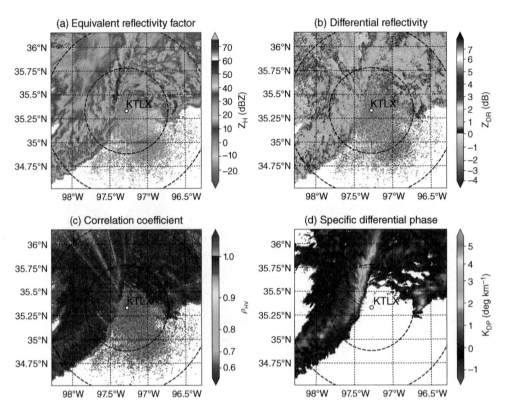

Figure 7.33   KTLX NEXRAD (Twin Lakes, Oklahoma) 0.5° elevation PPI S-band observations from October 17, 2016 at 10:52 UTC of the squall line shown in Figure 7.21. (a) Reflectivity at horizontal polarization, (b) differential reflectivity, (c) co-polar correlation coefficient, and (d) specific differential phase

7.5.9   Hydrometeor classification

We have seen the variables available from a dual-polarization radar covary in characteristic ways in certain hydrometeor types. We have also seen that this behavior is dependent on wavelength. From the discussion earlier in this chapter, it is clear that the effects of attenuation, differential attenuation, and non-Rayleigh scattering and its impacts on Z_H, Z_{DR}, ρ_{HV}, and K_{DP} can cause the behavior of each variable to change depending on the meteorological situation. In addition, effects related to the polarimetric information at each frequency change as well (i.e., sensitivity of K_{DP} increases with decreasing wavelength). In the late 1980s, early polarimetric radar research demonstrated that, in addition to improving rainfall retrievals, dual-polarization radar could be used to automate the process of identifying hydrometeor types within convective storms. In the 1990s and early 2000s, this effort was extended to mid-latitude cool season precipitation. Today, the WSR-88Ds have implemented a hydrometeor classification algorithm (HCA) following the dual-polarization upgrade in the early 2010s.

Hydrometeor classification algorithms, or sometimes called *hydrometeor identification* (or *HID*), have been designed in several ways, but the basic principle is to identify which hydrometeor type corresponds best to the observed radar variables at a location of interest. The classification is applied at each radar gate such that a full field of hydrometeor types is produced consistent with the radar data. Hydrometeor type can be mapped as a PPI or an RHI or examined in three dimensions to understand where particular hydrometeors are present. Such products are useful for identifying hazards such as hail, transitions from rain to snow, or in studies of cloud processes (such as understanding the production of hydrometeor types from convective storms).

In order to design and implement an HCA algorithm, there are two steps involved. The first is to examine the polarimetric signatures of every hydrometeor type to be classified. This can involve empirical thresholds of variables for each hydrometeor type, or using simulations of dual-polarization radar characteristics for various hydrometeors (i.e., drizzle, rain, hail, rain mixed with hail, snow, melting snow, ice crystals, and non-meteorological echo). The second step is to implement an algorithm to make the decision which hydrometeor type, given input data, to assign to each radar gate. This can be accomplished through hard thresholds or decision trees, or through supervised classification techniques such as **fuzzy logic** approaches.

Before beginning such an exercise, two primary issues must be dealt with. The first is to recognize that for HCA algorithms to be successful, the classification accuracy will only be as good as the data input to the classification. HCA classification schemes rely on good radar measurements that relate to the expected or assumed intrinsic dual-polarization characteristics of each hydrometeor type. That means, for example, that the radar data must be calibrated for Z_H, and Z_{DR} and be corrected for attenuation and differential attenuation and non-uniform beam filling must be identified. If attenuation lowers Z_H, and Z_{DR} relative to expected values, then an incorrect hydrometeor type will be assigned because the values of variables will not match up with the intrinsic values expected at that frequency. The second issue in HCA algorithms is that polarimetric signatures in drizzle and rain match up closely with those of frozen hydrometeor types (i.e., cloud ice, snow, and melting snow). Thus, most HCA approaches require vertical temperature profile information (i.e., from a nearby radiosonde or meteorological analysis) to incorporate into the algorithm. These data from the sounding or model are matched up with the height of the radar beam at each gate to aid in discriminating ice and liquid hydrometeor types. During the mid-latitude warm season or in the tropics, usually a single sounding is acceptable. However, in winter storms, precipitation is often times near fronts and strong temperature gradients, so the vertical position of the melting (and freezing) levels is non-uniform in space and time. In some cases, a fully 3-D meteorological analysis of temperature may be required. Finally, the height of the lowest scan of the radar beam is often a significant distance above the ground, such that the hydrometeor type at the surface is not necessarily what is determined from the lowest radar beam. Melting or re-freezing can often occur below the beam height, especially at mid to far ranges.

Here, we describe one example, an HCA algorithm developed using a fuzzy logic supervised learning approach.[9] Output from this algorithm is shown in Figure 7.34, illustrating a convective storm observed with the Vance Air Force Base, Oklahoma WSR-88D. A variety of hydrometeor types are identified in this cluster of convective cells, including rain, big drops (rain with large Z_{DR}), hail, and graupel. Near the

---

[9]To learn more about this approach, see Liu, H. and Chandrasekar, V. (2000) Classification of hydrometeors based on polarimetric radar measurements: development of fuzzy logic and neuro-fuzzy systems, and *in situ* verification. *J. Atmos. Oceanic Technol.*, **17**, 140–164.

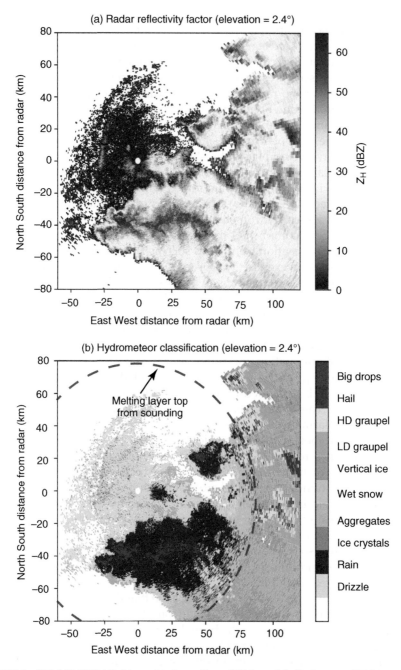

Figure 7.34 KVNX NEXRAD (Vance Air Force Base, Oklahoma) 2.4° elevation PPI S-band observations from May 23, 2011 at 22:15 UTC showing (a) radar reflectivity at horizontal polarization and (b) hydrometeor classification algorithm results (hydrometeor type indicated by the legend)

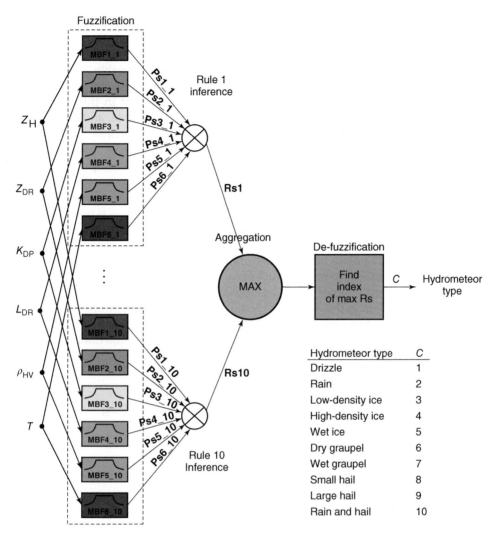

Figure 7.35 Illustration of a fuzzy logic hydrometeor classification algorithm (Adapted from Liu, H. and Chandrasekar, V. (2000) Classification of hydrometeors based on polarimetric radar measurements: development of fuzzy logic and neuro-fuzzy systems, and *in situ* verification. *J. Atmos. Oceanic Technol.*, **17**, 140–164. © the American Meteorological Society, used with permission)

melting level, melting snow is assigned, along with graupel. Above the melting level indicated by data from a sounding, particle types such as hail, low- and high-density graupel, and aggregates are classified by this scheme.

An overview of the fuzzy logic HCA algorithm structure, implemented in Figure 7.34, is shown in Figure 7.35. Fuzzy logic algorithms use predefined **membership functions** to classify multidimensional input data into categories. Here, the multidimensional input data are the data from a dual-polarization radar, and the membership functions can be thought of as functions that represent the likelihood that a given range of a polarimetric variable could be observed for that hydrometeor type. The generation of these rules is called *fuzzification* of the variables to be classified. These membership functions, one for each hydrometeor type, contain values between 0 and 1 (here represented mathematically by three-parameter beta functions), indicating whether a particular value of each variable is associated with a variable of each class. Figures 7.36–7.38, and 7.39 show the membership functions for the classification shown in Figure 7.34. Note that in this algorithm, there are membership functions defined for S-, C-, and X-band data, and these nicely summarize some of the frequency-dependent differences in intrinsic polarimetric variables across hydrometeor types that were discussed earlier in this chapter. In addition, it is clear from examining the hydrometeor types that temperature plays an important role in discriminating the various hydrometeor types with significant overlap (i.e., rain and wet snow). L_{DR} membership functions are available in this algorithm, although they can be removed from the algorithm if L_{DR} is not available.

To apply the algorithm, polarimetric variables are input from each gate in the radar data, in addition to the local temperature at the radar gate. The observed data are then used to calculate values of membership functions to calculate a "score" for each hydrometeor type, which is simply a sum of each of the membership function values across all of the radar variables and temperature. This is the "inference" step. Finally, in the "aggregation" step, the maximum score is chosen as the hydrometeor type for that radar gate. An integer code is assigned, which corresponds to the hydrometeor type shown in Figure 7.35. This code can be stored alongside the radar data for display, or for later examination.

Important terms

alternate transmission and reception
amplitude
backscatter differential phase shift
canting angle
co-polar correlation coefficient
co-polar power
cross-polar power
depolarization
differential attenuation
differential phase
differential phase shift
differential reflectivity factor
disdrometer
droplet oscillations
dual-polarization retrievals
frequency
fuzzy logic classification
hydrometeor classification algorithm
intrinsic polarization variable
K_{DP} column
L_{DR} cap
linear depolarization ratio
maximum drop diameter
membership function

non-uniform beam filling
oblate
phase
phase shift
polarimetric variables
polarization agile
polarization plane
polarization radar
prolate
propagation differential phase shift
propagation polarization variable
quasi-vertical profile (QVP)
radar reflectivity at horizontal polarization
reciprocal media
reflectivity-weighted axis ratio
retrieve
simultaneous transmit and receive
specific differential phase shift
taurus
two-dimensional video disdrometer
wavelength
Z_{DR} column

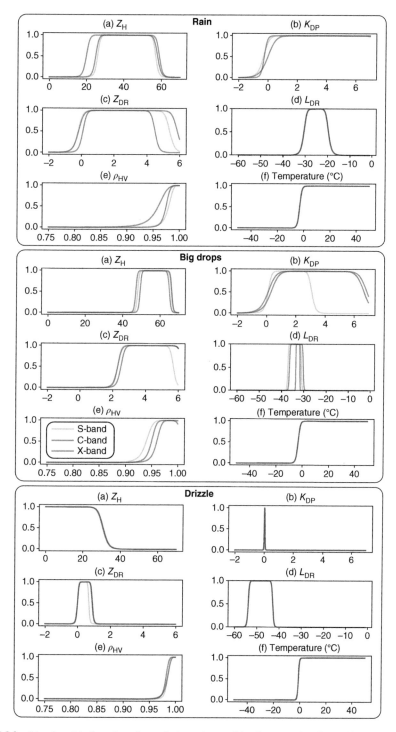

Figure 7.36   Membership functions for polarimetric variables for rain, drizzle, and big drops from the Colorado State University hydrometeor classification algorithm at S-band (green), C-band (blue), and X-band (red) data. These values were obtained from the CSU RadarTools software package available at https://github.com/CSU-Radarmet/CSU_RadarTools

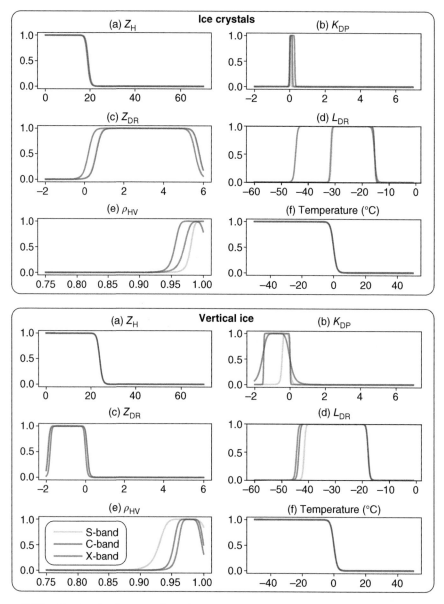

Figure 7.37 Membership functions for polarimetric variables for ice crystals and vertical ice from the Colorado State University hydrometeor classification algorithm at S-band (green), C-band (blue), and X-band (red) data. These values were obtained from the CSU RadarTools software package available at https://github.com/CSU-Radarmet/CSU_RadarTools

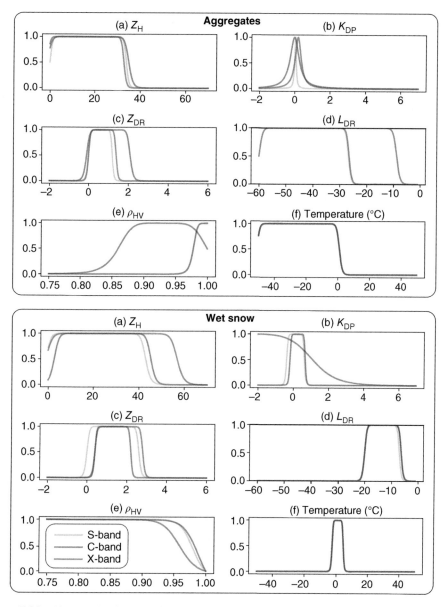

Figure 7.38 Membership functions for polarimetric variables for aggregates and wet snow from the Colorado State University hydrometeor classification algorithm at S-band (green), C-band (blue), and X-band (red) data. These values were obtained from the CSU RadarTools software package available at https://github.com/CSU-Radarmet/CSU_RadarTools

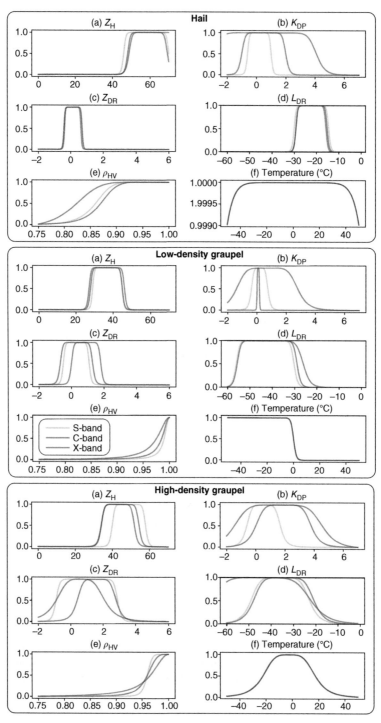

Figure 7.39 Membership functions for polarimetric variables for hail, low-density graupel, and high-density graupel from the Colorado State University hydrometeor classification algorithm at S-band (green), C-band (blue), and X-band (red) data. These values were obtained from the CSU RadarTools software package available at https://github.com/CSU-Radarmet/CSU_RadarTools

Review questions

1. Why are conventional radars (i.e., non-polarimetric radars) typically not vertically polarized?
2. Why are intrinsic values of specific differential phase and differential reflectivity usually positive?
3. List and describe three differences between S-band and C-band interpretation of intrinsic polarimetric variables in hailstorms.
4. Why are values of differential reflectivity near 0 dB in drizzle?
5. How can the co-polar correlation coefficient be used to remove non-meteorological radar echo?
6. Why can't a polarimetric radar directly measure specific differential phase?
7. What are the three common situations where co-polar correlation coefficient is reduced in convective storms? Why?
8. Why can't linear depolarization ratio be measured by a radar using simultaneous transmit and receive dual-polarization sampling?
9. How can a QVP be used to examine storm characteristics using dual-polarization radar?
10. How are the values of the intrinsic polarimetric variables different in rain and hail? Why?

Challenge problems

11. You are given the following disdrometer data from rain to perform calculations:

Maximum diameter raindrop size bin (mm)	Number distribution $(mm^{-1}\,m^{-3})$
0–1	1310
1–2	150
2–3	13
3–4	3
4–5	2
5–6	0.7

Use the following equation for axis ratio formula to estimate axis ratio A in each bin: $A = 1.012 - 0.144D - 1.03D^2$ where D is expressed in centimeters.

(continued)

Assume that there is no droplet canting or oscillation and that the number concentrations are uniform across each bin.

   a. What is the S-band reflectivity at horizontal polarization of this distribution?

   b. What is the S-band differential reflectivity of this distribution?

   c. You found out that the data from the next minute were the same, except it contained drops larger than 5–6 mm. Repeat (a) and (b) if a bin of diameter 7–8 mm is added and the concentration in that bin in $0.2\,mm^{-1}\,m^{-3}$.

12. Using the data in problem 11, and Eqs. (7.17) and (7.18)), what is the specific differential phase value for

   a. the size distribution in the table

   b. the size distribution with the extra 7–8 mm diameter size bin.

13. Consider a three-species hydrometeor classification algorithm comprised membership function values for hail, high-density graupel, and low-density graupel only in Figure 7.39 (i.e., ignore all other species). Assume that an S-band polarimetric radar measured the following parameters:

$$Z_H = 57\,dBZ; \quad K_{DP} = 2.5\,deg\,km^{-1}; \quad Z_{DR} = 0.5\,dB; \quad \rho_{HV} = 0.97,$$
$$temperature = -10°C$$

   a. Calculate the value of the sum of the membership functions for each of the three species for each measurement.

   b. Select a hydrometeor type based on these observations.

   c. Repeat (a) and (b) for a C-band radar. Assume that the attenuation in reflectivity is 5 dBZ and the differential attenuation is 2 dB. What is your selected hydrometeor type?

14. Assume that the initial differential phase of a radar system is 0°, and you can measure 360° of differential phase shift with your radar system before the values are folded. Assume that the mass-weighted mean axis ratio of droplets is 0.85 and that the distribution of raindrops in a cloud is infinitely uniform. Using the approximation of Eq. (7.17), what is the liquid water path ($kg\,m^{-2}$) along the radar beam that will cause 360° of phase shift (before observed differential phase gets folded) at (a) S-band, (b) C-band, and (c) X-band? Ignore backscatter differential phase and noise. If the liquid water content in a storm is uniformly $5\,g\,m^{-3}$, over what distance will this phase folding occur at (d) S-band, (e) C-band, (f) X-band?

8

Clear Air Echoes

Objectives

By the end of this chapter, you should understand the following:

- The primary sources of clear air echoes.
- What ground clutter appears like on radar, and how ground clutter patterns can change due to anomalous propagation.
- Basic techniques to mitigate the effects of ground clutter.
- The nature of biological echoes and how they evolve based on insect and animal behavior.
- How echoes from insects can be used to determine low-level winds.
- How dust and debris contribute to radar echoes, and how these echoes can be used to interpret meteorological phenomena.
- How aircraft appear on meteorological radar, and the effects that aircraft-released chaff can have on radar echoes.
- How noise appears in both reflectivity and radial velocity fields.
- The mechanism by which Bragg scattering occurs, and how Bragg scattering can be used to interpret meteorological processes.

8.1 Introduction

Radar meteorologists' primary interest is normally in radar echoes from **hydrometeors**, that is, cloud droplets, raindrops, ice crystals, snowflakes, graupel, and hail. The intensity and polarization state of hydrometeor echoes are used, for example, to determine the location, intensity, and type of precipitation, whereas the Doppler radial velocity associated with the movement of hydrometeors as they fall and blow along with the wind can be used to derive characteristics of the wind field.

Many echoes appearing on the radar display do not originate from hydrometeors. Meteorologists often refer to this echo as **clear air echo** to differentiate it from echo from hydrometeors, even though the air may not be "clear" in the sense of infinite visibility. Clear air echo can originate from a variety of sources. Targets on the surface of the earth, such as buildings, trees, vehicles, power transmission towers, poles, wind turbines, and water towers contribute to echo that collectively is called **ground**

Radar Meteorology: A First Course, First Edition. Robert M. Rauber and Stephen W. Nesbitt.
© 2018 John Wiley & Sons Ltd. Published 2018 by John Wiley & Sons Ltd.
Companion website: www.wiley.com/go/Rauber/RadarMeteorology

clutter. The earth itself can be a source of ground clutter in the case where anomalous propagation (AP) causes a beam to return to the ground, or a hill or mountain intercepts the beam as it propagates through the atmosphere. Waves on large bodies of water can also intercept radar energy and create **sea clutter**, another type of surface echo that can confound the meteorological signal. In some cases, surface debris and dust can be lofted producing radar targets that can range from blowing leaves to shingles to parts of houses. **Biological echo**—from insects, bats, and birds—represents another type of non-meteorological signal. Insects, in particular, can produce echoes that can cover large areas on a radar screen, especially when a radar antenna is pointed at a low elevation angle. Aircraft also produce radar echoes. An Aircraft is normally point source, but military aircraft sometimes releases **chaff**, small filaments of reflective material much like shards of Christmas tree tinsel, designed to confuse enemy aircraft. Clear air echo also comes from the sun, receiver noise, and, in some cases, local **radio interference** sources. such as cellular telephone or Wi-Fi transmissions.

Finally, a source of radar echo exists that *is* meteorological, but not from hydrometeors. Radar echoes can develop through a process called **Bragg scattering**. This occurs when there are significant variations in atmospheric density on a scale of half the radar wavelength, normally associated with turbulence. Constructive interference can occur in these situations that can lead to coherent backscattered wave energy. The process can lead to significant echoes with longer wavelength radars such as S-band. In fact, it is the primary source of echo for clear-air wind profiling radars.

Each of the echo sources described above has the potential to confound the meteorological signal. However, many of these echoes can also be used to extract meteorological information. In this chapter, we will consider each of these echo sources, examine how they appear on a radar display, how the echoes can be filtered when they are unwanted, and how they can be used productively to understand atmospheric processes.

8.2   Ground clutter

Ground clutter refers collectively to echoes from the ground or objects attached to the ground, regardless of what type of objects they are. The term *clutter* implies that the signal from these objects contaminates weather signals, biasing or making it impossible to derive any meteorological information that the weather signals may contain. Unless the signal from ground clutter can be appropriately filtered, biases in the derived radar reflectivity factor, radial velocity, spectral width, and polarization variables can occur. In general, ground clutter leads to increased reflectivity because ground targets are large, radial velocity values biased toward zero because most ground targets are near-stationary, and spectral widths that are shifted toward zero because of the strong contribution from targets with near-zero radial velocities.

8.2.1   Ground clutter characteristics

The distribution of ground clutter and the characteristics of ground clutter are unique to each radar location. Figure 8.1 shows an example of ground clutter in

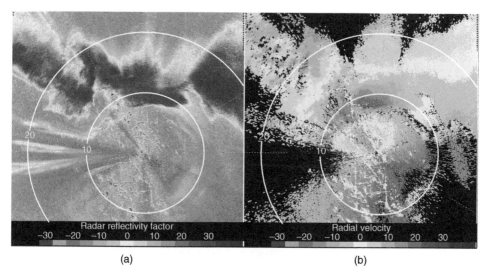

Figure 8.1   Ground clutter in the vicinity of convective echoes as it appears in the reflectivity (a) and radial velocity (b) fields of a Doppler on Wheels X-band radar. Note the grid pattern within the 10 km range ring where targets have zero radial velocity (Image courtesy Jeffrey Frame)

the vicinity of convective echoes as viewed by a Doppler on Wheels X-band radar. The clutter pattern to the south of the radar shows characteristic patterns found in rural areas—grids of clutter echo created by poles, silos, and farm buildings located along evenly spaced north–south and east–west highways. Note that the stronger echoes associated with clutter are marked by near-zero radial velocity. Near the radar, ground clutter blends with insect echo, so the radial velocities are nonzero, a reflectivity-weighted combination of the insect and ground echo. To the north of the radar, there is likely some ground echo embedded within the meteorological echo. Here, the effects of ground echo, if any, are difficult to separate from the meteorological echo. In general, ground contamination will bias reflectivity higher and radial velocity toward zero.

Ground clutter can also appear on range-height indicator (RHI) scans. Figure 8.2 shows an example of an RHI taken by the NCAR CP2 S-band radar in Florida at the Kennedy Space Center. As might be expected, ground clutter appears near the radar as strong low-level echo. However, at approximately 22 km from the radar, a long narrow arc of echo can be seen extending upward to the top of the scan. This feature is also ground clutter, in this case caused by the radar's main beam and *sidelobes*, intersecting the Space Center's Vehicle Assembly Building (Figure 8.3). Energy from sidelobes located far off the radar's main beam was reflected from the building surface facing the radar. The sidelobe returns were sufficiently strong that they appear at the same range along an arc as the antenna was swept upward above the horizon.

8.2.2   Sea clutter

Sea clutter is the generic name given to radar echoes generated by reflection of radar energy by waves on a large body of water. Sea clutter is normally confined to the

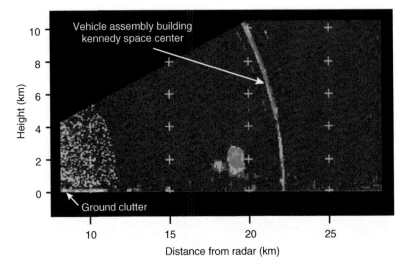

Figure 8.2 Ground clutter echo on an RHI caused by radar sidelobe energy striking the Vehicle Assembly Building at Cape Canaveral, Florida (see Figure 8.3)

Figure 8.3 The Vehicle Assembly Building at Cape Canaveral (Courtesy NASA)

lowest elevation beam of a radar unless AP is occurring. Water waves move, so sea clutter typically does not exhibit zero radial velocity like ground clutter. Figure 8.4 shows an example of sea clutter on Lake Ontario as observed by a Doppler on Wheels X-band radar during a lake-effect storm. A large band of precipitation straddles the middle of the lake, marked by $20\,\mathrm{m\,s^{-1}}$ radial velocities. On the north side of the band, closer to the radar is a region of weaker echo, marked by $5\,\mathrm{m\,s^{-1}}$ radial velocity and

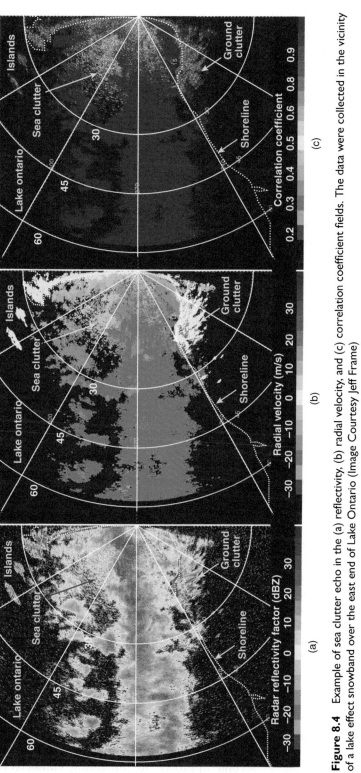

Figure 8.4 Example of sea clutter echo in the (a) reflectivity, (b) radial velocity, and (c) correlation coefficient fields. The data were collected in the vicinity of a lake effect snowband over the east end of Lake Ontario (Image Courtesy Jeff Frame)

correlation coefficients in the range of 0.4–0.6. This is sea clutter. The radial velocities are characteristic of the wave speed toward the eastern shore of the lake, the waves forced by the strong winds blowing down the axis of the lake.

8.2.3   *Effects of anomalous propagation*

The distribution and characteristics of ground clutter, unfortunately, can vary with time because of AP. Figure 8.5 shows an example from the KILX WSR-88D on an early winter morning. The data were taken in mid-February when an inversion was present and AP was occurring. The first image, taken before sunrise at 5:04 a.m. local standard time (LST), shows clusters of echo associated with turbines at wind farms. You can also see clutter from objects along major highways. Sunrise occurred at 6:53 a.m. Following sunrise, the inversion broke, AP ceased, and the wind farms and clutter along major highways rapidly disappeared.

The effects of AP complicate the problem of ground clutter filtering. In the absence of AP, a map of known ground clutter sources could be generated and filtering applied only in those regions. With AP, the map changes depending on weather conditions. Wind farms are particularly difficult to deal with because the large blades, if rotating parallel to a radar radial, have appreciable nonzero velocity.

8.2.4   *Ground clutter mitigation*

From the above discussion, we can see that there are two categories of ground clutter, normal propagation (NP) clutter and AP clutter. The former is typically always present, and therefore is easier to remove, whereas the latter comes and goes depending on the vertical gradient of the refractive index and is typically harder to identify and remove.

There are four fundamental approaches to mitigating clutter. The first is to avoid it in the first place through careful radar siting. The second is to consider pulse-to-pulse variations in power that occur with ground and meteorological targets. The third is through post-processing of derived radar moments, for example, by eliminating regions where radar echoes are stationary and have near-zero radial velocity, or using polarimetric variables. The final approach is to employ clutter filters during signal processing. This latter approach allows the possibility of retaining the weather signal while reducing or eliminating the contribution from clutter.

Regarding the first approach, advances in geographic information systems have made detailed information available from which ground clutter mapping can be done in advance of siting a radar. Terrain and land use data, as well as tools such as Google Earth® and Google Street View®, now allow potential sites to be virtually visited and analyzed well in advance of siting. Detailed ground clutter maps can be constructed for each site under consideration and the best site chosen based on the results. It is also possible to calculate clutter patterns under various conditions of AP based on an understanding of the meteorology of the site.

Weather echoes typically exhibit much larger pulse-to-pulse power fluctuations than ground echoes because hydrometeors shuffle from pulse to pulse, whereas ground targets do not. A second approach to mitigating clutter considers the

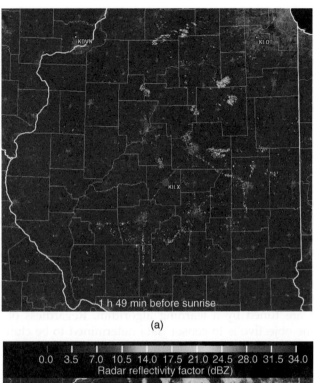

(a)

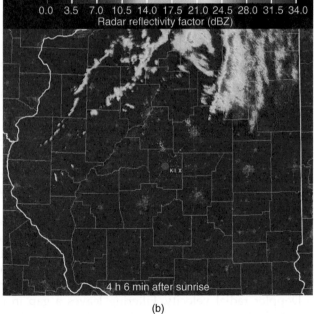

(b)

Figure 8.5 Example of anomalous propagation due to a nocturnal inversion. The larger echoes in panel a are from wind farms, whereas the thin lines are from cell phone towers along interstates. In panel b, most of these echoes have disappeared after daytime heating destroys the inversion. The larger echoes at the top of panel b are from precipitation that moved in during the intervening time between a and b

statistics of these variations. One can, for example, measure sample statistics over a specified sufficiently large number of pulses, statistics like the difference from peak to mean power, or the mean value of the power difference from pulse to pulse. These statistics can then be compared to similar statistics for targets known to be clutter or meteorological, and when the statistics match the statistics of clutter, the radar data at that point can be censored.

Post-processing of radar data to eliminate clutter has been a common approach used in clutter mitigation. Researchers use primary characteristics of clutter to identify clutter: (1) near-zero mean radial velocity, (2) narrow spectrum width, (3) shallow echoes in the vertical, (4) sharp changes in power in range and azimuth, and (5) stationarity of the echoes in time. A number of techniques have been developed that employ various aspects of these properties to eliminate clutter. These techniques typically employ logical decision processes based on rules using one or more of the five characteristics of clutter described above. Some employ decision criteria that are simple empirical rules (e.g., delete all data with reflectivity >40 dBZ and radial velocities with $|v_r| < 0.25\ \mathrm{m\ s^{-1}}$, or delete all data with co-polar correlation coefficient less than a threshold such as 0.6–0.7). Others apply rule-based logic such as **fuzzy logic** (see Chapter 7), or more complicated logic approaches such as a **neural network**, a statistically based model that consist of sets of adaptive weights or numerical parameters that are tuned by a learning algorithm. Regardless of the degree of sophistication, the objective is to censor data determined to be clutter. There is no attempt to retain the weather signal that might be confounded with the clutter signal.

In past decades, **clutter filtering** during signal processing was not possible operationally because of limits in computation speed, storage, and bandwidth. Recently, however, ground clutter filters have been developed and employed to suppress the clutter signal while retaining as much of the meteorological signal as possible. These filters take advantage of the fact that ground echoes are characterized by near-zero radial velocities and narrow spectrum width. The filters are normally applied at the signal-processing level, operating on the time series of received complex voltages—the in-phase (I) and quadrature (Q) time series data (see Chapters 4 and 6). The challenge in ground clutter filtering is twofold. First, if ground echoes are mistaken for precipitation echoes, the reflectivity will be overestimated and the radial velocity will be biased. Conversely, if precipitation echoes are mistaken for ground echoes, they will be reduced or eliminated when they should have been retained. Both problems can lead to errors in precipitation estimates.

Clutter filters can be applied in the time domain or the frequency domain. Figure 8.6a shows an example of a Doppler spectrum where a weather signal with a nonzero mean Doppler radial velocity is contaminated with ground clutter signal that leads to a secondary peak in the spectra at zero radial velocity. Clutter filters use a technique called **windowing** to filter selected bands of frequencies corresponding to the near-zero Doppler radial velocity. Filtering leaves a gap in the spectra, as illustrated in Figure 8.6b. Adaptive frequency domain filters exist that not only filter the signal from the ground clutter but can also interpolate across the gap, correcting for the lost signal eliminated by the filter (Figure 8.6c). This technique works well except when the weather signal itself has zero Doppler radial velocity, particularly when it is also characterized by a narrow spectral width. In that case, the filter wipes out both the clutter and the weather signal, making it difficult to recover the reflectivity and the true Doppler radial velocity.

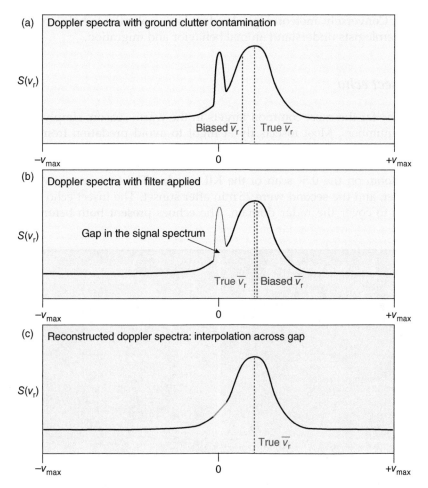

Figure 8.6 Graphic depiction of removal of ground clutter contamination. The Doppler spectra in (a) has a secondary peak near zero velocity associated with ground clutter. A clutter filter removes this echo (b), resulting in a radial velocity closer to the true value (c)

With continued advances in computing capabilities, and more widespread use of polarization technology (i.e., using correlation coefficient to identify clutter as discussed in Chapter 7), clutter filters have become more and more sophisticated and now have been developed to employ more than one of the techniques described above. For example, fuzzy logic algorithms are now employed to identify only those radar range gates that are potentially contaminated by ground clutter signal and then a clutter filter is applied to just those range gates, reducing the overall computation required to correct a field of radar data.

8.3 Echoes from biological sources

Birds, bats, and many species of insect spend part of their life airborne and become targets for meteorological radars. **Insect echo** indeed has a meteorological application. Conversely, meteorological radars have potential to help ornithologists and chiropterologists understand animal behavior and migration.

8.3.1   Insect echo

Insects are by far the most common targets in the warm season, simply because of their sheer numbers. Most insects fly at night to avoid predation from birds and many take flight near sunset. Swarms of insects simultaneously taking flight appear on radar as a widespread, weak echo "bloom." Figure 8.7 shows an example of an insect bloom on the 0.5° scan of the KILX radar. The first images were 53 min before sunset, and the second were 25 min after sunset. The insect echo during this time grows to cover the radar domain. The echoes present both before and after

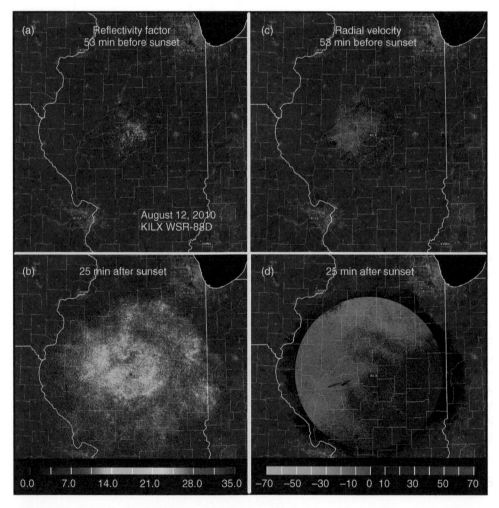

Figure 8.7   An insect echo bloom in the reflectivity field (a, b) and the radial velocity field (c, d) occurring at sunset at the KILX radar on a summer evening

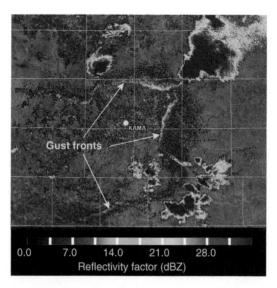

Figure 8.8   The gust front echoes in this reflectivity image from the KAMA radar are the result of insects being lofted by updrafts along the gust fronts

sunrise are from insects, not ground clutter. We know this because there are coherent nonzero radial velocities associated with the echo. Individually, insects fly in random directions while being carried along by the wind. The radar observes large numbers of insects within a pulse volume, so that their individual motions cancel, and their collective motion along the direction of the beam, that is, the radial wind velocity, is recorded by the radar. Insects therefore serve a meteorological purpose—they allow measurement of the low-level wind field in the absence of meteorological scatterers. The large axis ratios of most insects make them easy to identify on polarization radars. Insect echo is typically characterized by low values of reflectivity, but high values of differential reflectivity.

Another benefit of insect echo is the identification of gust fronts. Air ahead of an advancing gust front is lifted as the leading edge of the cool air moves forward. Insects are often lofted by the updraft and become detectable with radar. Figure 8.8 shows an example of several gust fronts in the vicinity of supercell thunderstorms near Amarillo, TX. The reflectivity scale in the figure is set to saturate for strong echoes so as to bring out the weaker echoes associated with the gust fronts. The insect echoes in this case clearly help map out the gust front boundaries. In the High Plains region of the USA, many gust fronts are dry, and insects constitute the entire echo. More generally, insect echo contributes along with echo from arcus clouds and/or dust lofted along the cold air outflow. Insect echo along gust fronts can be very advantageous while forecasting where deep convection may trigger, particularly where gust fronts collide with each other or with a synoptic scale frontal boundary.

8.3.2   Birds and bats

Birds and bats are normally point targets. However, flocks of migrating birds, as well as birds departing from roosts, or bats from caves can produce substantial echoes.

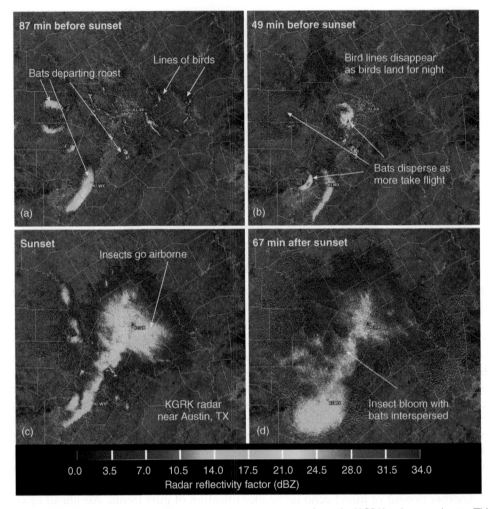

Figure 8.9   Radar echoes from birds, bats, and insects at sunset from the KGRK radar near Austin, TX

Figure 8.9 shows an example of bird, bat, and insect echoes, which illustrates all of these features. The data were collected near Austin, TX, near sunset. At 87 min before sunset, echoes from lines of birds are evident as well as circular echoes marking the departure of bats from caves (and the Congress Avenue Bridge in Austin). As the birds land, more bats emerge as sunset approaches. At sunset, insect echo emerges, eventually merging with the bat echo after sunset, as the bats feast on the insects.

Echoes near sunrise from birds and sunset from bats are easy to recognize because they tend to be circular and expand away from the roosts as the animals depart for their feeding forays. Echoes from migrating birds tend to appear as thin somewhat irregular lines of echo, corresponding to the V-formations that the birds often organize into when they fly. In Chapter 7, it was shown that birds show up as high differential reflectivity regions often within the eye of tropical cyclones.

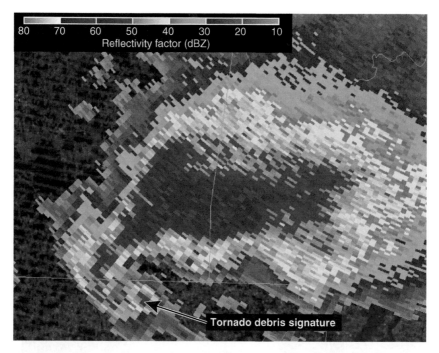

Figure 8.10 The echo at the end of the hook on the southwest side of the supercell is caused by debris from a tornado

8.4 Debris, dust, and smoke

Airborne debris, dust, and smoke can contribute to the echo patterns observed with meteorological radars. The most important debris signature is that produced by a tornado (Figure 8.10). Tornadoes can loft a significant amount of debris in the air, often producing a high-reflectivity debris signature at the end of the hook echo of tornadic supercells. We will defer our discussion of **tornado debris signatures** to Chapter 14.

Aside from tornadoes, it is unusual for debris to be lofted to altitudes where it can become a target for radar. Exceptions can occur with strong winds that can loft leaves or other light surface material in the air. One example of lofting of surface dust is the haboob (an Arabic term for an intense dust storm). On October 17, 2011, for example, a strong, dry cold front with wind gusts approaching 75 miles per hour passed across the Texas Panhandle. The resulting haboob is shown in Figure 8.11, along with the radar return from the KLBB WSR-88D radar at Lubbock, TX. The echo from the lofted dust clearly defined the frontal boundary. Haboobs also occur along gust fronts from thunderstorms, the echo from the dust (and lofted insects) defining the leading edge of the expanding cold pool.

Another source of airborne particles is fire. Large grass fires and forest fires loft a significant amount of material into the atmosphere in the form of smoke. Figure 8.12, for example, shows the smoke plumes from several grass fires in Oklahoma on March 12, 2008. The smoke echoes can be helpful in identifying the source of the fires and where new fires might be breaking out.

Figure 8.11 The line of radar echoes passing the KLBB radar was associated with the advance of a wall of dust called a *haboob* (Photo Courtesy of John M. Holsenbeck, Jr and the National Weather Service, Lubbock, Texas)

8.5   Aircraft echoes and chaff

An aircraft represents a point target, generally flies at high altitudes except when approaching or departing airports, and normally presents no issues when interpreting meteorological radar data. The exception is in research, especially when the research goal involves comparing data collected by a research aircraft with collocated data collected with a radar. In this case, aircraft echo so dominates any meteorological echo that direct comparison is not possible (Figure 8.13). The echo from an aircraft is sometimes referred to as a **skin paint** by research meteorologists, a reference to the radiation "painting" the skin of the aircraft. The approach used to compare datasets normally involves interpolating uncontaminated radar data close to the aircraft to the aircraft position and then performing the comparison. This approach works in cloud systems with uniform gradients, particularly stratiform clouds, but can lead to errors or fail altogether in systems with strong gradients such as convective clouds.

Another source of echo is aircraft-released chaff. Chaff consists of thin pieces of aluminum foil, metalized glass fiber, or plastic that is used by the military as

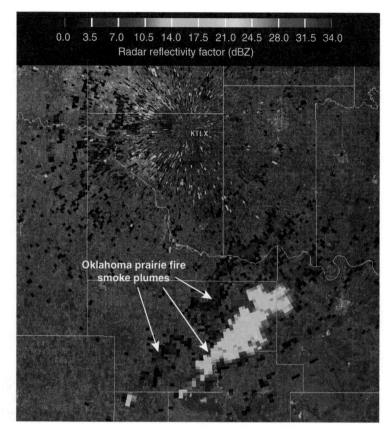

Figure 8.12 The radar return is associated with a plume of smoke originating from two fires near the southwest end of the echoes

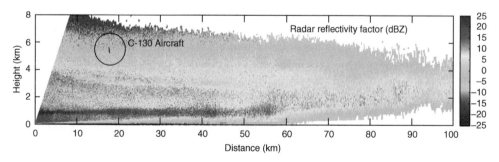

Figure 8.13 An example of an RHI scan with an aircraft echo embedded within the stratiform echo of a winter storm

a radar countermeasure for self-defense. The pilot launches a package into the aircraft's slipstream, where it releases millions of small fibers that are typically a few centimeters in length. The chaff confuses a missile's radar guidance system by creating a new, larger target than the aircraft. Several aircraft often lay out a chaff curtain.

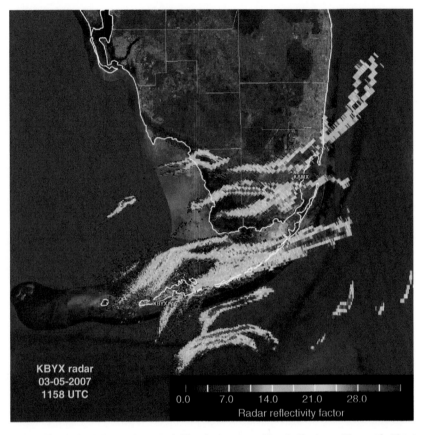

Figure 8.14   The radar echoes over south Florida are caused by chaff released from a military aircraft during training exercises

Chaff typically falls at a rate of 10–20 cm s^{-1}, so dispersal from aircraft altitudes can create a curtain that lasts for hours. Chaff echoes are rare but can mysteriously appear near military areas, generally as part of military exercises. For example, Figure 8.14 shows several curtains of chaff released by military aircraft from the Naval Air Station at Key West. The forecast discussion that day from the National Weather Service stated that "local radars are not detecting any precipitation echoes, but false echoes in the form of chaff is smothering the island chain and surrounding waters."

Chaff has also been used, albeit rarely, in meteorological experiments. In these cases, chaff has been released from aircraft as an aid to map out complex wind fields in air where no other echo sources, such as insects, are present.

8.6   Other non-meteorological echo sources

Microwaves at radar frequencies can be generated from other sources. These contribute to the background from which a meteorological signal must be extracted.

8.6.1   The sun

The sun, as a blackbody, is a source of microwaves. Normally, the sun does not inter-fere with radar signals unless the antenna is pointed very close to or directly at the sun. When this occurs (it typically happens with most WSR-88D radars shortly after sunrise and just before sunset), one or two beams will typically be filled with echo. The echo is weak and is most evident on days where the air is clear and no meteorological echo exists. These features, called **sun spikes** (Figure 8.15), can sometimes be seen appearing at radars from east to west as the earth rotates and the sun progressively rises or sets further to the west.

As the sun is a source of microwaves, it can be used as a source to calibrate radar antenna-pointing angles. The precise position of the sun in the sky can easily be cal-culated from astronomical equations, given a radar's position and the exact time of day. By focusing the antenna on the maximum return from the sun and noting the time and azimuth and elevation of the antenna, any error in the pointing angles can be identified and corrected.

8.6.2   Receiver noise

Noise is a random fluctuation in an electrical signal and is generated to some degree by all electronic circuitry. Receiver noise represents an ever present background input

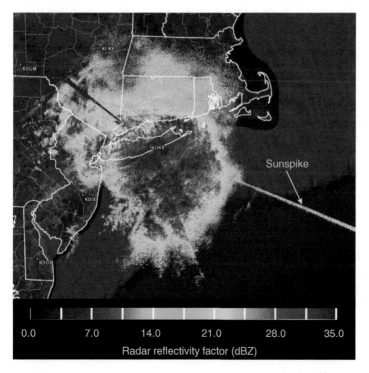

Figure 8.15   The echo labeled "sun spike" on the KOKX radar is caused by microwaves emitted from the sun. Sun spikes are most common near sunrise and sunset when the radar antenna points directly at the sun

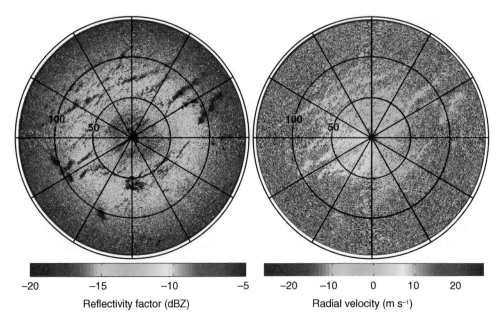

Figure 8.16   Receiver noise appears as an increase in the reflectivity factor with range and as random velocities (the speckled values) in the radial velocity field

to the radar. The signal from a target must sufficiently exceed the noise level to be detectable. In the absence of a noise filter, noise appears on a plan position indicator (PPI) display of the radar reflectivity factor as a background value that increases with the square of the range. The increase occurs because the radar equation used to calculate the reflectivity factor (Eq. (5.43)) has an r^2 in the numerator. Although the noise power is the same for each range, the r^2 correction leads to an increase of the reflectivity factor associated with noise with distance. Figure 8.16 shows an example of a radar display of reflectivity and radial velocity without a noise filter applied. Receiver noise has random phase, so on the radial velocity display, each range gate in a noise field displays a random velocity within the Nyquist range. The random phase of noise allows us to use low values of the co-polar correlation coefficient in a polarimetric radar to efficiently remove noise.

8.6.3   Radio interference

Some radar frequencies, particularly higher frequencies, are close to frequencies used for commercial and communication activities. These external sources, particularly if the bandwidth of the external source is not well calibrated, can lead to interference for meteorological radar operations. These sources are generally rare, may be intermittent or constant, and generally are a nuisance when attempting to interpret the meteorological component of a radar signal.

8.7   Bragg scattering

The term *Bragg scattering*, also referred to as **Bragg diffraction**, derives its name from a father and son team of physicists, William Lawrence Bragg and William Henry

Bragg, who together won the Nobel Prize in 1915 for their discovery of a mechanism of coherent scattering of radiation from a crystal lattice. Their discovery led to the understanding of the structure of atoms in crystals, hence the Nobel Prize. Bragg scattering occurs when the spacing of scattering elements (in the Braggs' case, atoms) is half the wavelength of impinging radiation, such that the waves of radiation from the scattering elements are in phase, that is, undergo **constructive interference** and are **coherent**, as they propagate away from the scattering material.

The atmosphere is not a crystal lattice, so one might naturally question what the scattering elements could be for radiation propagating outward from a radar antenna. To understand this, we need to return to Chapter 1 and the discussion of the refractive index. Recall from Eq. (4.6) that the index of refraction depends on pressure, temperature, and water vapor pressure. Pressure and temperature vary slowly in most places in the atmosphere, but water vapor is highly variable, as one can easily see by watching clouds form and evaporate on high-speed videos. The atmosphere is turbulent, particularly in regions of wind shear. **Turbulent eddies** occupy a wide range of scales, with the energy contained in eddies at larger scales transferred down in a cascade to energy in eddies at smaller scales until at the smallest scales energy is dissipated as heat. In turbulence theory, the **inertial subrange** is the range of scales where the net energy transfer from larger energy-containing eddies is in equilibrium with the net energy cascading through to smaller eddies; in other words, the inertial subrange is the range of eddy sizes that is in equilibrium in terms of energy transfer within a turbulent environment. **Turbulence** is said to be **isotropic** if eddies appear the same regardless of viewing angle and **anisotropic** if they have angular dependence.

If turbulence occurs in the presence of strong gradients of water vapor, temperature, or both, the turbulent eddies will create strong gradients in the index of refraction. These gradients occur on all scales within the inertial subrange. Some of these eddies will be on a scale of half the wavelength of the impinging radar waves. A component of the impinging waves undergoes specular reflection when encountering strong index of refraction gradients created by the eddies. If the turbulence is isotropic, only the scattered waves from eddies with spacing of half the radar wavelength will undergo constructive interference and therefore be coherent, a phenomena similar to that which the Braggs' observed with X-rays in their crystal experiments. This coherent return is called *Bragg scattering*. Bragg scattering is dependent on wavelength because larger eddies are more likely to be isotropic and remain coherent in the turbulent environment than smaller eddies. For this reason, Bragg scattering is often observed at S-band and longer wavelengths but is rare at C-band and essentially not observed at shorter wavelengths.

Turbulence tends to manifest itself in layers in the atmosphere because vertical shear, as well as moisture and temperature gradients, tend to be strongest in the vertical. In fact, strong gradients in temperature and moisture are often coincident with the shear zones because all three occur along airmass boundaries. For this reason, **Bragg echo** often tends to appear as layers. An example of Bragg echo manifested in layers on a PPI display is shown in Figure 8.17. These data were taken over the Caribbean Sea near the Island of Barbuda by the S-Pol radar in clear air. The layers of echo developed in locations coincident with strong gradients in relative humidity. The Bragg scattering from the layers was persistent over time so that the humidity layers could be tracked for hours while they maintained their structure as the trade winds moved across the radar site (Figure 8.18). Bragg echo can also develop in

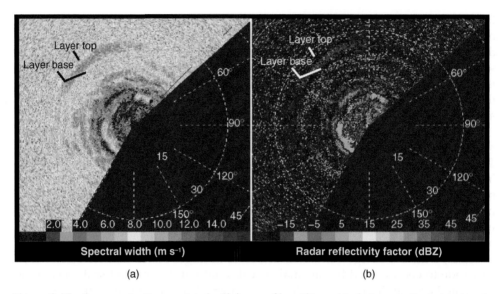

(a)   (b)

Figure 8.17   Bragg scattering associated with layers of humidity and turbulence in the tropical trade wind layer leads to a reduction in spectral width (green/blue) from noise values (yellow) in panel a and an enhancement in the radar reflectivity factor (b) (From Davison, J.L., Rauber, R.M., and Di Girolamo, L. (2013) A revised conceptual model of the tropical marine boundary layer. Part II: detecting relative humidity layers using Bragg scattering from S-band radar. *J. Atmos. Sci.*, **70**, 3025–3046. © American Meteorological Society, used with permission)

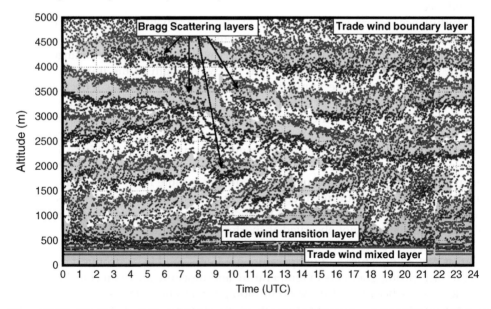

Figure 8.18   The time evolution of moisture layers (green) and the trade wind transition and mixed layers (blue) over a day, as mapped out by the Bragg scattering layers appearing in Figure 8.17. Each red and blue dot denotes the top and bottom of a layer, respectively, on one PPI scan. The data were obtained with the S-Pol 10 cm radar, which was located on the Island of Barbuda in the Caribbean (From Davison, J.L., Rauber, R.M., and Di Girolamo, L. (2013) A revised conceptual model of the tropical marine boundary layer. Part II: detecting relative humidity layers using Bragg scattering from S-band radar. *J. Atmos. Sci.*, **70**, 3025–3046. © American Meteorological Society, used with permission)

the convective boundary layer. In summertime, however, it is sometimes difficult to separate the effects of Bragg scattering from scattering associated with insects.

Important terms

anisotropic turbulence
biological echo
Bragg diffraction
Bragg echo
Bragg scattering
chaff
clear air echo
clutter filtering
coherent
constructive interference
fuzzy logic
ground clutter
inertial subrange

insect echo
isotropic turbulence
hydrometeors
neural network
noise
radio interference
sea clutter
skin paint
sun spike
tornado debris signature
turbulence
turbulent eddies
windowing

Review questions

1. What are the primary sources of clear air echo, and how and where might each appear on a radar screen?

2. When describing ground clutter, this text often uses the term *near-zero* radial velocity. What characteristics of some ground clutter sources would lead to near-zero rather than zero radial velocity?

3. How might differential reflectivity be used in clutter identification?

4. What are four methods used for ground clutter mitigation, and how do they differ from each other? Which has the potential of retaining the meteorological signal while discarding the ground clutter signal?

5. Why are nocturnal low-level winds less likely to be recovered with a Doppler radar in winter, compared to summer?

6. What polarization characteristic makes it easy to separate insect echo from meteorological echo? Why?

7. Why are haboobs detectable on radar?

8. Why is it difficult to directly compare microphysical data collected with research aircraft with simultaneous data collected with a radar scanning the aircraft position from the ground?

9. How does Bragg scattering differ from Rayleigh scattering and Mie scattering?

10. Based on the information presented in this chapter, why would it be relatively easy to precisely define the boundary of a cumulus cloud with a K_a-band radar, but not with an S-band radar? Assume that both radars have the same beamwidth and along beam resolution.

Challenge problems

11. On a very hot (40 °C), stiflingly humid day (90% RH) when the surface pressure was 1013 hPa, the position of ground clutter associated with a particular building 30 km from the radar shifts slightly from its position on a bitterly cold (−30 °C), exceptionally dry (5% RH) day when the pressure was 1030 hPa 6 months earlier. The radar is calibrated and is working properly on both days. What happened to cause the position of the ground clutter to shift? How far did the ground clutter shift and in which direction? Would such a shift be detectable with a WSR-88D?

    (Hint: An equation relating saturation vapor pressure (hPa) and temperature (°C) is

$$e_s = a + T(b + T(c + T(d + T(f + T(g + hT)))))$$

    where

$a = 6.107799961$ $b = 4.436518521 \times 10^{-1}$ $c = 1.428945805 \times 10^{-2}$
$d = 2.650648471 \times 10^{-4}$ $f = 3.031240396 \times 10^{-6}$ $g = 2.0340880948 \times 10^{-8}$
$h = 6.136820929 \times 10^{-11}$

---

12. Assume that a typical nocturnal airborne insect has a radar cross section equivalent to a sphere of diameter 3 mm. How many insects would need to be present in the lowest kilometer of the atmosphere over a football field to produce the stronger echo (~14 dBZ) seen at 25 min after sunset in Figure 8.9? Assume that the insects are distributed uniformly in the volume. (Note: Including end zones, an American football field is 360×160 ft.)

13. Identify a region of the country where there are a large number of wind farms close to a National Weather Service rawinsonde launch site and WSR-88D radar. Examine the radar data from the WSR-88D around the time of sunrise and obtain the 12:00 UTC sounding, which should characterize the atmosphere near sunrise. Pick a day when the wind farms are visible to long range in the lowest elevation scan. Use equations in Chapter 4 to calculate the path of the 0.5° radar beam through each low-level sublayer in the rawinsonde data to confirm that anomalous propagation was indeed occurring at the time the wind farms were visible as ground clutter.

14. At 45°N latitude, at what azimuth will a sun spike appear on a radar display at sunrise on the spring equinox? At what latitude and date will a sun spike appear at an azimuth of 0° on a radar display at sunrise?

15. Wind profilers are Doppler radars that most often operate in the VHF (30–300 MHz) or UHF (300–1000 MHz) frequency bands. How large are the turbulent eddies that produce strong coherent echoes for a profiler operating in clear air at 404 MHz? How about 915 MHz?

9

Propagation Effects: Attenuation and Refractivity

Objectives

By the end of this chapter, you should understand how

- attenuation affects a radar signal.
- gaseous attenuation can be used to measure water vapor.
- attenuation by cloud droplets can be used to retrieve cloud liquid water content.
- corrections for attenuation can be applied to retrieve rainfall rates.
- hail impacts attenuation and how corrections for attenuation can be applied.
- supercooled water can be identified in clouds based on attenuation at W-band.
- satellite precipitation retrievals depend on surface echoes to correct attenuation.
- refractivity is measured, what it represents, and its relation to moisture and temperature fields.

9.1   Introduction

We implicitly assumed while deriving the radar reflectivity factor that the intervening atmosphere between the radar and the target did not interact with the radiation. The validity of this assumption depends on both the radar wavelength and the characteristics of the intervening medium, which contains gases, and may also contain cloud droplets, raindrops, snow, and/or hail. Two processes, energy absorption and scattering of energy out of the radar beam, can contribute to energy loss, or **attenuation**, of the radar signal when radiation from the radar interacts with the intervening atmosphere. As discussed in Chapter 7 in the context of interpreting dual-polarization radar measurements, attenuation introduces uncertainty in quantitative interpretation of radar measurements and, in almost all cases, is an undesirable characteristic. However, in some limited applications, scientists have employed dual-wavelength radar systems and measured the **differential attenuation** between radars operating at different wavelengths to extract meteorological information about the atmosphere. They have also developed methods using dual-polarization

Radar Meteorology: A First Course, First Edition. Robert M. Rauber and Stephen W. Nesbitt.
© 2018 John Wiley & Sons Ltd. Published 2018 by John Wiley & Sons Ltd.
Companion website: www.wiley.com/go/Rauber/RadarMeteorology

radars to correct for attenuation. Attenuation can also give us information about the characteristics of the medium in which the attenuation is occurring.

Aside from attenuation, a second influence of the intervening medium is to alter the speed at which radar waves propagate through the medium. As we learned in Chapter 1, the speed of light is a function of the refractive index of the atmosphere, which, in turn, is a function of atmospheric temperature, pressure, and water vapor concentration. Characteristics of **radio refractivity**, that is, changes in the refractive index along the path of a radar ray, influence the propagation of radar waves and can be used under certain conditions to determine the local distribution of water vapor near the surface in the vicinity of a radar.

This chapter focuses on attenuation, its impact on radar measurements, potential uses of attenuation to extract meteorological information about the atmosphere, and ways to correct for its effects. Much work has been done in this regard—we will selectively concentrate on only a few broad highlights from the body of research on attenuation. We also explore methods that use radio refractivity to measure atmospheric properties.

9.2 Attenuation

Attenuation can be described mathematically using the **Beer–Lambert law**, which relates the attenuation of electromagnetic radiation to the properties of the material through which the radiation travels. The law arises by considering the reduction of backscattered power, dP, that occurs either due to absorption or due to scattering out of the beam as the radiation passes through a small distance, dr, through the atmosphere. We can write this formally as

$$dP_w = -2k_L P_w dr \tag{9.1}$$

where k_L is the **attenuation coefficient** that has units of inverse length. The factor 2 accounts for the fact that the radiation transits the atmospheric path twice during the round trip between the radar and the target. We can integrate Eq. 9.1:

$$\int_{P_{r_0}}^{P_r} \frac{dP_w}{P_w} = -2 \int_0^r k_L dr \tag{9.2}$$

where the limits of integration are P_r, the actual power received at the antenna, and P_{r_0}, the power that would have been received at the antenna in the absence of attenuation, with the result.

$$\frac{P_r}{P_{r_0}} = \exp\left(-2\int_0^r k_L dr\right) \tag{9.3}$$

The reduction in power in decibels can be obtained by writing this equation as

$$\Delta P_r(\text{dB}) = 10 \log\left(\frac{P_r}{P_{r_0}}\right) = -2\int_0^r k dr \tag{9.4}$$

where the attenuation coefficient now has units of decibels per unit length and is equal to

$$k = 10 \; \log(e)k_L \tag{9.5}$$

Attenuation of radar wave energy in the atmosphere is caused by atmospheric gases and hydrometeors, which take the form of cloud droplets, raindrops, ice crystals, snowflakes, or hail. Gases and cloud droplets act as absorbers, whereas rain, snow, and hail both absorb radiation and scatter it out of the radar beam.

With the advent of dual-polarization radars, several terms are now used while discussing attenuation. **Total attenuation** is the attenuation integrated along the path between the radar and the target and can be expressed as **one-way attenuation** (from the radar to the target) or **two-way attenuation** (from the radar to the target and back). Attenuation is a reduction in power and is normally expressed in decibels (dB). **Specific attenuation** is defined as the attenuation per unit distance and is normally expressed in units of decibels per kilometer. For dual-polarization radars, additional attenuation terms arise. **Differential attenuation** is the difference in total attenuation that occurs between the horizontally and vertically polarized signals and can be expressed as **one-way differential attenuation** or **two-way differential attenuation**. The term differential attenuation can also be applied to the difference in total attenuation that occurs between two radars of different wavelength, provided the radars have beams that are aligned and have the same beamwidth. **Specific differential attenuation** is the differential attenuation per unit distance, again expressed in units of decibels per kilometer.

9.2.1   Attenuation by atmospheric gases and measurement of water vapor

Water vapor molecules and oxygen molecules in the atmosphere have permanent dipole moments, the former associated with the electric field and the latter with the magnetic field. Both these molecules possess quantized vibrational and rotational states that are altered by absorption of electromagnetic energy at specific wavelengths. When radar energy is absorbed by a molecule, it attains a higher energy state. This energy is subsequently lost to the radar wave as it is dissipated to the atmosphere as heat. Within the range of meteorological radar wavelengths, water vapor molecules have a group of **absorption lines** centered at a wavelength of 1.35 cm, and another group at 0.2 cm. Because of molecular collisions, or **pressure broadening**, individual lines merge into wider **absorption bands**. Oxygen absorption lines, centered near 0.5 cm, also merge into a wider absorption band centered at this wavelength. Figure 9.1 shows the attenuation that results from absorption by water vapor and oxygen in an atmosphere at sea level pressure (1013.25 hPa), a temperature of 20 °C, and a **water vapor density**[1] of 7.75 g m^{-3}, the mean value for Washington, DC. Common radar wavelengths are superimposed. Figure 9.1 shows that for wavelengths 3 cm and greater, attenuation by gases is negligible

---

[1] Water vapor density, also called **absolute humidity**, is the mass of water vapor (typically measured in grams) contained in a volume of air (typically given in cubic meters).

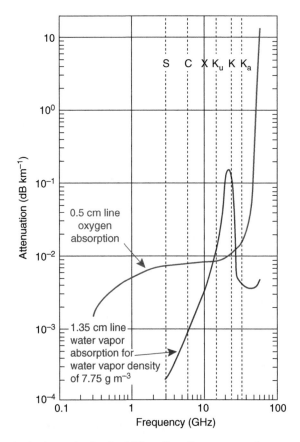

Figure 9.1   Atmospheric absorption by the 1.35 cm line of water vapor for a mean absolute humidity of 7.75 g m^{-3} (blue line) and by the 0.5 cm line of oxygen at a temperature of 20 °C and a pressure of 1 atm (After Bean, B.R. and Dutton, E.J. (1966) *Radio Meteorology*, National Bureau of Standards Monograph 92, U.S. Government Printing Office)

except over very long ranges. For example, an X-band radar would have the signal attenuated only 0.008 dB km^{-1}, or ~1 dB for a target located ~62 km from the radar. Note that radars using a K-band wavelength are particularly susceptible to water vapor absorption. Water vapor absorption is proportional to water vapor density, so in humid atmospheres, such as found in the tropics, attenuation at K band can be severe. At even shorter wavelengths such as W-band, both oxygen and water vapor contribute to attenuation. Because of attenuation, the primary use of short wavelength radars has been for applications where short distances are acceptable. Short-wavelength radars are most often used in vertically pointing mode for this reason. Because atmospheric density always decreases with altitude, and water vapor concentration generally decreases with altitude as well, the effect of attenuation by gases has its greatest impact on measurements in the lower atmosphere. This is illustrated in Figure 9.2, which compares gaseous attenuation in similar atmospheric conditions at S-band and X-band for beams at 0° and 5° elevation.

Can atmospheric attenuation by water vapor be turned into an advantage? Is it possible to use measurements of attenuation to extract information about the distribution

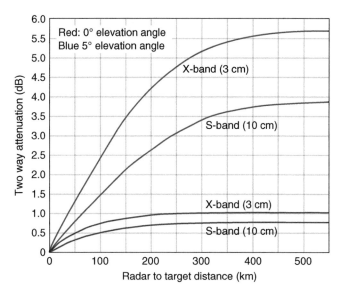

Figure 9.2   Gaseous two-way attenuation in similar atmospheric conditions at S-band and X-band for beams at 0° and 5° elevation (Data from Skolnik, M., *Introduction to Radar Systems*, 2nd edn, McGraw-Hill)

of water vapor in the atmosphere? The answer in some situations is yes. Techniques to measure water vapor in the atmosphere rely on measuring the along-path differential attenuation that occurs when different wavelength radars view the atmosphere with their beams aligned. One strategy,[2] for example, uses aligned beams of non-attenuating (S-band) and attenuating (K_a-band) wavelength radars to recover *vertical profiles* of water vapor density. The technique requires that scattered isolated meteorological echoes exist, such as those that might be produced by a field of towering cumulus near the radar. Water vapor density is assumed to be a function only of altitude over the area of the radar analysis. To apply the technique, radar data are either collected over an azimuth sector, or over complete 360° sweeps, at elevation angles covering the cloud depth.

To understand how the method works, consider Figure 9.3 that shows a radar ray segment between the NCAR SPOL-K_a dual-wavelength S- and K_a-band radar system located at r_0 and the leading edge of an echo associated with a cloud at range, r_1. This ray segment (and any ray segment originating at the radar) is called a *primary ray*. Now consider a ray at the same elevation angle, but at an azimuth that misses the first cloud, yet hits a second, more distant cloud, whose leading edge is located at range r_2. The segment along this ray from range r_1 to r_2 is called a *secondary ray*. Secondary rays can also extend from the back edge of a cloud outward to the leading edge of a more distant cloud, as shown between r_3 and r_4 in Figure 9.3. If scattering particles at the leading edge of the clouds at r_1 and r_2 are sufficiently small that they satisfy the Rayleigh criteria, the difference in the radar reflectivity factor measured by the radars

[2]The technique is described in Ellis, S.M. and Vivekanandan J. (2010) Water vapor estimates using simultaneous dual-wavelength radar observations. *Radio Sci.*, **45**, RS5002, doi:10.1029/2009RS004280.

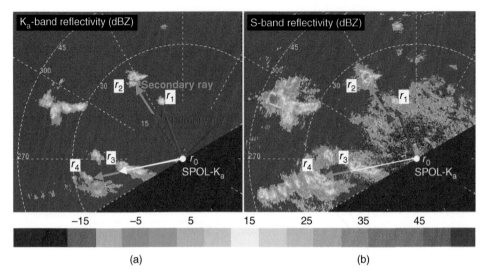

Figure 9.3   PPI plots of (a) K_a-band and (b) S-band reflectivity values. The arrows are meant to illustrate two methods of creating secondary rays for attenuation estimation as described in the text (From Ellis, S.M. and Vivekanandan, J. (2010) Water vapor estimates using simultaneous dual-wavelength radar observations. *Radio Sci.*, **45**, RS5002, doi: 10.1029/2009RS004280, © American Geophysical Union, used with permission)

at the leading edge of each cloud can be related to the attenuation that occurred due to absorption by the intervening atmosphere between each cloud and the radar.

Radar rays increase in altitude with distance from the radar, so attenuation between r_1 and r_0 is caused by absorption by water vapor in the lowest layer of the atmosphere, between altitudes z_1 and z_0. Note that z_1 is not the point where the radar ray crosses r_1, but rather is located at the top of the radar beam, as the beam has a finite width. Similarly, attenuation between r_2 and r_0 is caused by water vapor in the deeper layer between z_2 and z_0. If many clouds are present, the layer-mean attenuation in a number of layers can be estimated. Simple subtraction of the attenuation within a shallow layer $(z_1 - z_0)$, from that measured in a deeper layer $(z_2 - z_0)$ allows one to determine the attenuation in the elevated layer $(z_2 - z_1)$. Theoretical relationships[3] between radar wave attenuation and water vapor density (Figure 9.4) can be used to convert layer-mean attenuation values to water vapor density, allowing a vertical profile of water vapor density to be obtained.

Figure 9.5 shows an example of an actual sounding of water vapor density from a tropical location (black line), the same sounding, but averaged to reflect the resolution obtainable by radar analysis, and water vapor density derived directly from measurements of atmospheric attenuation using the dual-wavelength technique. In this example, the remotely-sensed technique retrieved a reasonable estimate of the average water vapor density profile compared with the profile obtained from a radiosonde.

---

[3]The theoretical model relating radar wave propagation to water vapor density is described in Liebe, H.J. (1985). An updated model for millimeter wave propagation in moist air. *Radio Sci.*, **20**, 1069–1089.

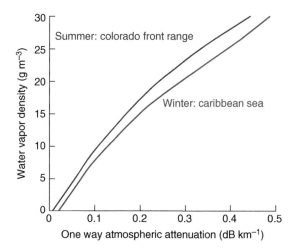

Figure 9.4 One-way attenuation due to water vapor (dB km^{-1}) for conditions characteristic of air over the Caribbean Sea in winter and over the High Plains of Colorado just east of the Rockies in summer (Adapted from Ellis, S.M. and Vivekanandan, J. (2011) Liquid water content estimates using simultaneous S and K$_a$ band radar measurements. *Radio Sci.*, **46**, RS2021, doi: 10.1029/2010RS004361, © American Geophysical Union, used with permission)

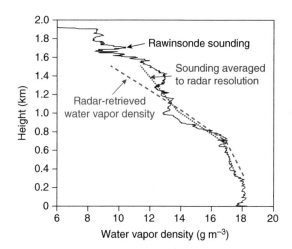

Figure 9.5 An example of radar-retrieved water vapor density (red) from the S-Pol radar when it was located on an island in the Caribbean. The black line is the water vapor density from a nearby sounding and the blue line is the same sounding averaged to match the radar resolution (Adapted from Ellis, S.M. and Vivekanandan, J. (2010) Water vapor estimates using simultaneous dual-wavelength radar observations. *Radio Sci.*, **45**, RS5002, doi: 10.1029/2009RS004280, © American Geophysical Union, used with permission)

A number of caveats and considerations were glossed over in the discussion above that must be considered in application of this (or other) techniques that employ differential attenuation to retrieve information about water vapor. With that caution noted, the technique nicely illustrates how an undesirable characteristic of radar

Table 9.1   Values of a_L in Eq. (9.6)

Temperature (°C)	Wavelength (cm)			
	0.9	1.24	1.8	3.2
20	0.647	0.311	0.128	0.0483
10	0.681	0.406	0.179	0.0630
0	0.99	0.532	0.267	0.0858

measurements, attenuation, can be turned into an advantage to retrieve atmospheric properties.

9.2.2   Attenuation by cloud droplets and measurement of liquid water content

Many clouds contain only tiny water droplets whose radii remain smaller than about 100 μm. These small droplets, which fall at such slow terminal velocity that they visually appear suspended in air, are called **cloud droplets**. Larger drops with appreciable fall speeds are called **raindrops**. Cloud droplets, and raindrops as large as 1 mm in diameter, are spherical in shape and are small compared to radar wavelengths and therefore fall into the Rayleigh scattering regime for common radar wavelengths. Rayleigh scattering theory can be used to calculate the attenuation caused by a field of cloud droplets as a function of wavelength and cloud temperature. From Rayleigh theory, a simple result emerges that the attenuation coefficient is given by

$$k_L = a_L(\text{LWC}) \tag{9.6}$$

where a_L is the **liquid attenuation coefficient** in units of $(\text{dB km}^{-1}\ (\text{g m}^{-3})^{-1})$ and LWC is the liquid water content of the cloud in units of g m^{-3}. Example values of a_L for four wavelengths and four temperatures are given in Table 9.1.

To estimate the distribution of liquid water content in actual clouds, it is necessary to use a dual-wavelength radar system. An important criterion for dual-wavelength retrieval of liquid water content is that the hydrometeors are sufficiently small that they conform to the Rayleigh criteria at the shorter wavelength. For example, if the shorter wavelength is K_a-band, the largest raindrops in the cloud should have diameters less than 1 mm. This criterion is satisfied, for example, by most oceanic and continental cumulus and stratocumulus clouds, provided temperatures within the clouds are sufficiently warm so that no ice is present. Unfortunately, the technique cannot be extended to measure ice water content because ice hydrometeors are normally not spherical and the requirement of Rayleigh scattering is therefore not met in mixed phase or ice clouds.

Retrieval of liquid water content is accomplished by measuring the differential attenuation that occurs along a radar ray as it passes through a cloud.[4] The measurements must first be adjusted to account for any attenuation from atmospheric gases.

---

[4] A review of methods to retrieve liquid water content in clouds can be found in Ellis, S.M. and Vivekanandan, J. (2011) Liquid water content estimates using simultaneous S and K_a band radar measurements. *Radio Sci.*, **46**, RS2021, doi: 10.1029/2010RS004361.

Using the dual-wavelength technique, liquid water content in clouds is determined from

$$LWC = \frac{\left[\frac{\partial}{\partial r}(DWR) - (a_g(\lambda_2, r) - a_g(\lambda_1, r))\right]}{(a_L(\lambda_2, r) - ka_L(\lambda_1, r))} \qquad (9.7)$$

where a_g is the gaseous attenuation in units of dB km^{-1} and DWR is the dual-wavelength ratio in dB, defined as

$$DWR(dB) = Z(\lambda_2) - Z(\lambda_1) \qquad (9.8)$$

Users of liquid water retrieval techniques must be cognizant of a number of error sources. For example, the geometry of the beams of the two radars must match sufficiently well or the sample volumes will be different. The Rayleigh scattering criteria must be met. The temperature in the sampled region must be determined using an independent measurement technique (such as a sounding) so that the liquid water attenuation coefficients can be determined. The measured DWR must also be large relative to the measurement noise level at each wavelength. Other factors may come into play depending on the meteorology of the region to be sampled and the characteristics of the environment (e.g., accounting for the influence of non-meteorological echoes such as insects or birds).

Figure 9.6 shows the S-band radar reflectivity factor measured by the NCAR SPOL-K$_a$ radar system in trade cumulus clouds over the Caribbean. Note the echo close to the radar near the edge of the sector scan. This cloud was sampled at several azimuth angles. Figure 9.7 shows vertical cross sections along the line through the circled echo of the S-band radar reflectivity factor and the dual-wavelength-derived liquid water content. The analysis shows the liquid water content increasing with altitude to about 0.5 g m^{-3} near the top of the radar reflectivity column. Above that level, the liquid water content decreases, likely due to entrainment of dry air at the cloud boundaries and top. The radar reflectivity increases downward through the

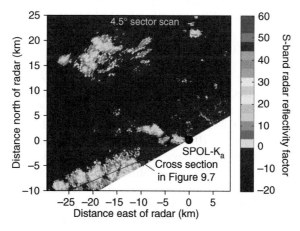

Figure 9.6 Trade wind cumulus clouds sampled using a 4.5° sector scan of the S-Pol radar. The cross sections in Figure 9.7 are along the red line (Adapted from Ellis, S.M. and Vivekanandan, J. (2011) Liquid water content estimates using simultaneous S and K$_a$ band radar measurements. *Radio Sci.*, **46**, RS2021, doi: 10.1029/2010RS004361, © American Geophysical Union, used with permission)

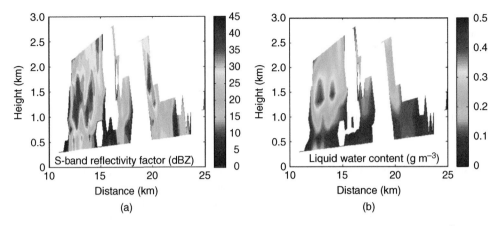

Figure 9.7 Vertical cross sections of (a) reflectivity (dBZ) and (b) liquid water content (g m⁻³) within clouds along the cross section in Figure 9.6 (Adapted from Ellis, S.M. and Vivekanandan, J. (2011) Liquid water content estimates using simultaneous S and K$_a$ band radar measurements. *Radio Sci.*, **46**, RS2021, doi: 10.1029/2010RS004361, © American Geophysical Union, used with permission)

liquid water column, consistent with growth of drizzle and rain in the cloud. The high values of LWC do not necessarily correspond to the high values of reflectivity. This is because the larger drizzle droplets contribute most to the reflectivity, but the smaller cloud droplets may dominate the LWC and thus the attenuation. Thus, below cloud base, the falling drizzle has relatively high reflectivity, but the LWC is lower in the absence of cloud drops.

9.2.3   Attenuation by rain and its correction

Attenuation of a radar signal by rain is dependent on both wavelength and temperature. Figure 9.8 shows a quantity we will refer to here as the **rainfall attenuation ratio**:

$$K(\mathrm{dB\,km^{-1}(mm\,h^{-1})^{-1}}) = k_r R^{-1} \tag{9.9}$$

defined as the ratio of the rainfall attenuation coefficient (k_r) to the rainfall rate (R), for three wavelengths. Based on Figure 9.8, at S-band, at the light rainfall rate of 1 mm h⁻¹, the specific attenuation is about 0.0008 dB km⁻¹, whereas at an extreme rate of 100 mm h⁻¹, the specific attenuation is 0.0007 dB km⁻¹ (mm h⁻¹)⁻¹ × (100 mm h⁻¹) = 0.07 dB km⁻¹. Based on these values, it should be evident that very heavy rainfall would have to occur over a very long path along a radar ray at S-band for attenuation to be noticeable. These data show that for practical purposes, except in specific research applications, attenuation can be ignored *in pure rain* at S-band. The emphasis is used here because in many thunderstorms, hail falling from aloft starts to melt when passing downward through the 0 °C isotherm. Raindrops with ice cores can exist well below the melting level as the melting proceeds. Wind tunnel studies show that these hybrid raindrops can remain stable and become quite large, growing to diameters approaching 13–14 mm with axis ratios of 0.5–0.7. Attenuation at S-band can become significant in these circumstances, which

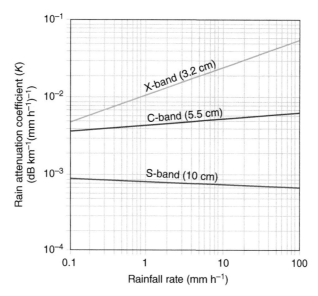

Figure 9.8 The rain attenuation coefficient for three wavelengths as a function of rainfall rate (Based on data in Wexler, W. and Atlas, D. (1963) Radar reflectivity and attenuation of rain. *J. Appl. Meteor.*, **2**, 276–280)

are not at all uncommon, for example, in mid-latitude squall lines. We will consider hail impacts on attenuation in the next section.

Rainfall estimation with radar is more challenging at shorter wavelengths, where attenuation can be serious. Consider again Figure 9.8. The specific attenuation at a moderately heavy rainfall rate of $50\,\text{mm}\,\text{h}^{-1}$ would be $0.25\,\text{dB}\,\text{km}^{-1}$ at C-band, and over $2\,\text{dB}\,\text{km}^{-1}$ at X-band. The situation for shorter wavelengths (not shown in Figure 9.8) is even worse. C-band radars form the backbone of the European radar network (see Chapter 10) and are used by the FAA at airports in their Terminal Doppler Weather Radar (TWSR) network (see also Chapter 10). They are also commonly used by television stations and at some universities.

Figure 9.9 shows an example from nearly collocated and coordinated scans from two radars, the WSR-88D KOUN S-band radar and the OU-PRIME C-band radar, both located in Norman, Oklahoma. The OU-PRIME radar has higher resolution than the WSR-88D. The C-band radar is located at $(x,y) = (0,0)$ to southeast of the region shown in Figure 9.9. The effects of attenuation of the radar reflectivity factor are obvious by comparing the top panels. For example, the regions labeled A and B on the panels, both located behind a line of heavy echo, show a marked reduction in reflectivity at C-band as compared to S-band. The effects of differential attenuation on Z_{DR} are even worse, with values of Z_{DR} as low as $-8.5\,\text{dB}$ at C-band in regions where the S-band Z_{DR} values are between 0.5 and 1.5 dB. The effect of attenuation and differential attenuation can be seen more clearly in Figure 9.10 on a cross section through points A and B (see line in Figure 9.9). Much of the cell west of the main line can no longer be seen by the C-band radar, and the reflectivity values in detectable regions are much weaker. The Z_{DR} fields look nothing alike in the second cell and are only similar at the leading edge of the cell close to the radar and aloft where the echo was weaker in S-band. The effects of differential attenuation on Z_{DR} can be understood by

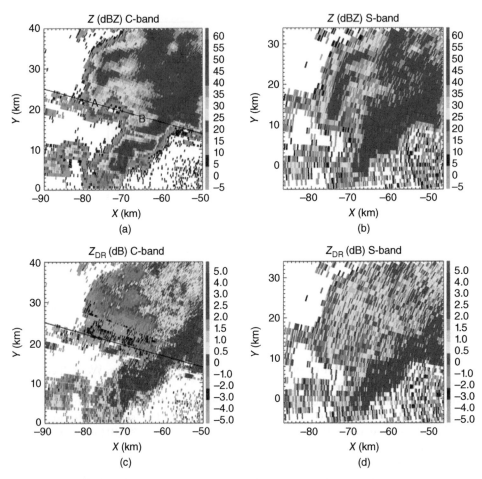

Figure 9.9 Reflectivity factor at C-band (a) and S-band (b) and corresponding Z_{DR} fields at C- and S-band (c,d) for a thunderstorm complex. Areas of negative bias in reflectivity caused by attenuation at C-band are marked as A and B in panel (a). In panels (a) and (c), a line indicates the azimuthal direction for the RHI plot in Figure 9.10 (From Gu, J.-I., Ryzhkov, A., Zhang, P., Neilley, P., Knight, M., Wolf, B., and Lee, D.-I. (2011) Polarimetric attenuation correction in heavy rain at C-band. *J. Appl. Meteor. Climatology*, **50**, 39–58. © American Meteorological Society, used with permission)

considering the definition of $Z_{DR} = 10 \log(Z_{HH}/Z_{VV})$. As large raindrops are oblate, horizontally polarized radiation is attenuated to a greater degree than vertically polarized radiation, leading to a reduction of the value of the ratio Z_{HH}/Z_{VV} with distance and a corresponding reduction in the value of Z_{DR}.

Under conditions of moderate to heavy rainfall, attenuation can pose serious problems for C-band radar systems. Unfortunately, for radars operating with a single polarization, there is no good way to estimate the accumulative effects of attenuation along the propagation path of a radar ray. However, dual-polarization radars, under some conditions, permit correction for attenuation using differential phase (Φ_{DP}) and specific differential phase (K_{DP}) measurements, both of which are immune to attenuation effects (because they are measurements of phase).

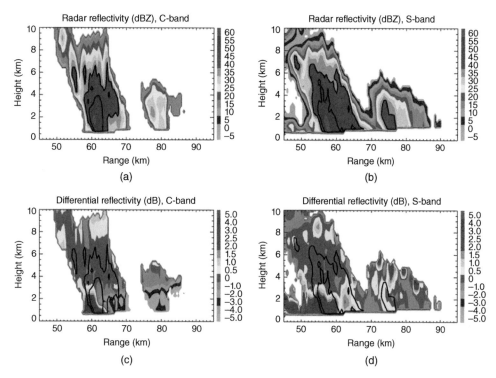

Figure 9.10   Composite RHI of the reflectivity factor at C-band (a) and S-band (b) and corresponding Z_{DR} fields at C- and S-band (c,d) along the line in Figure 9.9a,c (From Gu, J.-I., Ryzhkov, A., Zhang, P., Neilley, P., Knight, M., Wolf, B., and Lee, D.-I. (2011) Polarimetric attenuation correction in heavy rain at C-band. *J. Appl. Meteor. Climatology*, **50**, 39–58. © American Meteorological Society, used with permission)

The corrections are based on simple formulas:

$$\Delta Z = \alpha \Phi_{DP} \tag{9.10}$$

$$\Delta Z_{DR} = \beta \Phi_{DP} \tag{9.11}$$

where the coefficient α is the ratio of the specific attenuation A_h and K_{DP} and the coefficient β is the ratio of the specific differential attenuation A_{DP} and K_{DP}. Figure 9.11 shows the behavior of these ratios, which were calculated from a large number of raindrop size distributions during rain events in Oklahoma. For Z_{DR} values less than 3 dB, and reflectivity factors <45 dBZ, the data appear well behaved, aligning along a straight line with a small amount of spread about the line, allowing for good estimation of α and β from K_{DP}. However, when large raindrops were present ($Z_{DR} > 3$ dB), a large amount of scatter appears in the data shown in Figure 9.11. These larger raindrops occur in heavier rain regions, so-called "hot spots" in the reflectivity field. The scatter is caused by resonant effects associated with Mie Scattering, which maximize at specific drop sizes. Effort has been made to develop algorithms to correct C-band attenuation throughout a scan by estimating α and β across hot spot regions.[5]

---

[5]See Gu, J.-I., Ryzhkov, A., Zhang, P., Neilley, P., Knight, M., Wolf, B., and Lee, D.-I. (2011) Polarimetric attenuation correction in heavy rain at C-Band. *J. Clim. Appl. Meteor.*, **50**, 39–58.

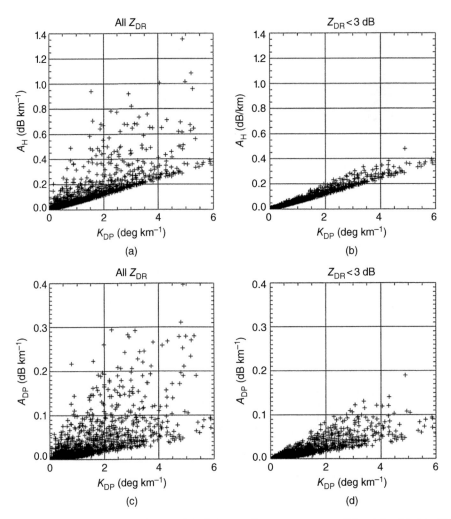

Figure 9.11   Scatterplots of (a,b) A_h and (c,d) A_{DP} vs K_{DP} in pure rain at C-band for (a), (c) all Z_{DR} and (b), (d) $Z_{DR} < 3$ dB. Radar variables are computed from 25,920 drop size distributions measured in Oklahoma (From Gu, J.-I., Ryzhkov, A., Zhang, P., Neilley, P., Knight, M., Wolf, B., and Lee, D.-I. (2011) Polarimetric attenuation correction in heavy rain at C-band. *J. Appl. Meteor. Climatology*, **50**, 39–58. © American Meteorological Society, used with permission)

The corrections depend on temperature and the nature of the drop size distributions that characterize the rainfall (e.g., stratiform, convective, and tropical). An example of results of these corrections appears in Figure 9.12, which is the same data as in Figure 9.9, but corrected for the effects of attenuation.

X-band radars are coming into more common use. For example, mobile X-band radars have been employed extensively during the past two decades for severe thunderstorm and hurricane research in the USA. X-band radars suffer severe attenuation in heavy rain (Figure 9.13). For this reason, they are not commonly used to measure rainfall. It is, however, possible to correct for attenuation in some situations, provided the X-band radar has dual-polarization capability (Figure 9.13). The attenuation

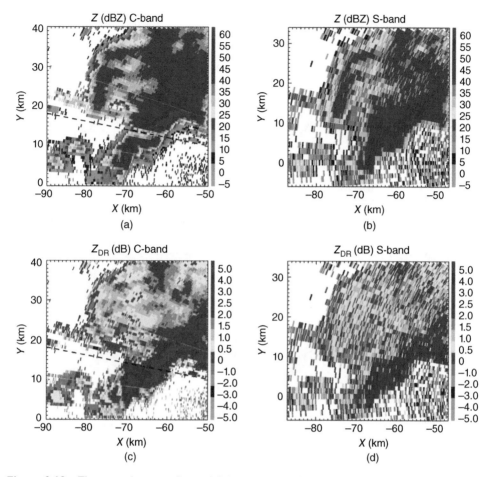

Figure 9.12   The same data as in Figure 9.9, but corrected for the effects of attenuation (From Gu, J.-I., Ryzhkov, A., Zhang, P., Neilley, P., Knight, M., Wolf, B., and Lee, D.-I. (2011) Polarimetric attenuation correction in heavy rain at C-band. *J. Appl. Meteor. Climatology*, 50, 39–58. © American Meteorological Society, used with permission)

correction procedure is generally similar to that used for C-band and requires determining the coefficients α and β for the specific type of rainfall and temperature at which the rain is observed. At X-band wavelengths, Mie scattering, and resonant effects are much more likely, particularly during heavy rainfall from mid-latitude thunderstorms where raindrops aloft may contain ice cores and grow to diameters approaching a quarter to a third of the wavelength.

9.2.4   Attenuation by hail

Exceptional challenges must be faced to correct for attenuation in hail-bearing storms, primarily because of complexities associated with the size, shape, composition, and microphysical evolution of hail. Hail forms and grows by accreting supercooled liquid water droplets at elevations above the melting level. Depending on temperature and

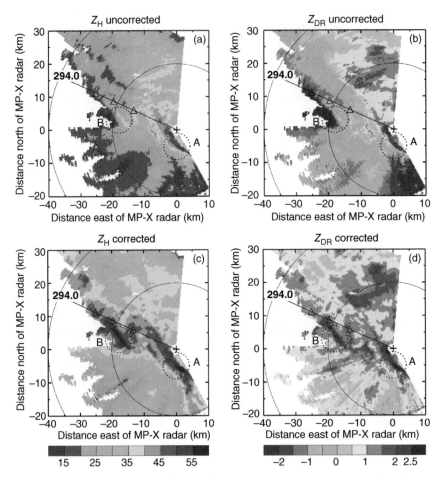

Figure 9.13 PPI images from an X-band radar of the reflectivity factor (a,c) and differential reflectivity (b,d) during a heavy rain event, uncorrected (a,b) and corrected (c,d) for attenuation. The delta symbols denote the locations of disdrometers (From Park, S.-G., Maki, M., Iwanami, K., Bringi, V.N., and Chandrasekar, V. (2005) Correction of radar reflectivity and differential reflectivity for rain attenuation at X band. Part II: Evaluation and application. *J. Atmos. Oceanic Technol.*, **22**, 1633–1655. © American Meteorological Society, used with permission)

the accretion rate of droplets, the surface of a hailstone can either be dry (i.e., the droplets immediately freeze on impact) or wet (i.e., the droplets impact on the stone and spread across the stone's surface, with the water freezing slowly). Microwave radiation is absorbed and scattered differently by hailstones under these conditions because of the different dielectric constant of water and ice, as well as the different shape and porosity of hailstones that result from the different growth modes.

The problem becomes more complicated as hailstones fall toward the ground and pass through the melting level. As hailstones fall, most transition from pure ice, to water drops containing ice cores, to pure rain as they descend. Larger hailstones with high fall velocities may develop a thin water shell before reaching the ground, or may not melt at all. During the time period of melting, drops with ice cores can become

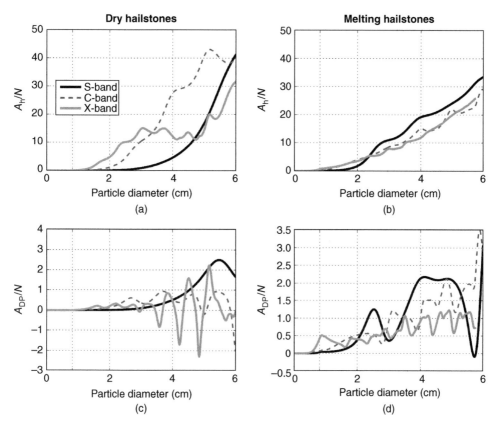

Figure 9.14 Normalized specific attenuation (a,b) and normalized specific differential attenuation (c,d) of dry (a,c) and melting (b,d) hail as a function of size at three radar wavelengths: 11.0 cm (S-band; black curves), 5.45 cm (C-band; dashed dark gray curves), and 3.2 cm (X-band; light gray curves). For the melting hailstones, the vertical dotted line at particle size 0.8 cm represents the cutoff between fully melted raindrops and melting hailstones (From Ryzhkov, A.V., Kumjian, M.R., Ganson, S.M., and Khain, A.P. (2013) Polarimetric radar characteristics of melting hail. Part I: theoretical simulations using spectral microphysical modeling. *J. Appl. Meteor. Climatology.*, **52**, 2849–2869. © American Meteorological Society, used with permission)

quite large, and Mie scattering strongly influences attenuation, particularly at shorter wavelengths and particular drop sizes where **resonance scattering** occurs.

Figure 9.14 shows A_h and A_{DP} determined from a numerical simulation of scattering at three wavelengths for monodisperse distributions of hydrometeors (i.e., all have the same size). The values are normalized by the hydrometeor concentration. All hydrometeors in the model were started as hail at the melting level 4 km above the surface and were allowed to fall to the surface so that smaller stones were completely melted but larger ones still had ice cores. The data shown in Figure 9.14 apply to the hydrometeors just before reaching the ground.

At diameters less than 1 cm, the values of A_h and A_{DP} are larger for raindrops compared to dry hail. However, both A_h and A_{DP} increase rapidly for both dry and wet hail at larger diameters. For small hail, the value of A_h is largest for X-band and smallest for S-band, but for large hail, the values are comparable at all three wavelengths.

Absorption is dominant for hydrometeors smaller than the radar wavelength, while scattering in directions other than the direction of wave propagation dominates for larger particles such as big hailstones. The value of A_h is much higher in hail compared to rain and is strongly affected by resonance scattering. Interestingly, A_{DP} exhibits oscillatory behavior and does not increase significantly in large hail.

The complexities of attenuation in hail become clearer by examining Figures 9.15 and 9.16. Figures 9.15 and 9.16 show the results of scattering computations of A_h and A_{DP} for a model simulation of falling hydrometeors including large, high-density hail. The hydrometeors begin their descent from the melting level at 4 km. There are six categories of hydrometeors, small raindrops (0–4 mm), larger raindrops (4–9 mm), and four size categories of hail (9–14, 14–19, 19–25, and >25 mm). The initial hydrometeor concentrations are based on the so-called **gamma size distribution**, which has been shown in previous studies to best represent hydrometeor populations in many types

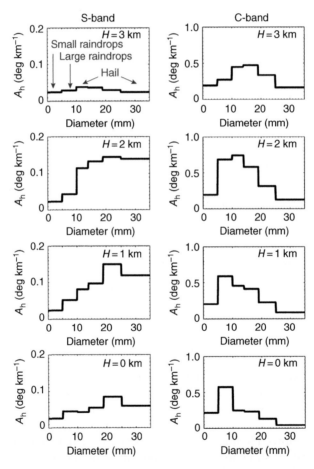

Figure 9.15   Relative contributions of different parts of particle size spectrum to S-band (a,c,e,g) and C-band (b,d,f,h) A_h at four height levels for large, high-density hail (From Ryzhkov, A.V., Kumjian, M.R., Ganson, and Khain, A.P., (2013) Polarimetric radar characteristics of melting hail. Part I: theoretical simulations using spectral microphysical modeling. *J. Appl. Meteor. Climatology*, **52**, 2849–2869. © American Meteorological Society, used with permission)

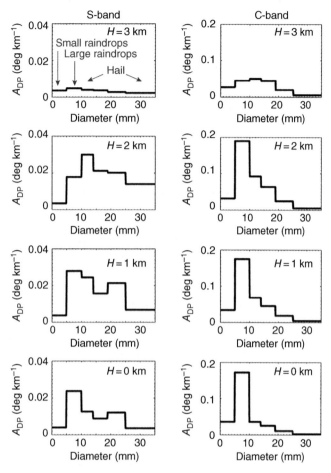

Figure 9.16 Relative contributions of different parts of particle size spectrum to S-band (a,c,e,g) and C-band (b,d,f,h) A_{DP} at four height levels for large, high-density hail (From Ryzhkov, A.V., Kumjian, M.R., Ganson, S.M., and Khain, A.P. (2013) Polarimetric radar characteristics of melting hail. Part I: theoretical simulations using spectral microphysical modeling. *J. Appl. Meteor. Climatology*, 52, 2849–2869. © American Meteorological Society, used with permission)

of weather systems. The hydrometeors are allowed to fall to the surface without interacting, although they do undergo melting. The first thing to note on Figures 9.15 and 9.16 is the changes that occur at a given wavelength in both variables with decreasing elevation. These changes are influenced by both melting and temperature changes as the hydrometeors approach the surface. The graphs would appear different for a different initial distribution of hydrometeors. Recall that a radar pulse volume has a finite depth and encompasses a range of elevations, and a radar ray increases elevation with distance from a radar. It should be obvious that the attenuation that occurs in natural hail-bearing storms can be quite complicated indeed. It is also notable that A_h and A_{DP} do not respond the same at different wavelengths. At S-band, large hail dominates raindrops for A_h, whereas the opposite is true at C-band. A_{DP} is relatively insensitive to large hail and is dominated by large water drops at C-band.

All of these factors make attenuation correction in hail more difficult than in pure rain. The current best approach used to correct for hail attenuation follows a similar procedure to that with rain, in that hot spots of large attenuation are identified using polarization variables (e.g., large K_{DP} and rapidly decreasing Z_{DR} along a radar ray.) An algorithm is applied that takes into account information outside the hot spot to estimate the values of Z_H and Z_{DR} behind the hot spot, from which a correction across the hot spot is applied. The goal in this effort is to obtain an estimate of precipitation intensity and information about hail size, which is important for severe storm forecasting.

9.2.5   Short-wavelength radars and attenuation

Airborne, spaceborne, and ground-based W-band and K-band radars are commonly used in atmospheric research. These radars have particular application for studies of clouds and, in the case of the spaceborne radars, retrieval of precipitation, including snowfall rates. Examples of airborne radars as of 2017 include the Wyoming Cloud Radar (WCR), which is operated by the University of Wyoming on both its own aircraft and the NCAR C-130 aircraft, the National Center for Atmospheric Research HAIPER Cloud Radar [HCR], flown on its Gulfstream-V research aircraft, and the W-band Cloud Radar system flown on the NASA ER-2 aircraft. In space, the Global Precipitation Mission (GPM) satellite has a dual-frequency K_u- and K_a-band radar system, the Tropical Rainfall Measuring Mission (TRMM) satellite (which ended its mission in 2015) employed a K_u-band radar, and the Cloud-Sat satellite flies the Cloud Profiling Radar, a W-band radar. Ground-based radars at K- and W-band wavelengths now operate at several universities and government facilities (such as the Atmospheric Radiation Measurement (ARM) sites).

As discussed earlier, short-wavelength radar radiation is attenuated by atmospheric gases and by cloud droplets. The radiation is also attenuated by wet snow, as occurs near the radar bright band at the melting level. Ice clouds, however, have minimal attenuation effects. For example, at a K_a-band wavelength of 0.9 cm, an ice cloud containing a reasonable ice water content of 1 g m^{-3} through its depth, assuming no liquid present, would attenuate the radar signal for a vertically pointing radar at a one-way rate of 0.0087 dB km^{-1} at 0 °C, 0.0029 dB km^{-1} at −10 °C and 0.0020 dB km^{-1} at −20 °C as the radiation passed upward or downward through the cloud. For practical purposes, attenuation of this magnitude would be undetectable and cloud attenuation by ice due to absorption can be safely neglected. Attenuation by scattering of radiation out the beam can be safely ignored in ice at wavelengths longer than W-band; however at W-band in ice the attenuation may be significant in heavy snow (2–4 dB km^{-1})

Unfortunately, short-wavelength radiation is strongly attenuated by cloud droplets, and even worse by raindrops and wet snow. For example, cloud droplets produce about two orders of magnitude more attenuation than ice particles in similar concentrations. Figure 9.17 shows an example of W-band reflectivity and vertical radial velocity obtained by the Wyoming cloud radar in a mid-lake lake-effect snowband over Lake Ontario. Note the strong attenuation coincident with the stronger convective updrafts. The updrafts contain supercooled water and are likely to be regions of graupel formation due to accretion of water droplets, accounting for

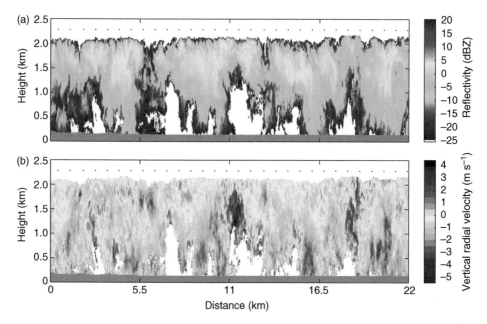

Figure 9.17   W-band reflectivity (a) and vertical radial velocity (b) obtained by the Wyoming cloud radar flying over a mid-lake lake-effect snowband over Lake Ontario. Note the strong attenuation coincident with the stronger convective updrafts, which likely contain supercooled water (Courtesy Jeffery Frame)

the increased attenuation. Although attenuation strongly affects the returned signal, attenuation can be interpreted as a signal of the presence of supercooled water. *In situ* aircraft studies of supercooled water in clouds are difficult because of aircraft icing, so remote detection of supercooled water through attenuation has utility, for example, in studies of aircraft icing or orographic cloud seeding, both of which require supercooled water to be present in clouds.

At short wavelengths, correction for attenuation is necessary to retrieve precipitation rates. This is a particularly important challenge for the spaceborne radars on the TRMM and GPM missions, as their purpose is to map global precipitation. TRMM and GPM algorithms to correct for attenuation take advantage of the radar echo from the surface of the Earth (dual wavelength in the case of GPM). Measurements of the surface return over cloud-free regions where the water vapor profile is known, and gaseous attenuation can be accounted for, are used to obtain an expected value of surface return over both water and land surfaces. Any change from the expected value of the surface return in the presence of precipitating clouds can be attributed to two-way attenuation of the signal. Algorithms designed to correct for attenuation based on the surface return can be further calibrated by coincident surface-based precipitation measurements.

9.3   Refractivity

We learned in Chapter 2 that the speed and direction of propagation of radar waves through the atmosphere are affected by both horizontal and vertical variations of the index of refraction, n, or equivalently the radio refractivity, N, along the path of a radar ray. The radio refractivity normally decreases with height and changes in the horizontal depending on the variations of dry air pressure, temperature, and water vapor pressure across the radar domain (see Eq. (4.6)). As refractivity depends on these variables, measurements of refractivity obtained using radar have the potential to provide information about the moisture and/or thermal structure of the atmosphere. Here, we overview a technique,[6] using ground targets, to retrieve the near-surface radio refractivity field. As we will see, the technique has proved useful in early detection of airmass boundaries such as drylines and front.

9.3.1   Basic principles

Measurement of near-surface refractivity requires that a large number of stationary ground targets are present within a radar's field of view. The phase of a stationary point ground target at range r with respect to a transmitted wave is given by

$$\phi(r) = 2\pi f_t t = \frac{4\pi f_t}{c} \int_0^r n[x(r'), y(r'), z(r'), t] dr' \tag{9.12}$$

where f_t is the transmitted frequency and t is the travel time. The integral formally states that the phase depends on the (continuously changing) value of n along the path, r', of the ray. If the phase of a stationary target varies with time, the phase variation must be related to variations in the refractive index between the radar and the target. These variations in n can cause a wave at one time to propagate more slowly, or more quickly, and/or along a different path that it would at some other time.

The refractivity retrieval technique does not depend on measuring phase alone, but rather a *phase change* from a reference phase, $\phi(t_0)_m$, determined for each of m ground targets in the field of view of the radar. The reference $\phi(t_0)$ for each target is measured at a time, t_0, when the reference value of N, N_0, is as homogeneous as possible across the radar domain (e.g., a dry sunny day with uniform temperatures and a deep mixed layer). The reference N_0 is determined independently from meteorological measurements of pressure, temperature, and dew point temperature, for example, from a meteorological mesonet. The reference phase information, $\phi(t_0)_m$, is then used to develop a base field of phase differences between targets aligned along various radar rays.

The phase change for a given target between the reference measurement and the phase measured at a different time is given by

$$\Delta\phi = \phi(t_1) - \phi(t_0) = \frac{4\pi f_t}{10^6 c} \int_0^r [N(x, y, z, t) - N_0(x, y, z, t_0)] dr' \tag{9.13}$$

If many targets are aligned, overlapping paths of the integrated N values can be dissected to determine N along shorter paths. The length of these paths, and therefore,

---

[6]Details concerning the near-surface refractivity retrieval technique can be found in Fabry, F., Frush, C., Zawadzki, I., and Kilambi, A. (1997) On the extraction of near-surface index of refraction using radar phase measurements from ground targets. *J. Atmos. Oceanic Tech.*, **14**, 978–987 and Fabry, F. (2004) Meteorological value of ground target measurements by radar. *J. Atmos. Oceanic Tech.*, **21**, 591–604.

the resolution of the recovered refractivity data, depends on the density of ground targets. In many locations, the density of these ground targets can be quite high. Across much of the US Midwest and Great Plains, for example, section roads are located every mile, and nearly all roads have lines of power poles along them, so well-spaced stationary ground targets are quite common. Because of earth curvature, the density of usable targets drops off with range, with only the tallest targets available at longer range. Depending on location, refractivity measurements can typically be made to a range of about 40–60 km. The retrieved refractivity data from each of the paths are normally objectively analyzed, usually with some smoothing to eliminate noise introduced because of variations in the altitude of the scatters; contamination from meteorological, biological, or other physical scatterers (e.g., a small shower, insects, and a passing truck); and then mapped to display the field of near-surface refractivity.

A key element for the technique to work is that each ground target must be stationary and unchanging in character. Swaying vegetation, such as large trees, are poor targets. For a 10 cm wavelength radar, for example, the movement of swaying tree branches only 5 cm results in a phase change of the returned echo of 180°. A mobile object such as a parked truck can change its position from day to day, resulting in large differences in phase. The best targets are sturdy poles and towers, as they remain stationary in the wind. A second criterion is that the landscape over which the retrieval of refractivity is done must be relatively flat. If it is not, the height variation of N will dominate any horizontal variation and airmass boundaries will appear diffuse or even impossible to identify.

9.3.2   Measurement of the water vapor field

Refractivity retrieval has now been tested in a number of environments ranging from a large city (Montreal) to the High Plains of Oklahoma and Colorado. Figure 9.18 shows an example of the evolution of the refractivity field measured during the International H_2O project (IHOP). The radar was located in the Oklahoma Panhandle, a common location for dryline formation. Figure 9.18 shows the evolution of the refractivity field between 15:00 UTC (10 a.m. LT) and 00:00 UTC (7 p.m. LT) on May 22, 2002 during which time a dryline forms and moves over the radar site. Although refractivity, at a given pressure level, is a function of both temperature and moisture, its sensitivity to moisture changes is greater than that for temperature. This is evident in Figure 9.19, which shows the derived refractivity field from the S-POL radar during IHOP, the calculated refractivity from a collocated meteorological mesonet station, and the temperature and mixing ratio at that station. The station was located in drier air west of a dryline before 00:30 UTC. The dryline passed over the station westbound so that the station was east of the dryline by 01:00 UTC. First note the close correspondence between N retrieved from the radar and N derived from surface station data. The high correlation between these independent datasets was typical of many other days, giving confidence in the radar retrieval. Note also the close correspondence between the evolution of the refractivity and moisture fields, and much weaker correspondence between refractivity and temperature. These data show that refractivity has great potential to monitor the evolution of surface moisture fields in real time. Researchers working with the US National Weather Service tested the use of refractivity in operational conditions in both the Colorado and Oklahoma regions

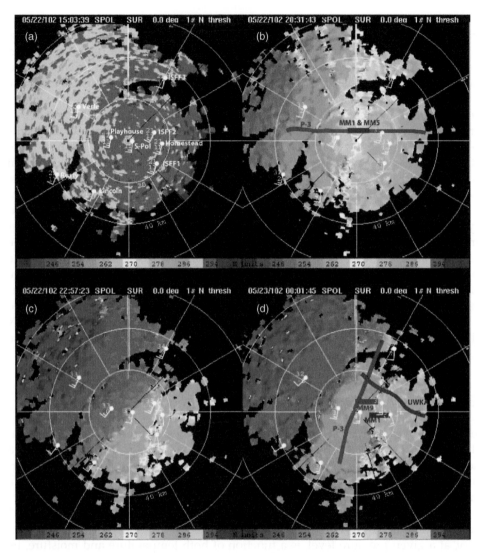

Figure 9.18 Radar refractivity retrievals from the S-Pol radar located in the vicinity of the western Oklahoma dryline. (a) 15:03 UTC: a relatively uniform field before dryline formation; (b) 20:30 UTC: after the dryline had formed, (running through the radar); (c) 22:57 UTC: when the dryline was 15 km east of S-Pol and a secondary dryline had formed running through the radar; and (d) with the dryline through the radar site at 00:01 UTC. Range rings are at every 20 km. Surface station names are shown with temperature (°C) plotted in the upper left and dew point temperatures (°C) in the lower left. Half barbs indicate wind speeds of 5 m s^{-1} and full barbs are 10 m s^{-1}. Red and blue lines denote research aircraft tracks (From Weckwerth, T.M. *et al.* (2005) Radar refractivity retrieval: validation and application to short-term forecasting. *J. Appl. Meteor.*, **44**, 285–300. © American Meteorological Society, used with permission)

in the period 2006–2008. Although refractivity was found to provide useful information to forecasters in many meteorological situations, at the time of this writing in 2017, refractivity had yet to be operationally implemented across the WSR-88D network.

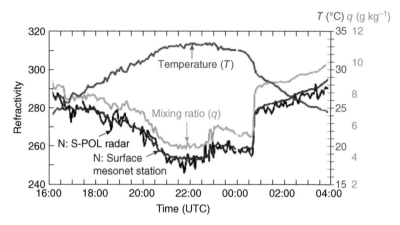

Figure 9.19   Surface station refractivity (blue) and corresponding S-Pol radar refractivity (black) comparisons from the same event as in Figure 9.18. Temperature (red; °C) and mixing ratio (green; g kg⁻¹) traces for each surface station are also shown (From Weckwerth, T.M. *et al.* (2005) Radar refractivity retrieval: validation and application to short-term forecasting. *J. Appl. Meteor.*, **44**, 285–300. © American Meteorological Society, used with permission)

Important terms

absolute humidity
absorption bands
absorption lines
attenuation
attenuation coefficient
Beer–Lambert law
cloud droplets
differential attenuation
liquid attenuation coefficient
gamma size distribution
one-way attenuation
one-way differential attenuation

pressure broadening
radio refractivity
raindrops
rainfall attenuation ratio
resonance scattering
specific attenuation
specific differential attenuation
total attenuation
two-way attenuation
two-way differential attenuation
water vapor density

Review questions

1. What is the difference between specific attenuation and specific differential attenuation?
2. Explain qualitatively, from a quantum physics viewpoint, why K-band radars are much more susceptible to attenuation by atmospheric gases than S-band radars.

(continued)

3. Contrast the two different approaches discussed in this chapter for measuring the distribution of water vapor in the atmosphere using radar.

4. What types of clouds would have suitable conditions for accurate recovery of liquid water content using the dual-wavelength approach discussed in this chapter?

5. Which two variables measured by dual-polarization radars are used to correct for rain attenuation at C-band? Under what meteorological conditions are these corrections likely to produce large errors?

6. You are viewing a line of thunderstorms on a display of a dual-polarization C-band radar. The display shows two panels, the reflectivity factor and the differential reflectivity. You suspect that there is attenuation associated with hail in the storms. What feature on the display might support your supposition?

7. Explain why attenuation by hail in a thunderstorm varies strongly with altitude.

8. Explain why K-band and W-band radars most often are used in vertically pointing mode for meteorological applications.

9. Which of the following ground targets are suitable for use in retrieval of radio refractivity and why: (a) a large tree, (b) a wind turbine, (c) a power transmission tower, (d) a tall building, (e) a grain silo, (f) a parked truck.

10. Why are hilly regions less suitable for refractivity retrieval than flat regions?

Challenge problems

11. Assume for this problem that gaseous attenuation at K-band varies linearly with water vapor density. How far from a radar would a target have to be for the returned power to be reduced by 1 dB due to attenuation if (a) the atmosphere was dry, (b) the atmosphere had a water vapor density of $5\,g\,m^{-3}$, and (c) the atmosphere had a water vapor density of $15\,g\,m^{-3}$? Assume that the radar is located at sea level where the pressure is 1013.25 hPa and the temperature is 20 °C.

12. Suppose a radar ray from an X-band radar first passes through a cloud-free region of the atmosphere 1 km wide at sea level characterized by a water vapor density of $7.75\,g\,m^{-3}$ and a temperature of 20 °C. Further along, the ray passes through a 1 km wide cloud at 10 °C with a liquid water content of $1\,g\,m^{-3}$. Further along still, the ray passes through a 1 km wide rain shower at 0 °C with a rain rate of $20\,mm\,h^{-1}$. Which event produces the greatest one-way attenuation and which the least? What is the value of one-way attenuation in dB in each of the three events?

13. Assume you have three co-located radars, each operating at a different wavelength (S-, C-, and X-band). A rain shower is located between the radars and a stationary target 20 km away. The shower has a dimension of 10 km along the direction of the beam, and the rain rate within the shower everywhere has the same value, 40 mm h^{-1}. How much would the signal from the target be attenuated at S-band? How about C-band and X-band? Assume the temperature is everywhere 0 °C and attenuation is solely due to rain.

14. Commercial aircraft have "weather avoidance radars" located in an aircraft's nose that monitor weather ahead of the aircraft so that the pilot has time to make strategic decisions to divert around dangerous weather conditions. Describe critical decisions that an aeronautical engineer might face in choosing wavelength and antenna size for such a radar and resulting problems pilots might face because of these choices in determining whether dangerous conditions are indeed ahead of the aircraft.

15. Measurements of radio refractivity require a reference value of N to be determined across the radar domain with independent measurements of pressure, temperature, and moisture content. It was stated in the chapter that at the time the reference value is measured, N should be as homogeneous as possible across the domain. What time of day, and under what meteorological conditions, might you expect to find the least variability in N?

10

Operational Radar Networks

Objectives

By the end of this chapter, you should understand the following:

- Characteristics of the WSR-88D network, including network coverage, radar components, radar specifications, and scanning strategies.
- The difference between Level-I, Level-II, and Level-III data.
- The volume coverage patterns of the WSR-88D, under what weather conditions they are used, and how they differ from one another.
- Basics of the processing algorithms used to correct range ambiguities and unfold radial velocities.
- The difference between legacy and super resolution WSR-88D products.
- Characteristics of the TDWR network, including network coverage, radar specifications, and scanning strategies.
- The differences in data collection strategies and data quality between the WSR-88Ds and TDWRs.
- Where international weather radar networks are located and the types of radars deployed in these networks.

10.1   Introduction

Operational weather radar networks are deployed across many parts of the world to detect and warn against weather hazards, measure precipitation, and provide basic weather information to the public, militaries, and governments. Operational radars differ from research radars in that they typically employ specific scan strategies targeted at accomplishing well-defined tasks, such as detecting a tornado, wind shear over an airport, or a flash flood. Operations are normally highly automated, with the user providing only basic decisions about scan sequences.

   The earliest network of S-band weather radars in the USA, called Weather Service Radar-1957 (WSR-57) radars, was first commissioned for operation in 1959 and deployed across the USA in the 1960s. These radars were supplemented by Weather Service Radar-1974 S-band (WSR-74S) and Weather Service Radar-1974 C-band (WSR-74C) radars in the 1970s and 1980s. These radars, in turn, were replaced by

---

Radar Meteorology: A First Course, First Edition. Robert M. Rauber and Stephen W. Nesbitt.
© 2018 John Wiley & Sons Ltd. Published 2018 by John Wiley & Sons Ltd.
Companion website: www.wiley.com/go/Rauber/RadarMeteorology

the current **Weather Service Radar-1988 Doppler** (WSR-88D) **network**, primarily in the 1990s. The radars, at their inception, were also called **Next-Generation Radars** (NEXRAD). The WSR-88Ds were upgraded to have dual-polarization capability beginning in 2011. A second network of C-band Doppler radars, called **Terminal Doppler Weather Radars** (TDWR), were deployed in the 1990s near major airports in the USA to detect and warn against low-level wind shear and microbursts that threaten aircraft departures and arrivals. In recent years, operational radar networks have been deployed across other regions of the world. These deployments continue today as the reach of technology spreads to less-developed parts of the world that are vulnerable to weather disasters.

10.2 The WSR-88D radar network

The WSR-88D radar network's primary purpose is to provide the National Weather Service (NWS), the Federal Aviation Administration (FAA), and the Department of Defense (DOD) with the capability to issue timely warnings of impending weather hazards.[1] The average lead time for tornado warnings, for example, increased to 13 min with the deployment of the WSR-88Ds. The radars also allow forecasters to pinpoint regions of severe hail and violent straight-line winds produced by thunderstorms and identify regions of greatest threat during landfalling hurricanes. In winter, the radars provide guidance concerning weather hazards such as freezing rain and heavy snow. A primary function of the WSR-88D network is to estimate precipitation and provide guidance for issuance of flash flood warnings. The radars also provide capability for short-term forecasting, for example, by detecting boundaries such as fronts and gust fronts that may trigger convection. Data from a single radar, composited with data from other radars across the network, provide large-scale coverage of weather systems, so their movement and intensity can be monitored and conveyed to the public. The WSR-88Ds also routinely use an algorithm called velocity–azimuth display (see Chapter 11) to measure the wind as a function of altitude over the radar site, permitting forecasters to monitor mesoscale features such as fronts as they pass over the radar. WSR-88D data are now assimilated into numerical forecast models. The data are also archived and are routinely used by researchers both within and beyond the atmospheric sciences.

10.2.1 Network coverage

The WSR-88D network consists of 122 radars operated by the Department of Commerce (the NWS), 12 radars operated by the Department of Transportation (the FAA), and 21 radars operated by the DOD.[2] Figure 10.1 shows the WSR-88D network configuration over the continental USA, including coverage at 4000, 6000, and 10,000 ft above ground level. Figure 10.2 shows the network over Alaska and Hawaii, US territories, and overseas US military bases. Coverage over the continental USA at the

---

[1]Students interested in WSR-88D forecaster training courses should visit http://www.wdtb .noaa.gov/courses/dloc/index.html.

[2]Information about WSR-88D and other networks discussed in this book is as of 2017.

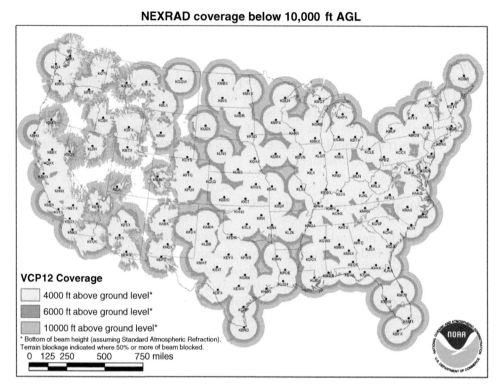

Figure 10.1   Coverage of the WSR-88D network at 4000, 6000, and 10,000 ft altitude above ground level across the continental USA (Courtesy NOAA)

3 km (10,000 ft) level is essentially complete east of the longitude corresponding to the eastern Montana border. However, noticeable gaps remain in coverage at the 1.2 km (4000 ft) level, particularly west of the Mississippi. Coverage west of the Rockies is not as dense, a result of beam blocking by mountains and fewer radars being deployed in sparsely populated regions.

10.2.2   Radar characteristics and data distribution

Figure 10.3 shows a block diagram of a WSR-88D radar system. WSR-88D radars have three basic units, each containing a number of components. The **Radar Data Acquisition Unit** (RDA) consists of the transmitter, antenna, receiver, and signal processor. A WSR-88D transmitter generates a coherent signal using a klystron amplifier with a precisely controlled amplitude, frequency, and phase, using a wavelength of 10 cm and a peak power of 750 kW. The transmitted signal is polarized linerally at a 45° angle. The receiver simultaneously receives the returned signal in the horizontal and vertical polarization planes. The signal processor then produces three base moments (reflectivity, velocity, and spectrum width) and three dual-polarization variables (differential reflectivity, correlation coefficient, and differential phase).

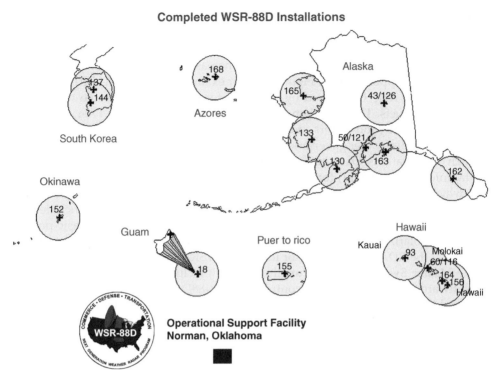

Figure 10.2   Locations of WSR-88D radars in Hawaii, Alaska, US Territories, and overseas US military bases (Courtesy NOAA)

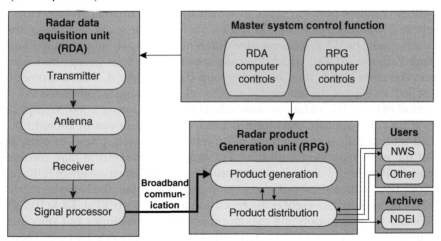

Figure 10.3   The components of a WSR-88D radar system

The **Radar Product Generator** (RPG) ingests and processes the base data received from the RDA. The RPG generates six base products (base reflectivity, base radial velocity, base spectrum width, differential reflectivity, correlation coefficient, and specific differential phase) at all elevations from the base data received from the RDA. The RPG also calculates derived products from these base data, such as the storm-relative mean radial velocity, composite reflectivity, tornado vortex signatures, and storm total precipitation. The data are displayed using different color scales determined by the data digital coding, which takes the form of either 3-bit, 4-bit, 8-bit, or 16-bit digital words. A 3-bit product has $2^3 = 8$ data levels, whereas a 16-bit product has $2^{16} = 65,536$ data levels. The upper panel of Figure 10.5, for example, shows the (4 bit) color coding of the radar reflectivity used by the NWS for publically assessable web-based reflectivity products. The RPG then transmits the products to NWS, FAA, and DOD computers for use by forecasters and distributes the products and the base data to outside users including other government agencies, private companies, universities, and research organizations. Data are also transmitted to the National Centers for Environmental Information (NCEI) for archiving purposes. Archived data include **Level-II data**, which consist of the data from the RDA at the highest resolution available (either legacy or super resolution, see below), and **Level-III products**,[3] which include any products generated by the RPG from the base data. **Level-I data**, the basic power, phase, and polarization measurements, are not archived for distribution but can be archived at the RDA signal processor for use by WSR-88D engineers.

The **Master System Control Function** (MSCF) includes the computer hardware and software that operators use to control the WSR-88D system and interface with the RDA and RPG through so-called **Human Control Interfaces** (HCIs). The RPG HCI, for example, is commonly used by operators to change a volume coverage pattern or edit parameters.

10.2.3   *Scanning strategies*

The WSR-88D radars perform scans in which the radar antenna sweeps clockwise through 360° at a series of elevation angles. Partial sector scans and RHI scans are never employed. The WSR-88Ds use nine different scanning strategies, called **Volume Coverage Patterns** (VCPs), which logically separate into four groups based on the common elevation angles used by each group (Figure 10.4). A VCP is a predefined series of 360° sweeps of the antenna at selected elevation angles completed in a specified period of time. The radar operator selects the VCP. During a 360° sweep, called an **elevation cut**, a specific **waveform** is used. A waveform can be thought of as a specified train of pulses in which the pulse repetition period, phase, or other properties of the wave may vary with time. For example, a waveform may consist of a few pulses with a long pulse repetition period for surveillance coverage followed by a burst of pulses with a short pulse repetition period to obtain the Doppler moments. The WSR-88D system uses eight pulse repetition frequencies (PRFs). Table 10.1 shows the PRFs and their corresponding maximum unambiguous ranges and velocities.

The first group of VCPs, VCP-11, and VCP-211, is recommended for use when convective storms are close to the radar, or when deep stratiform weather systems are present. Both VCPs use the same 14 elevation angles and take 5 min to complete. The second group, VCP-12 and VCP-212, is optimized for conditions when thunderstorms are located at long range or widespread convection is present. These VCPs also use 14

---

[3]A current list of products can be found at http://www.ncdc.noaa.gov/data-access/radar-data/nexrad-products.

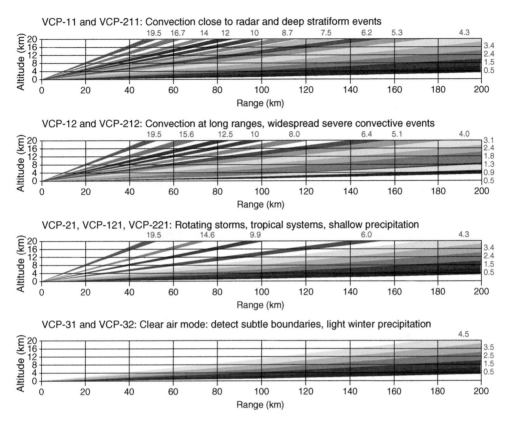

Figure 10.4 Volume coverage patterns used by the WSR-88D radar network

Table 10.1 WSR-88D pulse repetition frequencies (PRF) and corresponding maximum unambiguous ranges (R_{max}) and velocities (V_{max})

PRF number	1	2	3	4	5	6	7	8
PRF (Hz)	322	446	644	857	1014	1095	1181	1282
R_{max} (km)	466	336	233	175	148	137	127	117
V_{max} (m s^{-1})	8	11	16	21	25	27	30	32

elevation angles, but elevation cuts are packed at lower elevations to increase vertical resolution in distant storms. VCP-12 and VCP-212 take 4.5 min to complete. The third group, VCP-21, VCP-121, and VCP-221, use nine elevation cuts and take 6 min to complete. These VCPs are employed either when rotating thunderstorms are present or tropical storms make landfall, particularly VCP-121. The first three groups described collectively are referred to as **Precipitation Mode**, as they are used during precipitation events. The final group, called **clear air mode**, include VCP-31 and VCP-32. These are employed in conditions where precipitation and clouds are absent, nonprecipitating scattered clouds are present, or light winter precipitation is falling. These VCPs employ five elevation angles. VCP 31 differs from all others in that it uses a long pulse (4.7 μs), reducing range resolution, but increasing the radar's sensitivity in clear air

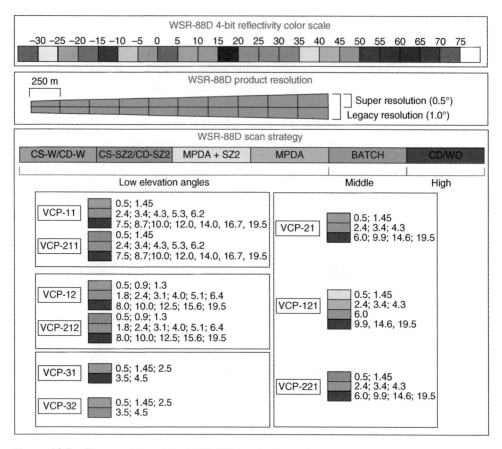

Figure 10.5   Characteristics of the WSR-88D resolution and scan strategies

conditions. The 4-bit scale of reflectivity used for precipitation mode and clear air mode differs, but the NWS has effectively combined them into a single scale used now for their reflectivity products (see top panel in Figure 10.5).

Five different sampling strategies with their associated processing algorithms are used, depending on the elevation cut and the VCP (Figure 10.5). Each of these strategies is summarized below. These strategies/algorithms are primarily designed to (1) resolve the Doppler dilemma, allowing precipitation echoes to be positioned accurately without range folding and radial velocities to be measured accurately over a wide range of velocities without velocity folding, (2) increase horizontal resolution to super resolution—half the actual beam width, and (3) minimize the impact of ground clutter on velocity estimates. The scan strategies and corresponding algorithms are described briefly below.

CS-W/CD-W (continuous surveillance/continuous Doppler-with unfolding)

This strategy, employed at low elevation angles of VCPs 11, 12, 21, 31, and 32, requires two complete 360° scans, called **split-cuts**, at the same elevation angle. The first scan (CS) uses a constant, low PRF, which permits accurate determination of the range of targets to a distance of 460 km (248 nm). The second scan (CD) uses a constant, high PRF, which more accurately measures radial velocity and spectrum width, but its range limited to 230 km (124 nm) with legacy resolution, and 300 km (162 nm) with super resolution (see below). Multiple-trip echoes frequently occur on the CD scan. Using techniques described in Chapter 6, data from the two scans are used together to eliminate multiple-trip echoes and unfold radial velocities in regions not influenced multiple-trip echoes. Unfortunately, the Doppler velocity spectrum within multiple-trip regions is a combination of the returns from both distant and near targets; as a result, the so-called **legacy unfolding algorithms** are unable to resolve velocities in these regions. If the CS-W/CD-W algorithm is used, radial velocity estimates in these ambiguous regions are coded as unknown and displayed as purple on radial velocity images. These purple regions, commonly referred to as **purple haze** by radar meteorologists, were common on velocity products before 2007.

CS-SZ2/CD-SZ2 (continuous surveillance using the Sachidananda-Zrnic (SZ-2) algorithm/continuous Doppler using the SZ algorithm)

This algorithm, employed at low elevation angles of VCPs 211, 212, 221, and in combination with a second algorithm (multi-PRF dealiasing algorithm (MPDA), see below) on VCP 121, also requires two complete 360° scans at the same elevation angle. The SZ-2 algorithm separates the velocities associated with multiple-trip echoes by varying the phase of multiple consecutive pulses. The phase information contained in the returned signal allows separation of the velocity information from the CS and CD scans and therefore permits determination of radial velocities in regions of multiple-trip echoes. SZ-2 has been proven to be extremely effective in mitigating purple haze. Velocities can be unfolded, and ground clutter filtered, effectively eliminating the large regions of ambiguous data common in pre-2007 data.[4] One problem that remains is a ring of ambiguous data caused by clutter that appears just beyond the edge of the first trip. This ring, however, can be eliminated by subsequently applying the MPDA.

MPDA (multi-PRF dealiasing algorithm)

The MPDA is implemented only within VCP-121. The MPDA is not a waveform but is rather software that uses data from scans employing the CS-SZ2, CD-SZ2, CS-W, CD-W, CS-WO, CD-WO, and/or BATCH waveforms, depending on the elevation angle. For example, at the lowest elevation angles, 0.5° and 1.5°, VCP-121 performs four sweeps at each elevation angle, the first using the CS-SZ2 waveform, the second CD-SZ2, the third CS-W, and the fourth CD-W, each with a different PRF. The initial velocity dealiasing is accomplished by the SZ technique; the MPDA algorithm then uses legacy unfolding techniques with the multiple PRFs to eliminate any remaining

---

[4]Students interested in learning more about purple haze mitigation can refer to http://www .roc.noaa.gov/wsr88d/PublicDocs/NNOW/NNautumn07b.pdf. Details of the SZ-2 algorithm are found in a series of reports at http://www.nssl.noaa.gov/publications/wsr88d_ reports/.

ambiguities in the radial velocity field. At middle elevation angles, data from scans using the BATCH, CS-WO, and CD-WO waveforms are used in the dealiasing algorithms. The MPDA largely eliminates ambiguous velocities in range-folded regions over a broad range of velocities, making VCP-121 ideal for situations involving severe tornado-producing convection and landfalling hurricanes.

BATCH

The BATCH waveform consists of a few pulses for surveillance coverage followed by tens of pulses to obtain the Doppler moments. This approach is used on each radial as the antenna conducts a sweep. This waveform is used at middle elevation angles (generally 1.8–6.2°, see Figure 10.5) on all VCPs except VCP-31. At these angles, clutter suppression is less stringent. This approach basically substitutes for two separate sweeps, with an overall faster scan time versus some minor degradation in the accuracy of reflectivity estimates at long range because of the reduced number of pulses.

CD/WO (continuous Doppler without unfolding)

The CD/WO scan uses a high PRF. As the CD/WO is only used at high elevation angles, the beam over-tops storms at relatively close range, so range folding is generally insignificant. Velocity aliasing is also somewhat reduced because of the beam angle with respect to the horizontal wind. Range and velocity unfolding are not performed for elevation cuts employing CD/WO.

10.2.4   Ground clutter suppression

The WSR-88D applies a sophisticated ground clutter suppression algorithm to remove ground clutter and still determine quality estimates of radial velocity. Returns from ground targets have little or no motion. Ground clutter also has other characteristics such as reflectivity texture, clutter phase alignment, and different standard deviations of Differential Reflectivity and Differential Phase that are used by the WSR-88D Clutter Mitigation Decision algorithm to detect areas of clutter on an elevation-by-elevation, bin-by-bin basis. Once clutter is identified, another algorithm uses spectrum width data to suppress the ground clutter-contaminated portion of the signal with near-zero velocity.

10.2.5   Super resolution

The beamwidth of the WSR-88D, defined by the −6 dB two-way power points of the gain function, varies from 0.87° to 0.96° depending on the radar, with an average of 0.89°. An estimate of reflectivity or radial velocity is determined not from a single pulse, but rather from a series of pulses that occur as the antenna sweeps in azimuth. This effectively smears the signal across a wider angle, called the **effective beamwidth**. Using an azimuthal sampling interval of 1°, the effective beamwidth for the various WSR-88Ds ranges between 1.36° and 1.41°. Legacy resolution products are provided with 1° azimuthal resolution, although they effectively include signal from this somewhat wider azimuthal range. Legacy reflectivity measurements are averaged in range from a gate spacing of 250 m to 1 km. Legacy resolution base data products from the WSR-88D network therefore consist of reflectivity on a 1 km by 1° polar grid and Doppler velocity and spectrum width on a 250 m by 1° grid.

In 2008, the WSR-88Ds began providing **super resolution** level-II data for the split-cut scans of the various VCPs. Super resolution data and products have an azimuthal resolution of 0.5°. The range resolution is kept at the range gate resolution of 250 m. To obtain the desired azimuthal resolution, the system uses pulses from overlapping 1° radials every 0.5° azimuth. Because of the antenna gain function, each pulse illuminates a slightly different region. The algorithm to extract finer resolution applies a mathematical filter, called a von Hann or Blackman window to the data. This has the mathematical effect of emphasizing the signal from the smaller azimuth sector. Super resolution requires an effective antenna beamwidth that is about 25% narrower than legacy resolution. Although narrowing the effective antenna beamwidth leads to better radar resolution, the price is somewhat reduced quality of the base data. However, experience has shown that super resolution reveals features such as hook echoes and tornado vortex signatures in much greater detail (Figure 10.6), so reduced accuracy of reflectivity and radial velocity estimates from a single range gate are an acceptable compromise.

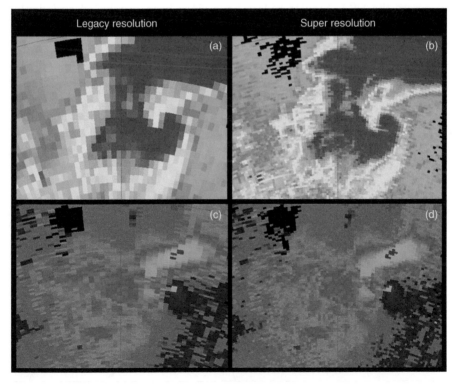

Figure 10.6 Illustration of legacy resolution reflectivity and super resolution reflectivity and radial velocity data (Courtesy NOAA)

10.2.6   *Additional features*

At the time of this writing, the NWS has implemented additional scan improvements to reduce the time between consecutive low-level scans, a feature particularly desirable during severe weather. The new algorithms include Automated Volume Scan Evaluation and Termination (AVSET), which essentially terminates a VCP early when higher elevation scans are topping all significant echo, Supplemental Adaptive Intra-Volume Low-Level Scan (SAILS), which inserts a low-level scan in the middle of a VCP volume, and Multiple Elevation Scan Option for SAILS (MESO-SAILS), which inserts more than one low-level scan into a VCP volume. These, no doubt, will be implemented by the time you are reading this book. Indeed, other improvements will continue to be developed as new ideas and technologies continue to emerge, and some strategies described in this chapter may eventually be retired. For the latest information, NWS websites will be your best source.

10.3   Terminal Doppler weather radars

TDWR were deployed by the FAA near major airports across the USA in the early 1990s. Their purpose is straightforward—to prevent aircraft accidents. The TDWRs provide warnings to air traffic control and pilots of wind shear and microbursts along runway approach and departure corridors. Figure 10.7 shows the locations of TDWR radars across the USA. Similar radars have since been purchased and deployed by other countries at international airports. The TDWR processing software has

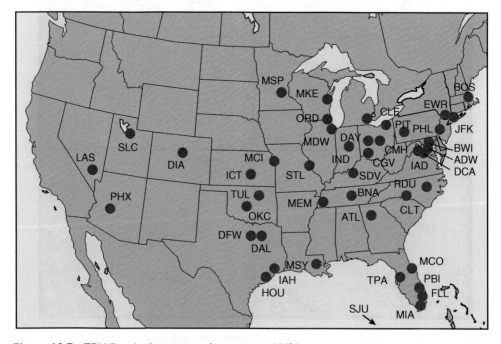

Figure 10.7   TDWR radar locations in the continental USA

sophisticated algorithms to detect wind shear and microbursts. When a microburst is detected, an alert sounds in the control tower and the controller sees on a ribbon display in the tower cab, a message like:

16A MBA 35 K- 2MF 240 25.

Based on this coded information, the controller immediately issues the following warning to arriving aircraft:

RUNWAY 16 ARRIVAL, MICROBURST ALERT, 35 KT LOSS 2 MILE FINAL, THRESHOLD WIND 240 AT 25.

In plain language, the controller is alerting approaching aircraft that on the approach to Runway 16, there is a potential microburst and to anticipate or expect a 35 knot loss of airspeed approximately 2 miles out on final approach, with runway winds currently from 240° at 25 knots. The crew, now forewarned, can prepare to apply wind shear/microburst escape procedures should they decide to continue the approach. As a second example, the ribbon display may read:

25D WSA 30 K+ RWY 210 40.

Here, the controller would advise departing aircraft on runway 25 that they could encounter wind shear on the runway with the possibility of a 30 knot gain in airspeed. The surface winds at the runway were 210° at 40 knot. The TDWR radars, along with other wind shear detection technologies, have virtually eliminated accidents associated with microbursts and wind shear at the airports where they are deployed.

In early 2009, the NWS completed connection to all 45 operational TDWR sites and developed a **Supplemental Products Generator** (SPG) for generation of TDWR products derived from the TDWR base data that are similar to WSR-88D products derived from the RPG.[5] Level III TDWR products generated by the SPG are archived at the NCEI. Level-II TDWR data, at the time of this writing, are not archived, but may be in the future. As TDWR products are now available to the NWS, the public, and for research, we will review basic characteristics of these radars here.

10.3.1   Radar characteristics and data distribution

TDWRs are C-band radars. The shorter wavelength implies that they are subject to significant attenuation in the presence of storms and will experience more significant range and velocity folding than WSR-88Ds. The TDWR's physical beamwidth is 0.6°, narrower than the WSR-88D, and its range resolution is smaller at 150 m. However, the NWS SPG generates products with 1° azimuthal resolution. The TDWR 150 m range resolution is retained. A TDWR uses different elevation angles compared to

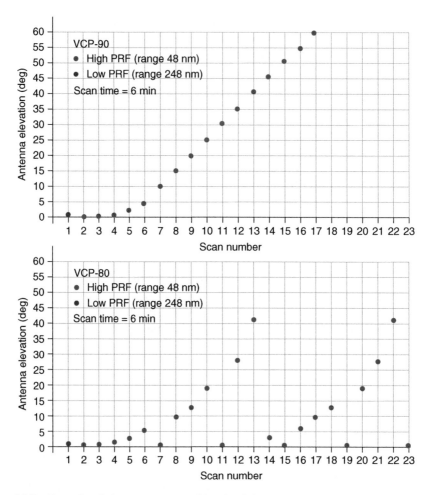

Figure 10.8 Example of elevation cuts used by the Baltimore – Washington Airport TDWR for VCP-90 and VCP-80

the WSR-88D. The elevation cuts for a specific TDWR depend on its distance from the airport and its relative altitude compared to the runway altitudes.

There are two VCPs used by TDWRs, termed VCP-90 and VCP-80. VCP-90 is called **monitor mode** and is used when wind shear threats are not detected near the airport. VCP-80 is called **hazardous mode** and is used when the TDWR detects a potential wind shear event that threatens the airport. Switching between these modes is controlled by automated algorithms that interrogate echo coverage, intensity, and range from the airport. The SPG produces products similar to the WSR-88D at the lowest elevation of VCP 80 every minute and at the remaining elevations every 3 min.

Figure 10.8 shows the sequence of elevation cuts for each VCP. The example is from the TDWR near the Baltimore – Washington International Airport. VCP-90 starts

---

[5]Students interested in examining TDWR training modules for NWS forecasters should visit http://wdtb.noaa.gov/buildTraining/SPG3/.

with a low PRF, long-range 460 km (248 nm) scan at 0.6° in which only reflectivity data are recorded. The remaining scans of VCP-90 use a high PRF corresponding to a maximum range of 89 km (48 nm), starting with the lowest elevation angle and stepping up to the highest elevation angle available for the particular TDWR. Short-range elevation angles can be as low as 0.1° and as high as 60°, depending on the radar. A volume is completed in 6 min.

VCP-80 differs substantially from WSR-88D VCPs. This VCP also starts with a long-range scan at 0.6° in which only reflectivity data are recorded. Following that, a low-elevation scan is performed every minute. Intermixed with these low-elevation scans, the TDWR performs two mini-volume scans using fewer elevation angles. A complete cycle containing both volumes takes 6 min.

The NWS SPG uses information from the long-range scan of either VCP-80 or VCP-90 to generate a reflectivity product with the range of 417 km (225 nm) and 1° × 300 m resolution, as well as precipitation accumulation products. The reflectivity data from the long-range scan are also used to flag range-folded data on short-range scans below 12°. Because of the short unambiguous range of the short-range scans, TDWRs have considerable potential for multiple-trip echoes. Attenuation is a major problem for TDWRs because of their C-band wavelength. Attenuation can lead to a poor depiction or complete loss of storm features behind regions of heavy rain or hail.

Despite these problems, the TDWR radars have proved to be an asset in severe storm monitoring, particularly when storms approach large cities having airports served by the TDWRs. The 1 min refresh scanning, higher resolution, and greater sensitivity of TDWRs can allow features such as supercell hook echoes and mesocyclones to be more clearly defined and more easily tracked, particularly in situations where the viewing angle between the radar and the storm is favorable (Figure 10.9).

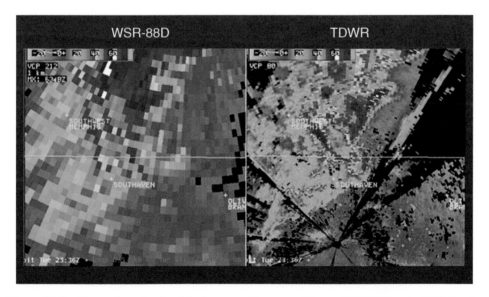

Figure 10.9 Comparison of reflectivity in a tornadic supercell from a WSR-88D and a TDWR radar where the TDWR was favorably located relative to the hook echo (Courtesy NOAA)

10.4   International operational radar networks

Many countries across the world have or are developing weather radar networks. For example, Canada's current weather radar network consists of 31 radars covering sections of Canada's more populated southern and eastern provinces (Figure 10.10). As of 2017, Environment Canada owned and operated 28 of these radars, the DOD two and McGill University one. All Canadian radars are C-band Doppler radars, except the McGill radar which is S-band. The network employs two separate scan sequences, similar to VCPs of the WSR-88Ds. At the time of writing, a new S-band dual-polarization network, with characteristics similar to the NEXRAD WSR-88D network is being implemented to replace the C-band radars across Canada.

The Australian government operates a network of S-band and C-band Doppler radars that ring the coastline of the continent where most of Australia's population resides and cover parts of the Australian interior (Figure 10.11). As of 2013, the Australian radar network continued a major upgrade called the Strategic Radar Enhancement Project to modernize its radar network to high-resolution Doppler capability and to supplement the severe weather warning capabilities of the Australian Bureau of Meteorology.

Other countries have invested significant resources in developing radar networks. China, for example, has invested in a network S-band and C-band Doppler radars that covers the populated regions of that country. China's S-band radars are similar to the WSR-88Ds in design. India has established a network of S-band radars

Figure 10.10   The Canadian operational radar network

Figure 10.11   The Australian meteorological radar network

called the *Cyclone Detection Radar Network* that monitors its entire coastline. Other countries such as Japan, New Zealand, and South Africa also have national radar networks.

The European weather radar network is unique in that it is a consortium of radars operated by Meteorological Services of many countries (Figure 10.12). Twenty-eight European governments have come together to establish the Operational Program for the Exchange of Weather Radar Information (OPERA), a coordination effort of the Network of European Meteorological Services. The goal of OPERA was to share data and technological expertise across the network of nearly 160 meteorological radars and develop products, such as composite reflectivity maps, that cover the continent and would serve the entire European community. Most European radars are C-band, and nearly 110 have Doppler capability. OPERA efforts over the last decade have led to the development of composite radar animations of European weather systems, now available in real time.

We can expect that upgrades of existing networks will continue, and weather radar technology will spread into the developing world at a rapid pace. Sharing of data across international borders will hopefully lead to economic advantages for all countries, as radar coverage of weather over the world's continents becomes ever more complete.

Figure 10.12   Locations of radars used in the European OPERA cooperative

Important terms

clear air mode

effective beamwidth

elevation cut

hazardous mode

Human Control Interface (HCI)

legacy unfolding algorithm

Level-I data

Level-II data

Level-III products

Master System Control Function (MSCF)

MPDA algorithm

monitor mode

Next Generation Radar (NEXRAD)
operational weather radar network
precipitation mode
purple haze
Radar Data Acquisition Unit (RDA)
Radar Product Generator (RPG)
split-cut
super resolution

Supplemental Products Generator (SPG)
Terminal Doppler Weather Radars (TDWR)
Volume Coverage Patterns (VCP)
waveform
Weather Service Radar-1988 (WSR-88D)
Doppler network

Review questions

1. What WSR-88D base variables can be obtained in Level-II format from the National Centers for Environmental Information?
2. What is the difference between Level-II and Level-III WSR-88D information provided by the NWS?
3. Which WSR-88D VCP has the greatest sensitivity?
4. Which WSR-88D VCP provides the best chance of correcting range ambiguities and unfolding radial velocities during a landfalling Category five hurricane?
5. Which WSR-88D VCP is best to use when a storm is a long distance from the radar?
6. Why is split-cut scanning used on low elevation angles of the WSR-88D in nearly all VCPs?
7. Why is it generally less problematic to correct for range and velocity folding on higher elevation scans of the WSR-88D?
8. What is purple haze?
9. Describe the primary difference in the VCP sequences for the monitor and hazardous modes of the TDWR. Which is least like a WSR-88D VCP?
10. What are some advantages to a forecaster of supplementing WSR-88D radar products with TDWR products when storms are in the vicinity of both radars?

Challenge problems

11. Using the WSR-88D beam height equation

$$h = s \, \sin \alpha + \frac{s^2}{15{,}417.8 \text{ km}}$$

determine the lowest altitude (above mean sea level) that the 0.5° beams of the Indianapolis (KIND) and Chicago (KLOT) WSR-88Ds intersect. In the

(continued)

equation, h is the altitude of the beam above ground level at the radar site, s is the slant range, and α is the elevation angle. KLOT is 231 m above sea level, KIND is 270 m above sea level, and the radars are 260 km apart.

12. If the physical beamwidth of a WSR-88D is 0.89°, the antenna rotates at $20°\,s^{-1}$, 40 pulses are used to calculate a reflectivity estimate, and the PRF is $446\,s^{-1}$, what is the effective beamwidth of the radar?

13. In past 4-bit resolution NWS depictions of radar reflectivity used for VCP-31 and VCP-32, the NWS used a color scale for reflectivity in which each color bin covered a range of 4 dBZ, except the lowest bin, which included all reflectivity values less than −28 dBZ, and the highest bin, which included all reflectivity values greater than X dBZ. What was X?

14. You are an NWS WSR-88D operator. During the early part of your work shift, insect echo is present on the radar. However, toward mid-afternoon, a squall line develops 150 km west of the radar site, moving toward the radar at $40\,km\,h^{-1}$. Two hours later, tornado-producing supercell thunderstorms develop 30 km ahead of the squall line. The worst tornado passes 20 km north of the radar 1 h later. Two hours after that, the convection associated with both the squall line and the supercells weakens and the precipitation evolves into a deep widespread stratiform rain. The back edge stratiform rain passes the radar site 2 h later and continues to move eastward at the speed of the system, $40\,km\,h^{-1}$. By the end of your shift, the back edge is 150 km east of the radar site. Describe a sequence of VCPs that you might choose to monitor the weather during your shift.

15. Shown below is the radar echo of a tornadic supercell from a WSR-88D at point A. Contrast the advantages/problems that a forecaster would have interrogating the storm if the radar instead was a TDWR. Do the same, assuming instead that the radar was located at point B.

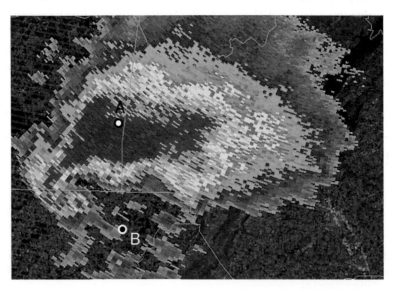

11

Doppler Velocity Patterns and Single-Radar Wind Retrieval

Objectives

By the end of this chapter, you should understand the following:

- The fundamental kinematic properties of a wind field: translation, divergence, vorticity, and deformation, and how they differ from one another.
- How to estimate wind speed and direction as a function of altitude by examining a Doppler radial velocity display.
- How to recognize frontal boundaries on a Doppler radial velocity display.
- The difference between Doppler radial velocity indications of rotation and divergence.
- How a wind profiler measures a profile of wind speed and direction.
- The principles behind the velocity–azimuth display technique for recovering vertical profiles of wind speed, wind direction, and properties of the deformation field over a scanning Doppler radar.
- How an extension of the velocity–azimuth display technique can be used to recover profiles of divergence, vertical air velocity, and hydrometeor terminal velocities over a radar.

11.1   Introduction

A Doppler radar measures the speed at which targets within a pulse volume are moving in the direction of the radar beam. Operational meteorologists need to know the horizontal wind speed and direction at the surface and above the radar site and use the radar to detect patterns in the wind field that can be associated with phenomena such as fronts, microbursts, and tornadoes. Scientists studying storm structure often seek to measure more complicated, derived properties of the horizontal wind, such as divergence, vorticity, and deformation, and obtain estimates of vertical air motion, as updrafts and downdrafts relate closely to storm intensity. In this and the next chapter, we focus on how Doppler radar radial velocity measurements can be

Radar Meteorology: A First Course, First Edition. Robert M. Rauber and Stephen W. Nesbitt.
© 2018 John Wiley & Sons Ltd. Published 2018 by John Wiley & Sons Ltd.
Companion website: www.wiley.com/go/Rauber/RadarMeteorology

interpreted qualitatively, and mathematically manipulated quantitatively, to obtain estimates of the wind field and its derivative properties.

11.2   Kinematic properties of the wind field

Before we examine wind retrieval techniques, it is worth spending a moment discussing the wind itself and its so-called **kinematic properties**. The word **kinematic** comes from the Greek word for motion. By kinematic, we mean a description of the motion of the wind without regard to how it came about or will evolve. Our goal in this section is to develop both a mathematical and an intuitive understanding of the basic kinematic properties of a wind field.

Consider the coordinate system shown in Figure 11.1. Suppose we want to estimate the wind field at an arbitrary point (x, y) given knowledge of the wind at a nearby point (x_0, y_0), which, for simplicity, we will consider the origin $(0, 0)$. A straightforward way to do this is to write a Taylor series in the form

$$u = u_0 + \frac{\partial u}{\partial x}x + \frac{\partial u}{\partial y}y + \text{higher order terms} \qquad (11.1)$$

and

$$v = v_0 + \frac{\partial v}{\partial x}x + \frac{\partial v}{\partial y}y + \text{higher order terms} \qquad (11.2)$$

where u is the wind component in the east–west direction (x), positive toward the east, and v is the wind component in the north–south direction (y), positive toward the north. If we assume that the wind field varies at most linearly over the area of interest, then we can ignore the higher order terms and write

$$u = u_0 + \frac{\partial u}{\partial x}x + \frac{\partial u}{\partial y}y \qquad (11.3)$$

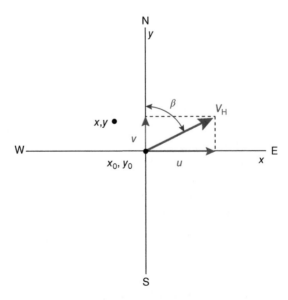

Figure 11.1   Wind components in a Cartesian coordinate system

$$v = v_0 + \frac{\partial v}{\partial x}x + \frac{\partial v}{\partial y}y \tag{11.4}$$

Let us now take a simple step and split each derivative term into two equal terms:

$$u = u_0 + \frac{1}{2}\frac{\partial u}{\partial x}x + \frac{1}{2}\frac{\partial u}{\partial x}x + \frac{1}{2}\frac{\partial u}{\partial y}y + \frac{1}{2}\frac{\partial u}{\partial y}y \tag{11.5}$$

$$v = v_0 + \frac{1}{2}\frac{\partial v}{\partial x}x + \frac{1}{2}\frac{\partial v}{\partial x}x + \frac{1}{2}\frac{\partial v}{\partial y}y + \frac{1}{2}\frac{\partial v}{\partial y}y \tag{11.6}$$

write two nonsense equations:

$$0 = \frac{1}{2}\left(\frac{\partial v}{\partial x} - \frac{\partial v}{\partial x}\right)y + \frac{1}{2}\left(\frac{\partial v}{\partial y} - \frac{\partial v}{\partial y}\right)x \tag{11.7}$$

$$0 = \frac{1}{2}\left(\frac{\partial u}{\partial x} - \frac{\partial u}{\partial x}\right)y + \frac{1}{2}\left(\frac{\partial u}{\partial y} - \frac{\partial u}{\partial y}\right)x \tag{11.8}$$

and add Eqs. (11.5) and (11.7), and separately add Eqs. (11.6) and (11.8) to get

$$\overset{A}{u = u_0} + \overset{B}{\frac{1}{2}\left(\frac{\partial u}{\partial x} + \frac{\partial v}{\partial y}\right)x} - \overset{C}{\frac{1}{2}\left(\frac{\partial v}{\partial x} - \frac{\partial u}{\partial y}\right)y} + \overset{D}{\frac{1}{2}\left(\frac{\partial u}{\partial x} - \frac{\partial v}{\partial y}\right)x} + \overset{E}{\frac{1}{2}\left(\frac{\partial v}{\partial x} + \frac{\partial u}{\partial y}\right)y}$$

$$\tag{11.9}$$

$$v = v_0 + \frac{1}{2}\left(\frac{\partial u}{\partial x} + \frac{\partial v}{\partial y}\right)y + \frac{1}{2}\left(\frac{\partial v}{\partial x} - \frac{\partial u}{\partial y}\right)x - \frac{1}{2}\left(\frac{\partial u}{\partial x} - \frac{\partial v}{\partial y}\right)y + \frac{1}{2}\left(\frac{\partial v}{\partial x} + \frac{\partial u}{\partial y}\right)x$$

$$\tag{11.10}$$

In Eqs. (11.9) and (11.10), the term A and the terms in parentheses (B–E) represent five distinct properties of atmospheric flows, called **translation** (A), **divergence** (B), **vorticity** (C), **stretching deformation** (D), and **shearing deformation** (E). These properties of the wind field have tremendous meteorological significance. We can understand these properties by considering how a fluid element (a unit area of fluid) is modified by each of these flows. Figure 11.2 shows the effects of each of the processes on a fluid element in the flow. Translation results in a change in the *location* of a fluid element, with no change in area, shape, or orientation. Translation may be thought of simply as the average wind. Divergence (or convergence) results in a change of the *area* of a fluid element, with no change in its location, shape, or orientation. Divergence and convergence in the atmosphere force air to rise or descend. At the surface, divergence is associated with outflows and microbursts. Vorticity results in a change of *orientation* of a fluid element, but no change in location, area, or shape. Vorticity is associated with rotation, as occurs with tornadoes and in the vicinity of low-pressure systems. Deformation (shearing and stretching) results in a change of the *shape* of a fluid element, with no change in location, area, or orientation. Deformation is associated with flow boundaries such as fronts.

The primary objective of Doppler radar processing algorithms used in meteorological research is to derive quantitative estimates of these properties of the wind field from Doppler radial velocity measurements so as to better understand the flows within storms, storm structure, and evolution. In operational meteorology, the objective is more straightforward: to understanding the basic character of the wind

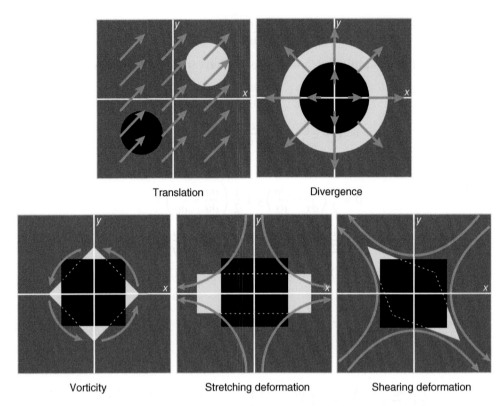

Figure 11.2 Kinematic properties of the wind field. The black square is the fluid element at the initial time and the white shape is the fluid element at a later time

field so as to detect and warn of impending hazardous weather. We begin by examining some common Doppler radial velocity patterns and their relationship to the wind field.

11.3   Doppler radial velocity patterns and the wind field

An important skill required of a radar meteorologist is to be able to examine **radial velocity patterns** on a radar display and from them deduce the wind fields that led to the patterns. The radial velocity represents only one component of wind—along the direction of the beam. Unfortunately, each beam on the display points in a different direction and, to complicate matters, distance radially outward on the display also represents increasing height above the radar. For example, in Figure 11.3, the elevation angle of the radar beam is 0.5°, and the edge of the display is at 12 km altitude, so the inner circle at the halfway point is about 6 km altitude (it will not be exactly 6 km because of earth curvature). Most radar displays mark range rings in term of distance from the radar rather than altitude, so independent information about the relationship between distance and altitude may be required to quantitatively interpret the radial velocity field on a specific display. For all these reasons, skillful interpretation of radial velocity displays takes some time.

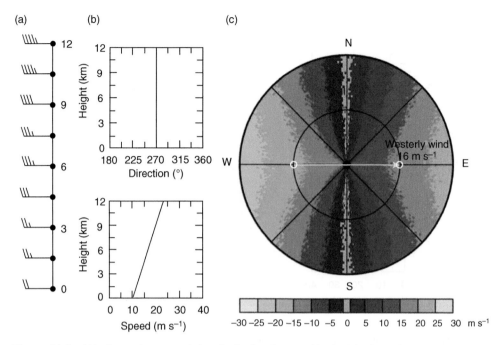

Figure 11.3   Winds as a function of altitude displayed as wind barbs (a), plots of wind direction and speed (b), and as appearing on the radial velocity display of a Doppler radar (c) (From Brown, R.A. and Wood, V.T. (2007) *A Guide for Interpreting Doppler Velocity Patterns*, 2nd edn, NOAA/NSSL)

Here, we will consider three families of Doppler velocity patterns that together represent the most common and important features found on Doppler radar displays.[1] These are (1) winds that increase with altitude, but are horizontally uniform across the region viewed by a radar, (2) local winds in the vicinity of severe convection, and (3) winds in the vicinity of fronts.

11.3.1   Large-scale flow patterns

A scanning Doppler radar typically views a circular region about 200–400 km in diameter. Relative to the dimensions of many large-scale weather systems, this area is sufficiently small that winds across the domain of the radar at any particular altitude may not change appreciably. This is particularly true during periods of stratiform precipitation, or clear weather when radar echoes are associated with flying insects. Figure 11.3 shows a radar display in such conditions. In this simple case, winds increase with altitude, but are from the west at all altitudes. Blue colors denote winds approaching the radar and red colors denote winds that are receding. Note that any circle centered on the radar display represents a constant altitude above the radar. If the winds are horizontally uniform at a particular altitude, then the radar will

---

[1]Brown and Wood (2007) provide a more extensive overview of Doppler velocity patterns. The images in this section were drawn from their work. See http://www.nssl.noaa.gov/publications/dopplerguide/.

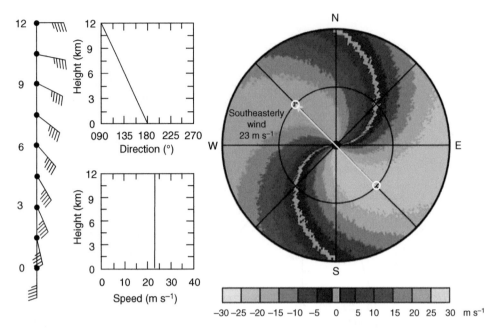

Figure 11.4 Radar display of radial velocity for winds that have constant speed but are backing (turning counterclockwise) with height (From Brown, R.A. and Wood, V.T. (2007) *A Guide for Interpreting Doppler Velocity Patterns*, 2nd edn, NOAA/NSSL)

observe the maximum negative radial velocity when the beam points directly upwind, and the maximum positive radial velocity when the beam points directly downwind, assuming that no folding of the radial velocities has occurred. The velocity will be near zero when the radar points normal to the wind direction, but not exactly zero because the beam also points upward and observes a component of the targets' fall speeds. To determine the wind direction at a specific altitude, simply draw an arrow with its tail at the minimum along the circle and its head at the maximum (see yellow arrow in Figure 11.3). The wind direction is the direction of the arrow, and the wind speed, at least at low elevation angles, will be approximately the value of the maximum radial velocity. At high elevation angles, the horizontal wind velocity will be a few meter per second different from the maximum radial velocity because the radial velocity includes a more significant component of the targets' fall speeds and the wind is viewed at an oblique angle.

Figure 11.4 shows a situation where winds are **backing**, that is, turning counterclockwise with altitude. In this example, the winds do not change speed. Note the "backward S" shape to the zero line. This is an immediate indication of backing winds with height, which, from dynamical principles, is associated with **cold air advection**. In this example, the winds back from southerly at the surface to easterly at 12 km. The same technique for determining wind speed and direction applies. Choose a specific circle to isolate a single altitude, draw an arrow from the minimum to the maximum to obtain the wind direction, and read the value of the maximum for the wind speed. In this case, the wind at 6 km altitude is from the southeast at approximately 23 m s^{-1}.

Contrast the last figure with Figure 11.5. Here, the zero line has a forward "S" shape. This is indicative of winds turning clockwise, or **veering** with height, which,

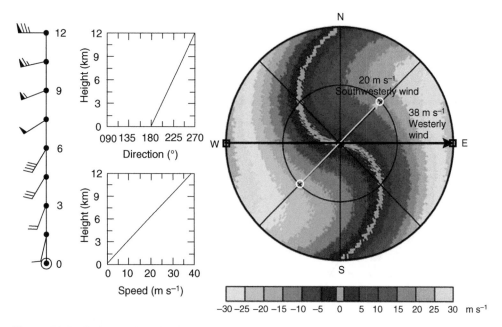

Figure 11.5   Radar display of radial velocity for winds that are increasing in speed and are veering (turning clockwise) with height. Note that radial velocities on the edge of the display are folded (From Brown, R.A. and Wood, V.T. (2007) *A Guide for Interpreting Doppler Velocity Patterns*, 2nd edn, NOAA/NSSL)

from principles of atmospheric dynamics, implies **warm air advection**. In this example, winds veer from southerly at the surface to westerly at 12 km. Applying the same technique on the inner circle, the winds at 6 km would be from the southwest at 20 m s^{-1}. But what if we apply the same technique at the perimeter of the radar display? Here, the technique would fail because the maximum and minimum velocities near the edge of the display are folded. We can see this because the radial velocities jump from one end of the velocity scale to the other near the edge of the display along, for example, the 90–270° radial. To get the correct wind direction on the display perimeter, one has to reverse the process and draw the arrow from the maximum to the minimum radial velocity on the circle. The wind speed is more complicated to determine, being twice the Nyquist velocity plus the value of the minimum radial velocity ($2v_{max} + v_r$), in this example [$2 \times 30 + (-22)] = 38$ m s^{-1}. If the wind velocities become even stronger, the arrow to find the wind direction would again have to be drawn from the minimum to the maximum along the circle. The same equation would apply until a second fold occurred.

11.3.2   Fronts

Fronts, the boundaries between airmasses with different thermal properties, are often associated with distinct wind shifts. The surface front may be passing directly through the radar domain or may have already passed so that the frontal boundary is overhead, as indicated by the three radars in Figure 11.6. Figure 11.7 shows a typical

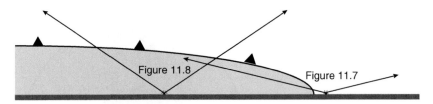

Figure 11.6 Radar beams passing through a front for radars positioned at two locations (see Figures 11.7 and 11.8)

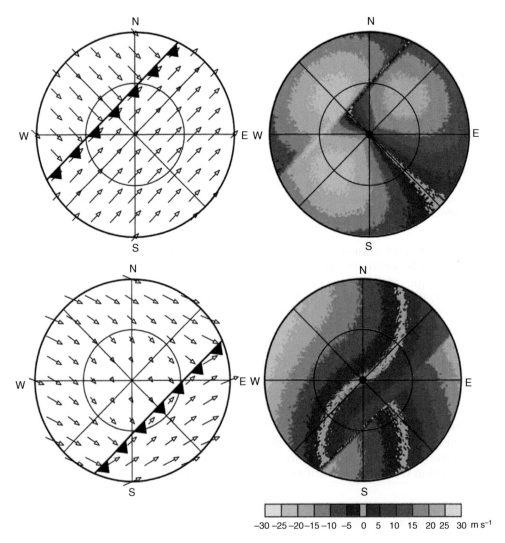

Figure 11.7 Radial velocity patterns for a front northwest of a radar (top) and southeast of a radar (bottom) (see Figure 11.6) (From Brown, R.A. and Wood, V.T. (2007) *A Guide for Interpreting Doppler Velocity Patterns*, 2nd edn, NOAA/NSSL)

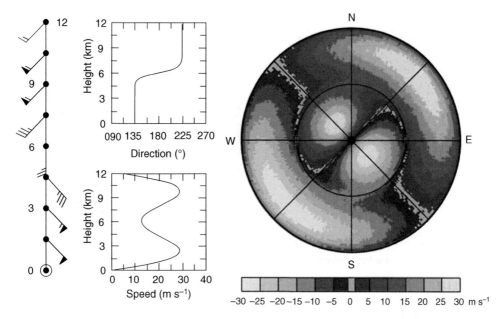

Figure 11.8 Radial velocity pattern for a radar beneath a warm front (see Figure 11.6) (From Brown, R.A. and Wood, V.T. (2007) *A Guide for Interpreting Doppler Velocity Patterns, 2nd edn*, NOAA/NSSL)

evolution of the radial velocity pattern on a display as a cold front passes across a radar site. The radial velocity pattern changes radically as the front moves across the site, although the position of the boundary itself is quite clear. As you can see, the technique for finding wind direction and speed in the previous section does not apply as the winds across the radar domain are not horizontally uniform. It may or may not be possible to determine the wind speed and direction in each airmass depending on how much radar coverage is within the airmass, and whether or not a clear maximum or minimum can be seen along an arc of a circle completely within one of the airmasses.

The pattern changes completely after the surface front departs the radar domain and the frontal boundary is overhead. A sharp wind shift often occurs above the radar between the lower airmass and the airmass aloft. Figure 11.8 shows the pattern a radar might observe if it were located beneath a warm front located northeast of a cyclone center (Figure 11.6). There winds often shift from southeasterly near the ground to southwesterly above the front. The wind shift in Figure 11.8 appears near 6 km altitude, with 30 m s^{-1} southeasterly winds below the front, and 30 m s^{-1} southwesterly winds aloft. The frontal boundary is quite distinct in the radial velocity field. In this case, the technique for estimating the winds can be applied on circles both above and below the front as winds are uniform over the radar domain at these altitudes.

11.3.3   Convective scale flow patterns

Individual convective cells typically occupy only a small part of the radar display. Two wind patterns associated with convection are of particular concern for the detection of severe weather, the first associated with storm rotation and the second with

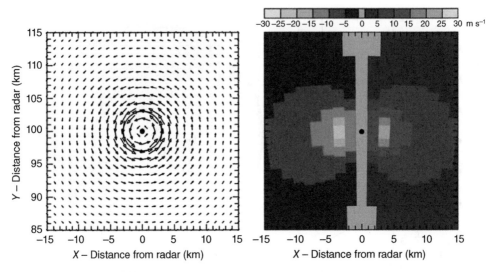

Figure 11.9 Radial velocity pattern associated with small-scale rotation similar to what would occur in a supercell thunderstorm. Here, the rotation is located 100 km north of the radar (From Brown, R.A. and Wood, V.T. (2007) *A Guide for Interpreting Doppler Velocity Patterns, 2nd edn*, NOAA/NSSL)

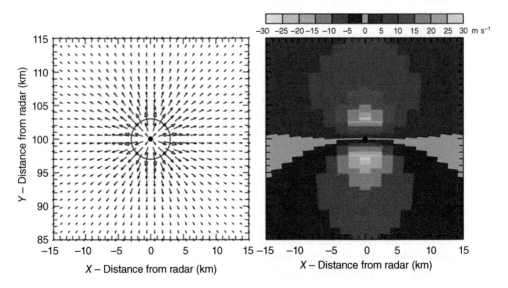

Figure 11.10 Radial velocity pattern associated with small-scale divergence similar to what would occur in a microburst. Here, the center of divergence is located 100 km north of the radar (From Brown, R.A. and Wood, V.T. (2007) *A Guide for Interpreting Doppler Velocity Patterns, 2nd edn*, NOAA/NSSL)

strong outflows associated with downbursts and strong straight-line winds. The way each of these patterns appear depends on the storm's location relative to the radar. In each case, the signature is a small **radial velocity couplet** with strong velocities of opposite sign. Figures 11.9 and 11.10 show these patterns for a situation where the circulations are directly north of a radar, with the center of circulation 100 km from the radar site. Figure 11.9 shows the signature pattern for cyclonic rotation, counterclockwise in the northern hemisphere—the normal way that tornadoes rotate. Note that the zero line dividing the radial velocity couplet is along the beam direction, which is directly north–south at the center of the figure. To the right of the beam, winds are strong and away from the radar, whereas to the left, winds are strong and toward the radar. We will explore how radial velocity patterns associated with rotation relate to the structure of supercell thunderstorms and squall lines when we examine convective storms in Chapter 14.

Figure 11.10 shows the pattern associated with strong outflows characteristic of a microburst. The pattern looks qualitatively similar to rotation, except that in this case, the zero line is east–west, perpendicular to the beam direction. On the "radar side" of the couplet, the winds are strong and toward the radar, whereas on the opposite side, the winds are strong and away from the radar. Again, we will examine how these patterns relate to the structure of convective cells in more detail in Chapter 14.

11.4 Wind retrieval with profiling radars

A **wind profile** consists of a series of measurements of wind speed and direction from the earth's surface upward through the troposphere and possibly into the stratosphere. Wind profiles are routinely made with balloon-borne devices called *rawinsondes*, typically every 12 h at locations around the world. Radars provide the opportunity to obtain wind profiles at much higher frequency, typically every few minutes. There are two very different approaches to obtaining wind profiles with radars, the first associated with a specific type of radar called a **wind profiler** and the second called the **velocity–azimuth display** (VAD) technique, which is used with standard scanning radars such as the WSR-88Ds. In this section, we explore each of these approaches.

11.4.1 Wind profilers

A wind profiler is a type of Doppler radar that operates in very high frequency (VHF) and ultra-high-frequency (UHF) radio bands. Unlike the typical scanning Doppler radar, which has a dish-type antenna, the antenna of a wind profiler is an array of cables (Figure 11.11). The antenna is called a **phased array antenna** (see Chapter 2) because the antenna transmits electromagnetic radiation with a slight time delay from one side to the other across the array. The delay has the effect of creating a beam of radiation that points in a specific direction. With traditional Doppler radar, energy is scattered primarily by raindrops and ice particles; with profilers, energy undergoes Bragg scattering, primarily by small variations in the index of refraction associated with turbulent regions of the atmosphere (see Chapters 1 and 8). These are manifested as small variations in the refractive index on a scale of half the radar wavelength. The wind profiler senses the motion of the air along the beam by determining the Doppler shift that occurs in the returned signal as these turbulent regions of air move toward or away from the profiler. Wind profilers work well in clear air.

By using three or sometimes five beams and trigonometry, a vertical profile of the wind (wind speed and direction at altitudes above the surface) can be obtained,

Figure 11.11   An example wind profiler. The phased-array antenna is the grid in the middle. The cylindrical devices with crowns are a part of a radio acoustic system used with the wind profiler to measure temperature profiles (Courtesy NOAA)

similar to the winds measured by a rawinsonde. Figure 11.12 shows a three-beam configuration, with one beam pointed east at an elevation angle, α, the second pointed north at the same elevation angle, and the third pointed vertically. A typical α might be 75°. The radial velocities measured by each beam of the profiler are related to the u, v, and W components by

$$v_{re} = u \, \cos \alpha + W \, \sin \alpha \tag{11.11}$$

$$v_{rn} = v \, \cos \alpha + W \, \sin \alpha \tag{11.12}$$

$$v_{rz} = W \tag{11.13}$$

where v_{re}, v_{rn}, and v_{rz} are the radial velocities along the east, north, and upward pointing beams, respectively, and W is targets' reflectivity-weighted mean total vertical velocity. In clear air, $W = w$, the vertical air velocity, whereas in precipitation, $W = w + \overline{w}_t$, where $\overline{w}_t$ is the reflectivity-weighted mean terminal velocity of the targets in still air (by convention in this text, $\overline{w}_t$ is negative, corresponding to targets falling toward the earth). Equations (11.11)–(11.13) can be easily solved for the horizontal wind components to obtain

$$u = v_{re} \, \sec \alpha - W \, \tan \alpha \tag{11.14}$$

$$v = v_{rn} \, \sec \alpha - W \, \tan \alpha \tag{11.15}$$

The wind speed is then given by $\sqrt{u^2 + v^2}$. The wind direction depends on the sign of u. If $u > 0$, the wind direction, θ, is given by $\theta = \frac{3\pi}{2} - \arctan(v/u)$, whereas if $u < 0$, $\theta = \frac{\pi}{2} - \arctan(v/u)$, where the arctangent is expressed in radians. Obviously, if $u = 0$, the wind is from the south ($v > 0$) or north ($v < 0$).

Figure 11.13 shows a series of wind profiles from a wind profiler that collected the data during the passage of a winter cyclone. A sharp boundary marking a passing

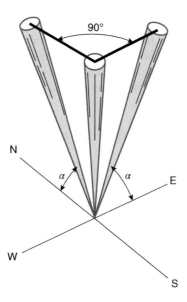

Figure 11.12   A three-beam profiling radar system

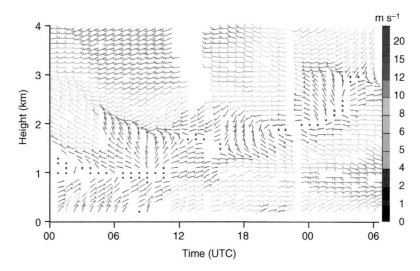

Figure 11.13   Winds recovered from a boundary layer wind profiler as a function of time and altitude

warm front can be seen in the wind field ascending from about 1 km altitude at 04:00 UTC to 3 km altitude at 00:00 UTC the next day. Winds increase with time from 2 to 15 m s^{-1} at 0.7 km altitude, while decreasing from 15 to 5 m s^{-1} at 3.5 km. Data such as these help delineate the passage of mesoscale features such as fronts and jets as they pass over a profiler site.

The recovery of winds by a wind profiler depends on the horizontal homogeneity of the scattering medium across the profiler's beams. If convection is occurring in the vicinity of the profiler, for example, such that a rain shaft is located in one beam, but not the other two, the recovery of u and v will contain significant errors. Like all radars, profiling radars have sidelobes that can influence the measurement. Many research profilers today are mobile. If parked near a highway, for example, passing trucks may intercept sidelobes, contributing to a spread in the Doppler spectra measured by the profiler, and more uncertainty in the recovered winds.

11.5   Velocity–azimuth display wind retrieval

The VAD technique is used routinely with WSR-88D radars, as well as research radars, to derive wind profiles above the radar site. The terminology "velocity–azimuth display" originates from the manner in which data are displayed: the radial velocity is plotted as a function of azimuth angle at a given elevation angle and distance from the radar. Extended velocity–azimuth display (EVAD) algorithms[2] exist to derive profiles of deformation, divergence, and even vertical air motion and particle terminal velocities above the radar.

11.5.1   VAD technique

Consider the geometry in Figure 11.14 that shows a Cartesian coordinate system with x positive to the east, y positive to the north, and z, positive upward. Consider a ground-based radar located at the origin, with a beam directed at elevation angle α rotating through 360° of azimuth angle β, where $\beta = 0$ is toward the north and coincides with the positive y axis. Assume that the radar receives a return around the entire scanned circle at a given altitude, measuring radial velocity v_r at range r. The radial velocity results from precipitation falling at a velocity, $W = w + \overline{w}_t$, and moving with the horizontal wind at speed V_h at a direction θ to the y axis. We will assume that positive v_r is away from the radar.

We can write the radial velocity as

$$v_r = u \sin \beta \cos \alpha + v \cos \beta \cos \alpha + W \sin \alpha \qquad (11.16)$$

Unfortunately, one cannot determine the components (u, v, W) from a single radial velocity measurement as there are three unknowns and only one equation. However, let us consider a ring of radial velocity measurements at a specific altitude as in Figure 11.14. We will assume in the discussion that follows that the velocity gradient across any ring varies at most linearly so that we can write

$$u = u_0 + \frac{\partial u}{\partial x} x + \frac{\partial u}{\partial y} y \qquad (11.17)$$

and

$$v = v_0 + \frac{\partial v}{\partial x} x + \frac{\partial v}{\partial y} y \qquad (11.18)$$

---

[2]The EVAD technique is described in Matejka, T. and Srivastava, R.C. (1991) An improved version of the extended velocity-azimuth display analysis of single-Doppler radar data. *J. Atmos. Oceanic Technol.*, **8**, 453–456). This section simplifies the analysis, summarizing the paper's essential points.

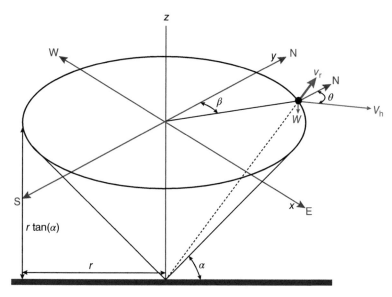

Figure 11.14   Geometry used for VAD scans (Adapted from Browning, K.A. and Wexler, R. (1968) The determination of kinematic properties of a wind field using Doppler radar. *J. Appl. Meteor.*, **7**, 105–113. © American Meteorological Society, used with permission)

Substituting (11.17) and (11.18) into (11.16), substituting $x = r \sin \beta$ and $y = r \cos \beta$ and applying the trigonometric identities

$$\sin^2 a = \frac{1}{2} - \frac{1}{2} \cos 2a \tag{11.19}$$

$$\cos^2 \alpha = \frac{1}{2} + \frac{1}{2} \cos 2\alpha \tag{11.20}$$

$$\sin a \cos a = \frac{1}{2} \sin 2a \tag{11.21}$$

We obtain

$$V_r = \frac{1}{2} r \cos \alpha \left(\frac{\partial u}{\partial x} + \frac{\partial v}{\partial y} \right) + W \sin \alpha$$

> No β dependence
> Contains divergence and fall speed

$$+ u_0 \cos \alpha \sin \beta$$

> Depends on sinβ
> Contains translational u wind component

$$+ v_0 \cos \alpha \cos \beta$$

> Depends on cosβ
> Contains translational v wind component

$$+ \frac{1}{2} r \cos \alpha \left(\frac{\partial v}{\partial x} + \frac{\partial u}{\partial y} \right) \sin 2\beta$$

> Depends on sin 2β
> Contains shearing deformation

$$- \frac{1}{2} r \cos \alpha \left(\frac{\partial u}{\partial x} - \frac{\partial v}{\partial y} \right) \cos 2\beta$$

> Depends on cos 2β
> Contains stretching deformation

$$\tag{11.22}$$

We can write Eq. (11.22) as a Fourier series where

$$V_r = \frac{1}{2}a_0 + \sum_{n=1}^{\infty}(a_n \sin n\beta + b_n \cos n\beta) \tag{11.23}$$

and the coefficients a_n and b_n are given by

$$a_0 = r \cos \alpha \left(\frac{\partial u}{\partial x} + \frac{\partial v}{\partial y} \right) + 2W \sin \alpha \tag{11.24}$$

$$a_1 = u_0 \cos \alpha \tag{11.25}$$

$$b_1 = v_0 \cos \alpha \tag{11.26}$$

$$a_2 = \frac{1}{2}r \cos \alpha \left(\frac{\partial v}{\partial x} + \frac{\partial u}{\partial y} \right) \tag{11.27}$$

$$b_2 = -\frac{1}{2}r \cos \alpha \left(\frac{\partial u}{\partial x} - \frac{\partial v}{\partial y} \right) \tag{11.28}$$

$$a_3 \text{ to } a_\infty, \quad b_3 \text{ to } b_\infty = 0 \tag{11.29}$$

As noted above, the Fourier coefficients are related to the kinematic properties of the wind field. We can understand why this is so by examining Figure 11.15. We see from Figure 11.15a that the coefficient a_0 is related to divergence and target fall speed, from Figure 11.15b that a_1 and b_1 are related to the translational wind components, and from Figure 11.15c that a_2 and b_2 are related to the deformation field. If we can determine through analysis of the radar data the Fourier coefficients, a_1 and b_1 and a_2 and b_2, we can determine the wind speed and direction and deformation field using

$$V_H = (u_0^2 + v_0^2)^{1/2} = \frac{-(a_1^2 + b_1^2)^{1/2}}{\cos \alpha} \quad \boxed{\text{Wind speed}} \tag{11.30}$$

$$\theta = \tan^{-1}\left(\frac{u_0}{v_0}\right) = \tan^{-1}\left(\frac{a_1}{b_1}\right) \text{ for } b_1 \text{ positive } \quad \boxed{\text{Wind direction}} \tag{11.31}$$

$$\theta = \pi + \tan^{-1}\left(\frac{u_0}{v_0}\right) = \pi + \tan^{-1}\left(\frac{a_1}{b_1}\right), \quad \text{for } b_1 \text{ negative} \tag{11.32}$$

$$\theta = \frac{\pi}{2}, \quad \text{for } b_1 = 0, \ a_1 > 0 \qquad \theta = \frac{3\pi}{2}, \quad \text{for } b_1 = 0, \ a_1 < 0 \tag{11.33}$$

$$\left[\left(\frac{\partial u}{\partial x} - \frac{\partial v}{\partial y} \right)^2 + \left(\frac{\partial u}{\partial y} + \frac{\partial v}{\partial x} \right)^2 \right]^{1/2} = -\frac{2(a_2^2 + b_2^2)^{1/2}}{r \cos \alpha} \quad \boxed{\text{Total deformation}} \tag{11.34}$$

$$\gamma = \frac{1}{2}\tan^{-1}\left[\frac{\left(\frac{\partial u}{\partial y} + \frac{\partial v}{\partial x} \right)}{\left(\frac{\partial u}{\partial x} - \frac{\partial v}{\partial y} \right)} \right] = \frac{1}{2}\tan^{-1}\frac{a_2}{b_2} \quad \boxed{\text{Axis of dilatation}} \quad \text{for } a_2 \text{ positive, } b_2 \text{ negative,} \tag{11.35}$$

$$\gamma = \frac{\pi}{2} - \frac{1}{2}\tan^{-1}\left[\frac{\left(\frac{\partial u}{\partial y} + \frac{\partial v}{\partial x}\right)}{\left(\frac{\partial u}{\partial x} - \frac{\partial v}{\partial y}\right)}\right] = \frac{\pi}{2} - \frac{1}{2}\tan^{-1}\frac{a_2}{b_2}, \quad \text{for } b_2 \text{ positive} \tag{11.36}$$

$$\gamma = \pi - \frac{1}{2}\tan^{-1}\left[\frac{\left(\frac{\partial u}{\partial y} + \frac{\partial v}{\partial x}\right)}{\left(\frac{\partial u}{\partial x} - \frac{\partial v}{\partial y}\right)}\right] = \pi - \frac{1}{2}\tan^{-1}\frac{a_2}{b_2}, \quad \text{for } a_2 \text{ negative, } b_2 \text{ negative}$$

$$\tag{11.37}$$

$$\gamma = \frac{\pi}{4}, \quad \text{for } b_2 = 0, \ a_2 > 0 \tag{11.38}$$

$$\gamma = \frac{3\pi}{4}, \quad \text{for } b_2 = 0, \ a_2 < 0 \tag{11.39}$$

where γ is the clockwise angle between the positive y axis and the axis of dilatation.

We determine the Fourier coefficients by analyzing radial velocity data on a ring. For example, the right panel in Figure 11.16 shows radial velocity data collected at a range of 5932 m, elevation angle of 17.7°, and altitude of 2126 m in a winter storm. The line through the data is the best-fit line using a regression technique. Each data point represents one radial velocity measurement. The panel on the left shows the derived spectral components of the Fourier series—the constant and sine and cosine functions in Eq. (11.23). In short, if you add up the five curves on the left panel, you get the right panel plus a small residual.

The data appearing in Figure 11.16 are well behaved, mapping out an easy-to-see wave function. Not all data are so clean. For example, Figure 11.17 shows radial velocity data from a ring located at a range of 2325 m, elevation angle of 0.5°, and an altitude of 257 m. Some data points have near-zero velocity and are likely to be ground clutter. Others have low signal-to-noise ratio and are likely from locations with weak or no echo. Although not appearing in Figure 11.17, it is also possible that the radial velocities can exist outside the Nyquist interval and are folded. Before determining the Fourier coefficients, data must be filtered and unfolded to correctly recover the winds. Automatic processing algorithms exist that can be applied to correct folding and remove many of the unrepresentative data points. Filtering was done in Figure 11.17 to remove unrepresentative points (the green points) before the Fourier components were calculated.

The Fourier coefficients are determined using a mathematical technique called "least squares regression," which uses matrix algebra. To extract the Fourier coefficients, a matrix is defined:

$$b_i = \sum_{k=1}^{N} f_i(\beta)v_r(\beta) \tag{11.40}$$

where $f_i(\beta)$ are either 1, $\sin(\beta)$, $\cos(\beta)$, $\sin(2\beta)$ or $\cos(2\beta)$, and $v_r(\beta)$ is a column vector consisting of the measurements of the radial velocity on a ring. We multiply b_i by the

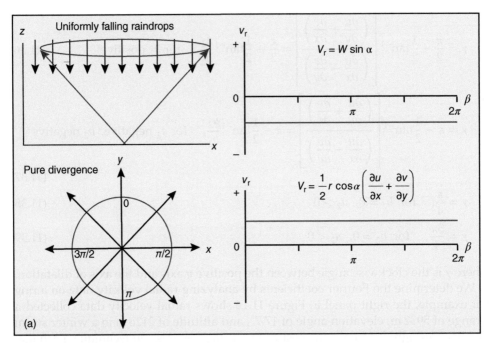

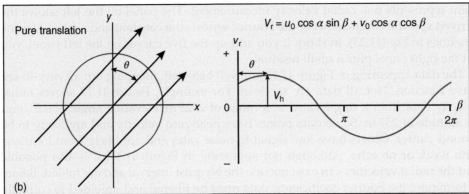

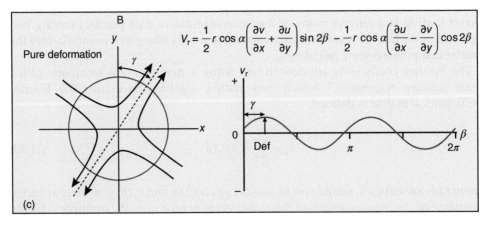

Figure 11.15 Relationship between the kinematic wind fields and the Fourier components derived from VAD analysis

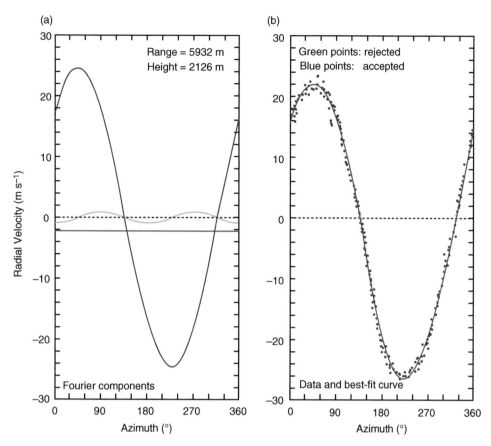

Figure 11.16 (b) Radial velocity data from a ring at constant elevation and the best-fit curve to the data. (a) The zeroth, first, and second harmonic components of the best-fit curve

inverse matrix A^{-1}, where

$$A_{i,j} = \sum_{k=1}^{N} f_i(\beta) f_j(\beta) \tag{11.41}$$

that is,

$$[A] = \begin{bmatrix} \sum 1 \times 1 & \sum \sin \beta & \sum \cos \beta & \sum \sin 2\beta & \sum \cos 2\beta \\ \sum \sin \beta & \sum \sin^2 \beta \ldots & & & \\ \sum \cos \beta \ldots & & & & \\ \sum \sin 2\beta \ldots & & & & \\ \sum \cos 2\beta & \ldots & & & \sum \cos^2 2\beta \end{bmatrix} \tag{11.42}$$

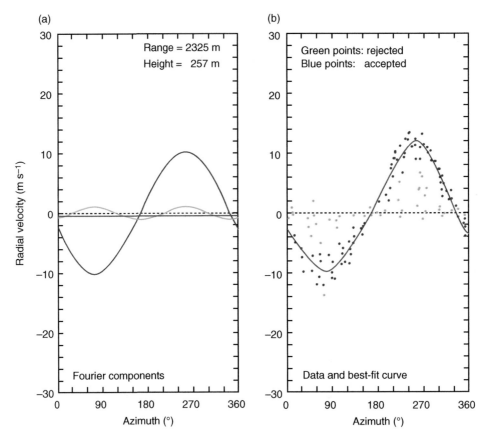

Figure 11.17   (a) The zeroth, first, and second harmonic components of the best-fit curve. (b) Radial velocity data from a ring at constant elevation and the best-fit curve to the data

such that

$$\begin{bmatrix} \frac{1}{2}a_0 \\ a_1 \\ b_1 \\ a_2 \\ b_2 \end{bmatrix} = [A^{-1}] \begin{bmatrix} \sum V_r \\ \sum V_r \sin \beta \\ \sum V_r \cos \beta \\ \sum V_r \sin 2\beta \\ \sum V_r \cos 2\beta \end{bmatrix} \tag{11.43}$$

Although this looks complicated, it is relatively simple arithmetic expressed concisely using matrix notation. The VAD technique to recover winds only requires a scan at one elevation angle. Lower elevation angles, because of their slow ascent, give high vertical resolution, but limited depth coverage, and measure the average wind over a broad region. In contrast, high elevation angles have lower vertical resolution, but cover the full depth of the troposphere over a circle with a smaller diameter centered on the radar site, and thus represent more closely the winds directly over the site.

Examples of products of VAD analysis from a winter ice storm over Illinois are shown in Figure 11.18. Each dot on each diagram represents one ring of data.

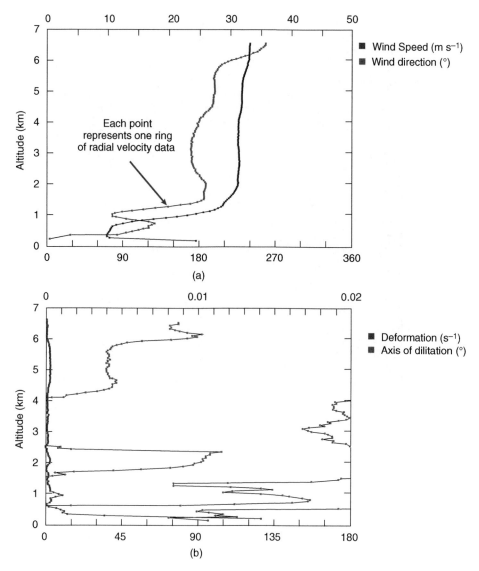

Figure 11.18   Profiles of (a) wind speed and direction, and (b) deformation and the axis of dilatation from a radar performing VAD scans in the vicinity of a warm front during an ice storm in Illinois

These data were collected at an elevation angle of 17.7°. The plots include (a) a vertical profile of the wind speed and direction and (b) the magnitude of the deformation and axis of dilatation. As can be seen in Figure 11.18b, the deformation fields over a radar are typically weak and rarely used except in limited research applications, as they tend to be noisy except when a strong front is located over the radar site. However, the National Weather Service routinely uses VAD analysis to obtain profiles of wind speed and direction over the radar at ~6 min time resolution (Figure 11.19). These data supplement the 12 h measurements of the wind made by rawinsondes.

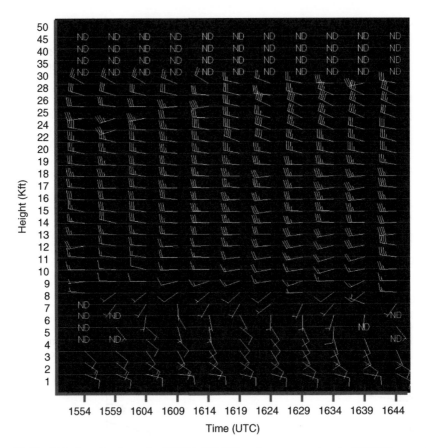

Figure 11.19 Wind profiles from the KIND WSR-88D radar recovered up to 25,000 ft. every 6 min, while the radar was routinely scanning during a rain event. In the figure, ND means no data are available at that level and time (Courtesy Paul Srivatka, College of DuPage)

11.5.2 Extended VAD analysis

The VAD products shown in Figure 11.18 do not include divergence. The reason can be understood by rearranging Eq. (11.24).

$$\left(\frac{\partial u}{\partial x} + \frac{\partial v}{\partial y}\right) = \frac{a_0}{r\cos\alpha} - \frac{2W}{r}\tan\alpha \tag{11.44}$$

Note that there are two unknowns in this equation, the divergence and W, the total particle vertical motion. To understand this more clearly, first consider a pure divergent flow centered directly over the radar (with no particle vertical motion) similar to the lower diagram in Figure 11.15a. With this flow pattern, the radar would measure receding radial velocities of the same, constant value at all azimuth angles. Now consider a calm atmosphere where precipitation particles were all falling at a constant speed (the upper diagram in Figure 11.15a). Here, the radar would measure constant approaching velocities at all azimuth angles. The two effects, divergence (or convergence) and fall velocity, produce similar looking VAD profiles. In general, both occur

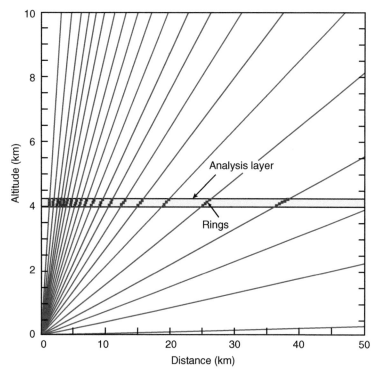

Figure 11.20 Extended VAD analysis is performed by analyzing data from many rings within a layer using scans at a large number of elevation angles. The technique allows recovery of particle fall velocity and divergence

simultaneously, so they cannot be easily sorted out with one equation. The **EVAD technique** provides a method to separate these quantities.

Let us rewrite Eq (11.44) as

$$\frac{a_0}{r \cos \alpha} = -W \frac{2}{r} \tan \alpha + D_v \tag{11.45}$$

where D_v is the divergence. Note that this equation has the form of a straight line $y = Wx + D_v$, where W is the slope of the line and D_v is the intercept. EVAD analysis requires volume scans consisting of a range of elevation angles (Figure 11.20). The measurements from all rings within predefined layers are then used to obtain a number of values of y and x. Each ring contributes one data point within a layer that is used in a least squares regression to obtain the equation of the straight line and thus D_v and W within the layer.

Once a vertical profile of divergence is obtained, appropriate boundary conditions can be specified to calculate vertical motion (e.g., zero vertical velocity at the base of the lowest elevation scan and at the top of the cloud) using the mass continuity equation

$$\frac{\partial(\rho w)}{\partial z} = -\rho \left(\frac{\partial u}{\partial x} + \frac{\partial v}{\partial y} \right) \tag{11.46}$$

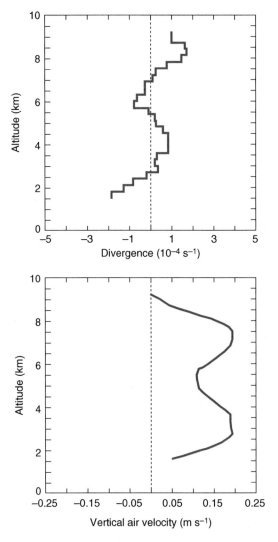

Figure 11.21 Profiles of divergence, vertical air velocity, and hydrometeor terminal velocity recovered from a radar scanning in EVAD mode during a winter ice storm in Illinois (From Rauber, R.M., Ramamurthy, M.K., and Tokay, A. (1994) Synoptic and mesoscale structure of a severe freezing rain event: The St. Valentine's Day ice storm. *Weather Forecasting*, **9**, 183–208. © American Meteorological Society, used with permission)

provided a profile of atmospheric density

$$\rho(z) = \rho_0 \exp(-z/H) \tag{11.47}$$

is available from a sounding. In this equation, H is a constant known as the scale height. We will discuss this procedure more thoroughly in the next chapter. Once one has an estimate of the vertical air motion, w, the particle fall velocity can be obtained from $w_t = W - w$. Figure 11.21 shows an example of these fields recovered using EVAD analysis during a disastrous ice storm over central Illinois.

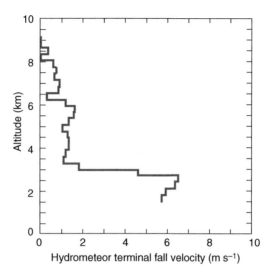

Figure 11.21   (*Continued*)

Summarizing, VAD analysis is a technique that uses the radial velocities in rings at constant elevation around a radar to determine profiles of kinematic properties of the wind over a radar site, such as the wind direction and speed, the magnitude of deformation, and the axis of dilatation. The area over which these variables are averaged is the diameter of the rings at the elevation where they occur. To do the calculations, the winds must vary no more than linearly across the radar domain, and there must be sufficient echo coverage for the ring to be coherent. Otherwise, VAD wind recovery will be less accurate. VAD only requires one elevation angle. Low elevation angles, because of their slow ascent, give high vertical resolution, but limited depth coverage. High elevation angles have lower vertical resolution but cover the full depth of a cloud.

EVAD analysis separates the divergence from the fall speed component in the zeroth harmonic equation. With EVAD, one can obtain a vertical profile of divergence. With this and assumed boundary conditions, one can integrate the continuity equation to obtain a profile of the vertical air motion. Subtracting that from the total particle fall velocity yields the terminal fall velocity of the precipitation particles in still air.

In the next chapter, we will expand our analysis of wind retrieval to include two or more radars. With multiple Doppler analysis, we will see that it is possible to map out the wind fields at very high resolution throughout a large volume within a storm.

Important terms

backing	kinematic properties
cold air advection	phased array antenna
divergence	radial velocity couplet
EVAD technique	radial velocity patterns
kinematic	shearing deformation

stretching deformation

vorticity

translation

warm air advection

veering

wind profile

velocity–azimuth display

wind profiler

Review questions

1. How do the five kinematic properties of the wind modify the fluid elements upon which they act?

2. The following equation shows that the u component of the wind can be expressed as a function of five basic kinematic properties:

$$u = -\frac{1}{2}\left(\frac{\partial v}{\partial x} - \frac{\partial u}{\partial y}\right)y + \frac{1}{2}\left(\frac{\partial u}{\partial x} - \frac{\partial v}{\partial y}\right)x + \frac{1}{2}\left(\frac{\partial u}{\partial x} + \frac{\partial v}{\partial y}\right)x + \frac{1}{2}\left(\frac{\partial v}{\partial x} + \frac{\partial u}{\partial y}\right)y + u_0$$

$$\quad\quad A \quad\quad\quad\quad\quad B \quad\quad\quad\quad\quad C \quad\quad\quad\quad\quad D \quad\quad E$$

Which terms are associated with frontogenesis? Which term will appear as a constant offset while applying the VAD technique (i.e., will not be a function of azimuth angle)? Which term is not calculated using VAD or EVAD? What term is calculated operationally by the National Weather Service using VAD?

3. The low-level jet is a common weather feature on the southern plains of the USA. Assume that within the scan range of a WSR-88D radar that winds are from the south at all levels, the surface wind is $10\,\text{m s}^{-1}$, the winds above 4 km are $15\,\text{m s}^{-1}$, and winds in the core of the low-level jet at 2 km are $30\,\text{m s}^{-1}$. Sketch with colored pencils how a WSR-88D 0.5° scan radial velocity display would appear in this situation, assuming 100% echo coverage.

4. Using colored pencils, draw a sketch of convective scale rotation as it would appear on a radar radial velocity display if (a) the radar was 100 km west of the center of rotation, (b) the radar was 100 km east of the center of rotation, (c) the radar was 100 km north of the center of rotation, and (d) if the radar was 100 km south of the rotation.

5. Search the internet for "wind profilers." Determine what common wavelengths are used for wind profilers in atmospheric research and operations.

6. If the antenna of a WSR-88D could be controlled so that it alternately pointed north at 75° elevation, then upward, then east at 75° elevation, could the resulting Doppler radial velocity data be used to derive a vertical wind profile?

7. Both wind profilers and WSR-88D radars derive vertical wind profiles. If the two radars were placed side by side, how would you expect the recovered wind profiles to differ on (a) a clear day in winter, (b) a clear day in summer, (c) a day with widespread precipitation, (d) a summer day with scattered showers, and (e) a winter day with a mid-level cloud layer?

8. Would the VAD technique for wind profile recovery work well when a strong cold front was passing directly across a WSR-88D radar site? Why or why not?

9. Data on constant elevation rings are used in VAD to derive wind profiles. What types of weather conditions would lead to poor wind recovery when using the VAD technique?

10. The EVAD scan sequence requires a number of elevation angles to independently estimate total particle fall velocity and divergence. Why are so many elevation angles required?

Challenge problems

11. Shown below is a WSR-88D radial velocity display of the 1.5° scan from KIWX during a winter storm. The outer edge of the echo return is at an altitude of 4 km. Draw a vertical wind profile similar to those appearing on Figure 11.13 based on this radial velocity display.

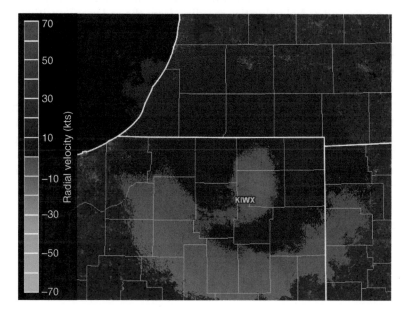

12. Assume that a $10\,\mathrm{m\,s^{-1}}$ westerly wind is present at 2 km altitude above a three-beam wind profiler, with the slant beams oriented at an elevation angle of 75° along north–south and east–west axes. Assume further that at 2 km, in the absence of rain, the vertical air velocity, $w = 0\,\mathrm{m\,s^{-1}}$.

   a. What is the radial velocity measured in each of the three beams at 2 km?

   b. A rain shower moves over the profiler. During the shower, rain is detected in all three beams, and the vertical beam measures $W = -5\,\mathrm{m\,s^{-1}}$. What is the radial velocity detected at 2 km in the other two beams.

c. Later, the shower moves off the profiler site such that rain is detected in the slant beams, but not the vertical beam. Calculate the error in the u and v wind components at 2 km that will occur because of the rain.

13. Wind profiles derived from VAD analysis or wind profilers are sometimes plotted such that time runs backwards; that is, the latest profiler is on the left side of the plot and the earliest on the right. What might be the advantage of such a plot?

14. A WSR-88D is conducting a VAD scan at an elevation angle of 45° on a rainy day. Measurements from another source indicate that the atmosphere is nondivergent in the area of the scan. There is also no evidence of deformation flow anywhere near the radar. In fact, the winds are constant at any altitude across the scanned region. On a ring at an altitude of 3 km, the measured radial velocity was found to conform to the following functional relationship:

$$v_r = -3 - 15 \cos(\beta)$$

where β is the azimuth angle, the radial velocity is in meter per second, $\beta = 0$ is north, and radial velocities receding from the radar are positive.

a. What is the horizontal wind speed at 3 km?

b. What is the horizontal wind direction at 3 km?

c. What is the total fall velocity (W) of the particles at 3 km?

d. Is the ring above or below the bright band?

15. Examine the radial velocity pattern from a WSR-88D radar that currently has active weather and widespread echo coverage. Determine, using the techniques described in Section 11.3, what the approximate wind profile is likely to be. Now use the internet to locate a series of VAD wind profiles from the same WSR-88D radar site at the same time. Does your qualitative wind profile agree with the quantitative VAD-derived profile?

12

Multiple Doppler Wind Retrieval

Objectives

By the end of this chapter, you should understand the following:

- Important considerations in network design for multiple Doppler wind retrieval.
- Sampling issues that must be considered to insure accurate wind recovery.
- Problematic issues with Doppler radial velocity data that must be dealt with before accurate retrieval of winds are possible.
- Basic methods of interpolation of radial velocity data from spherical to Cartesian coordinates.
- The method to calculate Cartesian wind components from radial velocity measurements.
- Errors associated with the recovery of Cartesian wind components and their relationship to the geometry of sampling.
- How dual-Doppler lobes are calculated and what they represent.
- How vertical air motion is retrieved from integration of the continuity equation and the consequences of choosing a particular integration method.

12.1 Introduction

Doppler radars are sometimes deployed in a network or on aircraft in such a way that the measured radial velocities can be used to derive three-dimensional wind fields (u, v, w) at very high resolution. At present, these **multiple Doppler wind syntheses** are largely limited to specialized field campaigns. Multiple Doppler wind syntheses can produce remarkable views of storm structure, such as the vertical cross section through a squall line shown in Figure 12.1, which can ultimately lead to a better understanding of storm processes. In this chapter, we examine basic principles of multiple Doppler analysis, focusing on both the methods and limitations associated with determining winds from a network of Doppler radars.

12.2 Network design and deployment

Networks of Doppler radars have been used to study a wide range of atmospheric phenomena that occur on spatial scales ranging from a few kilometers to about

Radar Meteorology: A First Course, First Edition. Robert M. Rauber and Stephen W. Nesbitt.
© 2018 John Wiley & Sons Ltd. Published 2018 by John Wiley & Sons Ltd.
Companion website: www.wiley.com/go/Rauber/RadarMeteorology

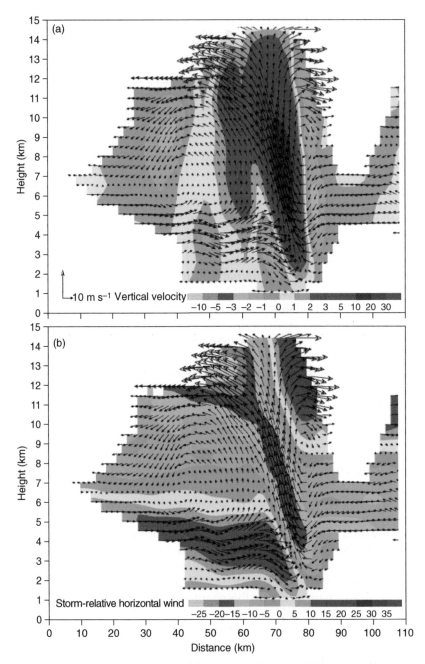

Figure 12.1 Winds (arrows), vertical air motion (a, colors), and horizontal air flow (b, colors) in a squall line that occurred on June 29, 2003 during the Bow Echo and Mesoscale Convective Vortex Experiment (BAMEX)

100–200 km. Radar network design and deployment must take both the spatial and temporal scales of these phenomena into careful consideration so that the retrieved wind fields represent the fundamental features that characterize the phenomena.

12.2.1   Meteorological considerations

The structure and temporal behavior of the atmospheric phenomena to be studied normally places strong constraints on network design. Consider, for example, a supercell thunderstorm. A supercell normally evolves rapidly (minutes), has rapid forward movement (10–25 m s^{-1}), is deep (~12–16 km), and occupies a small geographic area (~25 × 25 km). Air in the storm's updraft can rise from the surface to the tropopause in about 10 min, and the storm typically contains sharp gradients of reflectivity and velocity. The core updraft may have a horizontal scale as small as 1–4 km, and the tornado emerging from the storm's wall cloud may be only 100 m wide. These storms impose very severe challenges on observational strategies where updraft scales smaller than 1 km must be resolved, demanding extremely high spatial sampling resolution. The full volume of the storm must be scanned in under a minute for adequate temporal resolution. The storm must pass through a specific, relatively small area where the radars are positioned to observe it.

Contrast this with a local microburst emerging from a cumulus congestus cloud. Microbursts are strong outflows of wind at the surface that form when downdrafts reach the ground and spill out beneath convective clouds. Microbursts occur on very short time scales, usually a few minutes. The core of the downburst covers a small area (~5 × 5 km), and the outflow is generally quite shallow. Accurate, high-resolution measurements in the boundary layer must be available to adequately resolve the winds. Due to Earth curvature, the radars must be located close to the microburst to view boundary layer flows. Unfortunately, concentrating the radars near each other to observe the boundary layer significantly reduces the probability that a microburst will occur anywhere near the radars.

As a third example, consider a rainband embedded in a cyclonic storm. Rainbands can extend hundreds of kilometers along fronts, but only a few tens of kilometers or less across fronts. Rainbands generally have lifetimes of 1–2 h, although much shorter and longer lifetimes are common. Their depth is of the order of a few kilometers, but their strongest circulations are often above a frontal surface, with circulations tied to larger scale phenomena such as lower and upper tropospheric jet streams and fronts.

Clearly, distinctly different scanning strategies would be necessary to insure adequate data collection in each of the three phenomena discussed above to resolve their kinematic structure. In general, before designing a multiple Doppler radar network, it is essential to take into consideration the temporal and spatial scales associated with the atmospheric features to be studied.

12.2.2   Sampling limitations

Consider the simple sine wave in Figure 12.2. Let us suppose that the curve represents vertical velocity within a thunderstorm that contains two updrafts and downdraft

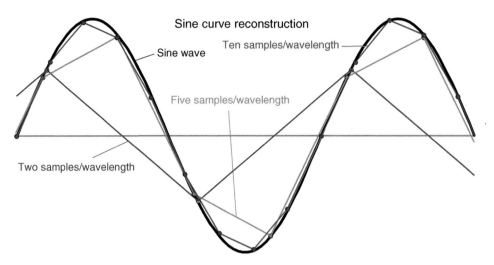

Figure 12.2 Sine wave (black) and a reconstruction of the sine wave using two samples per wavelength (blue), five samples per wavelength (orange), and ten samples per wavelength (red)

between them. Suppose further that we could directly measure the vertical velocity. How many evenly spaced samples would we need to obtain to resolve the updraft and downdraft amplitude and phase?

As shown in Figure 12.2 by the blue line, the phase, that is, whether it is an updraft–downdraft couplet, can be resolved with two samples per wavelength (the total width of one updraft and its neighboring downdraft). Two samples, however, give us no information whatsoever about the strength of the updraft and downdraft. To resolve the intensity, we obviously need many more samples. In fact, studies have shown that 10 independent samples per wavelength (the red line on Figure 12.2) are required to ensure approximately 90% recovery of flow amplitude.

With this in mind, consider a radar scanning two thunderstorms, each 15 km deep but located at different distances, roughly 20 and 70 km down range (Figure 12.3). Let us suppose that the closer storm's dimensions require scanning 100° of azimuth and 50° of elevation to cover the storm completely. Storms evolve rapidly, so to obtain a "snapshot" of the flow, the scanning must be completed quickly. Studies have shown that a reasonable time frame to complete a scan is no more than about 100 s. To cover

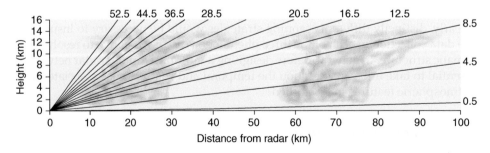

Figure 12.3 Two thunderstorms located at different distances from a radar. The black lines denote elevations required for the radar to sample the nearby storm within a 100 s time period

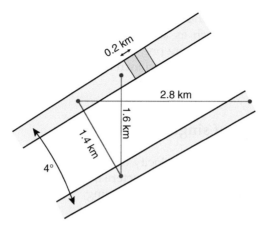

Figure 12.4   Radar beam and range gate spacing at 10 km height and 20 km slant range for a typical Doppler radar attempting to scan a thunderstorm over an elapsed time of 100 s. The angular separation between beams is 4° and the range gate spacing is 0.2 km (Adapted from Carbone, R.E., Carpenter, M.J., and Burghart, C.D. (1985) Doppler radar sampling limitations in convective storms. *J. Atmos. Oceanic Technol.*, **2**, 357–361. ©American Meteorological Society, used with permission)

the entire storm, given a $16°s^{-1}$ scan rate (a typical rate used in many radars) and allowance for scan turnaround time, at best a radar would have to use elevation angles separated by 4° (e.g., the beams shown in Figure 12.3) and cover only 25% of the solid angle of the closer storm, assuming the radar has a 1° beamwidth. Although the range gates (pulse volumes) may have 200 m spacing, the horizontal spacing between scan planes would be 2.8 km and the vertical spacing 1.6 km (Figure 12.4). If the storm updraft is 2 km wide, the horizontal spacing would not even meet the two-point per wavelength criteria, much less than the 10 points required to recover the updraft amplitude. The distant storm fares somewhat better because the small solid angle of scan required to contiguously sample the storm volume would allow more elevation angles in the same time frame. However, in this remote proximity situation, the radar beam acts as an effective low-pass filter. Scales of motion less than 4 km or so are suppressed. The filtering significantly alters the true values of the updraft, widening its structure and weakening its amplitude.

Sampling problems of a similar nature must be considered and understood for any type of phenomena that one wishes to observe with multiple Doppler radars, particularly when derivative and integral quantities such as vorticity, divergence and vertical velocity are to be calculated. Otherwise, a set of observations may yield poor, or worse yet, good looking, but completely erroneous results.

12.2.3   Siting and logistics

Some mesoscale meteorological phenomena are tied to geographic features, such as orographic clouds or lake-effect storms, and a network can be placed in advance with knowledge of the geography and topography. Many important phenomena, such as thunderstorms and rainbands, occur in random locations and require a rapidly deployable radar network to increase the probability of sampling. The presence of ground clutter, blockage, and radio interference can complicate deployments. Other logistical problems, such as access to roads, power, communications, and property, can ultimately limit where one can put radars. Careful planning again is required to minimize these problems and increase the probability of success during a deployment.

12.3   Characteristics of single Doppler data

Before considering fields of radial velocity data and how they are synthesized, it is worthwhile to step back and examine characteristics of raw data to get a better appreciation of problems encountered in data processing.

12.3.1   Geographic location of a range gate

A simple elementary question that must be answered before a multiple Doppler wind synthesis is possible: "is the antenna pointing where it is indicating that it is pointing?" The answer, unfortunately, is generally "no," particularly for research radars that are sited temporarily for a particular project. The pointing problem is not so serious if the purpose is to analyze data from a single radar. However, when calculating wind vectors from two or more radars, position errors can be quite serious, particularly in regions of strong wind velocity gradients. Unfortunately for the analyst, this is usually where all the interesting meteorology is occurring.

Azimuth, elevation, and range information must be calibrated by pointing the antenna at known ground targets and maximizing the return from those targets, and/or pointing at the sun (a strong source of microwaves due to blackbody radiation), as the position of the sun is known as a function of time. In temporary deployments, the ground may shift slightly under a radar's weight, causing the radar to misalign. This is usually checked several times during a project, and when it occurs, data must be adjusted after analysis for any misalignment that is discovered. Other problems such as antenna oscillation can occur, which can also lead to pointing ambiguity. Rapid deployments, such as those conducted by researchers studying tornadoes with radars, involve a quick setup, operation, and departure, often with strong winds against the antenna during the sampling period. It is particularly important to obtain an antenna position calibration if multiple Doppler analyses are to be performed.

12.3.2   Characteristics of raw data

Figure 12.5 shows the reflectivity and radial velocity data from a single beam from a Doppler radar viewing rain cells over the ocean. Note the behavior of the radial velocity data in the lower panel. In regions where echoes exceed the minimum detectable reflectivity, the velocity profile is slowly varying as would be expected. In regions of noise, the velocity is randomly distributed across the Nyquist interval. Note the edges of the signal associated with the meteorological target. Where are they? As the signal-to-noise ratio becomes smaller, the radial velocity becomes noisy. Edges

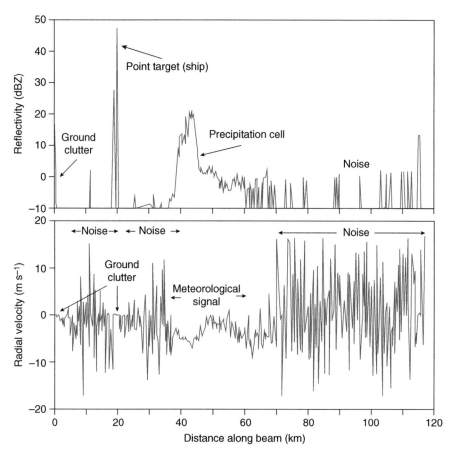

Figure 12.5 Reflectivity and radial velocity measurements from a single beam of data from a Doppler radar viewing rain cells over the ocean (Courtesy L. Jay Miller, with changes)

of storms are sometimes difficult to identify. Depending on how one thresholds the data, noise (random velocities) can creep into the analysis. On the other hand, too high of a threshold can mask the true meteorological signal. Data along a second beam in Figure 12.6 shows the types of sharp gradients one might encounter in reflectivity and velocity. This can be a true meteorological signal. One might note here that conversion of such data to a Cartesian or other coordinate system could weaken these true gradients considerably depending on the interpolation scheme, as data might be averaged across several range gates. Figure 12.6 also shows ground clutter (note the high reflectivity and zero radial velocity), a velocity fold, and strong shear in the vicinity of a precipitation cell. Note that a noise component exists throughout the data, even in regions of strong signal. These high-frequency fluctuations, if carried through the multiple Doppler analysis, will be amplified when a derivative quantity such as convergence is calculated. It is very important to filter these fluctuations while retaining the primary characteristics of the data. One way to filter the data is by range averaging—for example, a moving average of three range gates.

The magnitude of these fluctuations becomes apparent by applying a simple three-point "top-hat" filter to the unfolded velocity data and subtracting the filtered

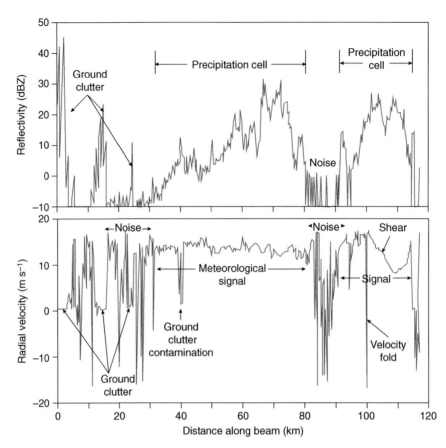

Figure 12.6   Reflectivity and radial velocity measurements from a second beam of data from the same Doppler radar as in Figure 12.5, but at a different location (Courtesy L. Jay Miller, with changes)

data from the original data to obtain the residual (Figure 12.7). The high-frequency noise component contained in the velocity measurements is of the order of $\pm 0.5\,\mathrm{m\,s^{-1}}$. The resulting velocity field after the weak filtering appears much smoother, but still retains nearly all of the amplitude of the original data. High-frequency variations can be expected in the raw data because of variations in the signal-to-noise ratio, variable distributions of velocity and reflectivity in the resolution volume, limited averaging times, and contributions to the total received power from sidelobes, non-meteorological targets, and targets beyond the maximum unambiguous range. Figure 12.7 suggests that a small amount of filtering of the original radial velocity data before synthesis is required to reduce noise in the derived wind fields. Averaging in range has another benefit: it creates averaging volumes that have the dimensions in range that more closely match those in azimuth and elevation. For these reasons, it is common to perform a small amount of filtering of the raw data in the range direction before synthesis of the wind fields.

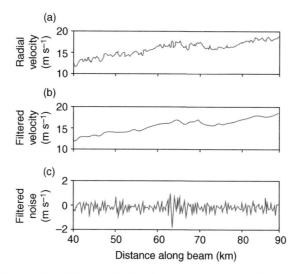

Figure 12.7   Unfiltered (a) and filtered (b) Doppler radial velocity data from a single beam in a rain cell using a three-point, top-hat filter. Panel (c) shows the residual obtained by subtracting the filtered data from the unfiltered data (Courtesy L. Jay Miller, with changes)

12.3.3   Ambiguities and Doppler radar data editing

Doppler radar data collected from field experiments are contaminated with artifacts and ambiguities. These must be removed before wind synthesis by editing the data. Examples of artifacts include (1) ground clutter, sea clutter, beam blockage, and noise; (2) range and velocity ambiguities; and (3) echoes associated with sidelobes and multiple scattering.

Ground clutter, sea clutter, beam blockage, and noise

The signal returned from ground objects, such as terrain, mountains, buildings, or trees, is called **ground clutter**. The signal returned from ocean waves and sea spray is called **sea clutter**. Ground clutter is usually characterized by a strong reflectivity factor and a near-zero Doppler velocity. At least this is true if the ground clutter is not mixed with weather echoes. When that happens, the Doppler velocity will be biased toward zero depending on the strength of the weather echo. It is generally not difficult to recognize ground clutter, particularly if one looks at animations of more than one volume. Meteorological echoes will move while ground clutter echoes remain fixed. Figure 12.8 shows an example of ground clutter and its effect on the radial velocity field, as well as examples of **beam blockage**. Ground clutter "maps" made by using the radar during fair weather can also help locate ground clutter. Sea clutter is more difficult to identify because the Doppler shift is generally nonzero and can often be close to the value of the true wind because of wave movement.

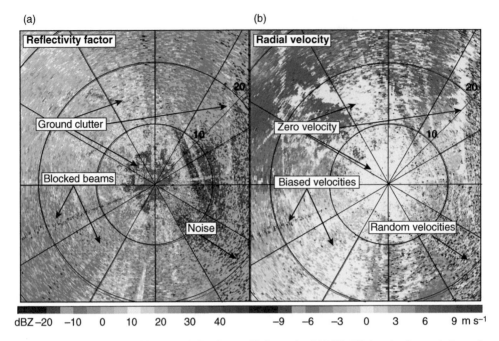

Figure 12.8 Reflectivity (a) and radial velocity (b) from the NCAR CP-4 radar located along the north Florida coast during the Convection and Precipitation/Electrification Experiment (CaPE) illustrating ground clutter, blocked beams, and noise (Image courtesy Nolan Atkins)

However, in short time periods, sea clutter areas tend to remain stationary relative to the moving weather echoes. Extended ground objects, such as groves of trees or mountains, can block radar beams. The energy transmitted past these objects is either severely reduced or totally lost and often characterized by streaks of velocities different from the surrounding weather returns.

Noise is contributed from several sources, such as the background radiation of the atmosphere and heat generated by the radar system. Noise limits the sensitivity of the radar to detecting signals. The return power of weather echoes is usually much higher than the noise level. However, the noise level imposes a problem in detecting very weak echo regions such as nonprecipitating clouds or clear air returns, where the returned power is often very weak. Noise occurs with random phase and the radial velocities measured in regions of noise are uniformly distributed across the Nyquist interval. Regions with radar echoes characteristic of noise appear in Figure 12.8. Regions of ground clutter, blocked beams, and noise all must be eliminated from the data before performing a wind synthesis using multiple Doppler techniques.

Range and velocity ambiguities

The sources of range and velocity ambiguities have been discussed in previous chapters. Here, we consider what the echoes look like on a radar display, and how they may influence multiple Doppler wind retrieval. As a radar interprets all return signals as located between the radar and the maximum unambiguous range, r_{max},

(a)   (b)

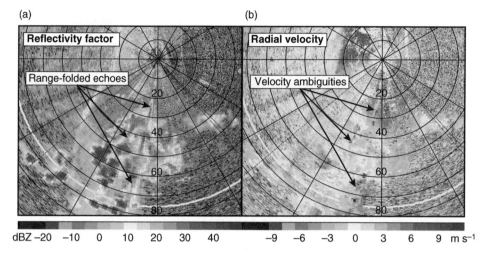

Reflectivity factor

Range-folded echoes

Radial velocity

Velocity ambiguities

dBZ −20 −10   0   10   20   30   40   −9   −6   −3   0   3   6   9 m s⁻¹

Figure 12.9   Reflectivity (a) and radial velocity (b) from the NCAR CP-4 radar located along the north Florida coast during the Convection and Precipitation/Electrification Experiment (CaPE) illustrating range folding effects (Image courtesy Nolan Atkins)

the second-trip echo at distance r beyond r_{max} will appear at distance $r - r_{max}$ on the radar display. The processor will correct the power of a returned signal by a factor of r^2 when calculating reflectivity; however, the second-trip echo will only be corrected by a factor of $(r - r_{max})^2$. The reflectivity factor from a second-trip echo will therefore be much weaker than its true intensity due to improper range correction. However, a second-trip echo from a distant thunderstorm can still have a significant reflectivity factor. With second-trip echoes, the solid angle of an echo is preserved. As the distance is changed from r to $r - r_{max}$, conservation of the solid angle in the azimuthal direction results in a narrow and elongated-looking echo on a PPI scan (see Figure 12.9), whereas conservation of the solid angle in the vertical direction results in a shallow appearance in an range-height indicator (RHI) scan. A second-trip echo should appear only in the lowest few degrees of scanning. The radial velocity of a second-trip echo is the radial velocity of the distant storm. Regions with these velocities must therefore be removed before multiple Doppler wind retrieval is carried out.

Ambiguity also occurs when the absolute value of the radial velocity of targets exceeds the Nyquist velocity. A general rule in recognizing velocity folding is to look for a sharp jump of velocity from one end of the spectrum into the other end. Figure 6.6 showed an example from the WSR-88D radar at Norman, Oklahoma. Folded radial velocities must be unfolded to their correct velocities before calculating wind components using multiple Doppler analysis techniques.

Echoes associated with sidelobes and multiple scattering

Antenna beam patterns and their associated sidelobes were discussed in Chapter 5. Recall that in most radars, the first sidelobes are ~20–25 dB lower in power than the main lobe and 2–6° away from the beam axis. The returned power from targets illuminated by the sidelobes will be interpreted as part of the main lobe. Problems

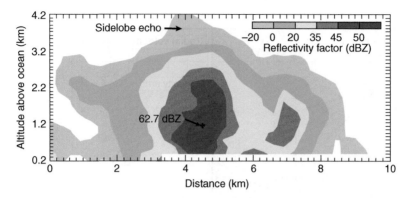

Figure 12.10 Reflectivity from the NCAR CP-4 radar illustrating a sidelobe echo above a high-reflectivity core within a rainband observed during the Hawaiian Rainband Project

arise when the main lobe is pointed toward a weak echo or echo-free region, while the first sidelobe is pointed at a strong reflectivity region. Such a situation occurs at the edges and tops of thunderstorm echoes. Echoes associated with sidelobes therefore tend to appear on the flanks of strong echo regions. The echoes are much weaker (40–50 dB down from the reflectivity measured by the main lobe when it is pointed toward the strong echo region because of two-way propagation), but the velocity remains unaffected—it has the same value as the strong echo region. Figure 12.10 shows the reflectivity pattern in a rainband off the Hawaiian coast. The vertically extended echo at the top of the cloud located above the strong echo region is entirely due to sidelobes striking the high-reflectivity region in the middle of the rainband.

Another type of anomalous echo occurs with hail. A spike-shaped radar reflectivity and Doppler velocity signature is sometimes observed that extends outward radially behind a high-reflectivity region (usually larger than 55 dBZ). Called a **three-body scatter spike** (TBSS) or **flare echo**, it results from a three-way scattering process, where hail scatters energy toward the ground, part of the signal is then reflected back to the hail, and then scattered back to the radar. Due to the added distance of travel, the echo appears behind the storm as shown in Figure 12.11. Strong negative velocities are associated with the hail spike due to the large fall velocity of hail.

12.4   Procedures for multiple Doppler syntheses

12.4.1   Interpolation of data from spherical to Cartesian coordinates

Radar data are collected in a **spherical coordinate system** (r, α, β). Analysis and interpretation of data, however, are best done in a **Cartesian coordinate system** (x, y, z), particularly the calculation of gradient quantities such as divergence and integral quantities such as vertical velocity. One must therefore transform the radar data from spherical to Cartesian coordinates. Although this might seem a straightforward task, there are different approaches to the problem of transforming the data that may produce different results. In general, interpolation of radar data to a point in Cartesian space from surrounding points containing data in spherical space involves two

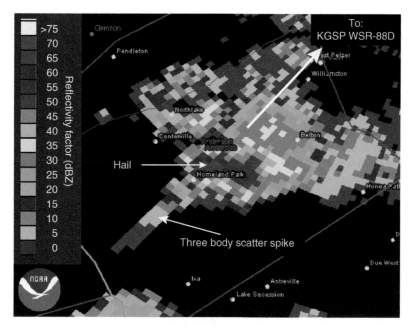

Figure 12.11   Reflectivity from the Greer, South Carolina KGSP WSR-88D radar illustrating a three-body scatter spike

steps: (1) choosing which points in spherical coordinates will be used in the interpolation and (2) determining the method, or weighting function which will be used to calculate the value at the Cartesian point. Also, one must decide if the analysis will be limited to interpolation (Cartesian point surrounded by radar data) or if extrapolation will be permitted (Cartesian point not surrounded by radar data in all quadrants).

The simplest, and least accurate, method is the "closest point" approach. With this approach, a Cartesian point is assigned the value of its nearest neighbor in spherical space. This approach may work with very high-density data in spherical space but degrades rapidly as the density of points available becomes smaller.

A more accurate approach is bilinear interpolation. With a bilinear interpolation, a maximum of eight data points in spherical space that surround a given Cartesian point can be used to estimate the value at the Cartesian point (see Figure 12.12).

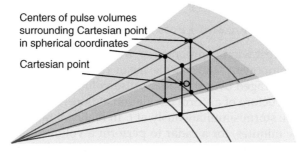

Figure 12.12   Illustration of bilinear interpolation, where eight points in a spherical coordinate system are used to calculate a value at a point in a Cartesian coordinate system

Bilinear interpolation to the Cartesian point is done in a series of steps, with linear interpolations first done in range and azimuth and then in elevation. Problems arise if not all eight points have valid data (e.g., near-echo boundaries and ground clutter). The algorithm has to decide how to assign values in these cases (e.g., use closest point or some other method). If all eight points are invalid, no number can be assigned to the Cartesian point.

Another approach is to use the concept of a **radius of influence**. The user selects a distance about the Cartesian point of interest. For example, one might choose all data points in a disc-shaped space within a distance from a Cartesian point to use in the interpolation scheme. The user, for example, may select $\Delta x = 2$ km, $\Delta y = 2$ km, and $\Delta z = 0.2$ km. Then all data points within

$$R = \left[\Delta x^2 + \Delta y^2\right]^{1/2} \tag{12.1}$$

within a 0.2 km thick layer will be used in the interpolation. The contribution of each point in the chosen space to the final value in Cartesian space is controlled by a **weighting function**. It is common to use weighting functions that give more weight to the spherical points close to the Cartesian point and less weight to points near the edge of the radius of influence. For example, a scheme called the **Cressman weighting function** takes the form

$$W_g = \frac{R^2 - r_p^2}{R^2 + r_p^2} \tag{12.2}$$

where r_p is the distance from the Cartesian point to the point in spherical space. At the edge of the radius of influence, $r_p = R$ and $W_g = 0$. At the Cartesian point, $r = 0$ and $W_g = 1$. One can see that extrapolation is possible by selecting the appropriate radius of influence. The approach using a radius of influence has the property of smoothing the field of data, effectively acting as a low-pass filter and further reducing the impact of noise.

12.4.2   Transformation of radial velocities to orthogonal particle motion components

As noted earlier, each radar collects data in a spherical coordinate system with coordinates (r, α, β). Once we have interpolated the data to a Cartesian system, each point in the system has a value of the radial velocity which can be related to the orthogonal velocity components $(u, v, w + \overline{w}_t)$ by the equation

$$v_r = u \sin(\beta) \cos(\alpha) + v \cos(\beta) \cos(\alpha) + [w + \overline{w}_t] \sin(\alpha) \tag{12.3}$$

where u is the east–west wind component, v, the north–south wind component, w, the vertical wind component, and $\overline{w}_t$, the mean reflectivity-weighted fall velocity of the particles, all averaged over the radar pulse volume. Strictly speaking, to analyze the kinematic structure of a storm circulation at time t, data should be collected at all locations in the storm simultaneously at t. In reality, it takes a finite amount of time, usually a few minutes, for a radar to perform a volume scan of a region of a storm. During that time, the storm is moving. For example, if a scan starts at $t = 0$ at the base of a storm moving at 10 m s^{-1} and reaches the top of the storm at $t = 120$ s, the storm will have moved 1.2 km during the volume scan. To compensate for storm motion during the time it takes to complete a scan, an advection scheme is required that accounts for the storm motion. This is routinely done in multiple Doppler analysis algorithms by advecting the scalar components of the radial velocity vector at each Cartesian point before synthesis. We will simplify the mathematical details here by assuming that the storm is steady during the volume scans of the radars participating in the multiple Doppler wind synthesis.

We can, in general, write an equation such as Eq. (12.3) in a simplified manner for each radar in our system as

$$v_{ri} = ua_i + vb_i + Wc_i \tag{12.4}$$

where $W = w + w_t$ and the subscripts i for the coefficients a, b, c, and v_r are different for each radar. The coefficients a, b, c are known geometric quantities, and v_r is the measured radial velocity, so in principle, one needs three equations to solve for the unknown u, v, W. In radar networks, there are often only two radars (dual-Doppler), although many experiments have used three or more radars. Let us first consider the general overdetermined solution to these equations where one has $n > 3$ radars observing the same volume. For n radars, we have a linear system of equations at time t in the form of Eq. (12.4) for $i = 1, 2, \ldots, n$.

Suppose, for example, we had seven radars viewing the same cloud system. We could choose any three of the $n = 7$ radars, find solutions to the three equations, and determine u, v, W. However, if we were to pick a different set of three from the seven and perform the same operation, would we get the exact same u, v, W at the same point in space? The answer, in general, is no. The u, v, W in Eq. (12.4) are derived from averages over different pulse volumes that are not the same for each radar (Figure 12.13).

Figure 12.13   The pulse volumes from radars viewing the same point in space from different locations will have different size and orientation, leading to differences in the solutions for the Cartesian wind components (Photo: Jeffrey Frame)

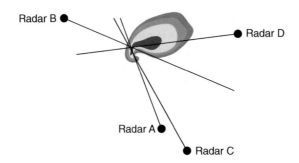

Figure 12.14 Geometry for four radars viewing a thunderstorm

In addition, the geometry of the radar triplets will be different and the ability of the radars to resolve the components of motion will change. For example, Figure 12.14 shows four radars viewing a storm. As A and C view the storm from nearly the same direction, the triplet A, B, C, would provide little more information than A and B alone. On the other hand, A, B, and D view the storm from very different angles and would provide better resolution of the individual velocity components.

For the case of more than three radars, a least squares approach to minimize the differences between the solutions for the triple radar set possibilities is used. It is rare in projects to have the luxury of more than three radars viewing a storm simultaneously. It is much more likely to have three, or two. We will consider situations where three or two radars are in the network.

For the **three-radar solution**, there is only one solution for the linear equations. The solution is

$$u = D^{-1}[v_{r1}(b_2c_3 - b_3c_2) - v_{r2}(b_1c_3 - b_3c_1) + v_{r3}(b_1c_2 - b_2c_1)] \tag{12.5}$$

$$v = D^{-1}[v_{r1}(a_3c_2 - a_2c_3) - v_{r2}(a_3c_1 - a_1c_3) + v_{r3}(a_2c_1 - a_1c_2)] \tag{12.6}$$

$$W = D^{-1}[v_{r1}(a_2b_3 - a_3b_2) - v_{r2}(a_1b_3 - a_3b_1) + v_{r3}(a_1b_2 - a_2b_1)] \tag{12.7}$$

Where the determinant, D, is given by

$$D = [a_1(b_2c_3 - b_3c_2) - a_2(b_1c_3 - b_3c_1) + a_3(b_1c_2 - b_2c_1)] \tag{12.8}$$

We have assumed above that we have three radars and that we are interested in the solution for the parameters u, v, and W. However, recall that $W = w + w_t$ is the average reflectivity-weighted total vertical velocity of the ensemble of particles within the resolution volume. In general, from the point of view of understanding the dynamics of the cloud system, we are generally more interested in the value of the vertical air motion, w. As we shall see momentarily, w can be obtained independently by invoking mass continuity and solving the continuity equation using values of u and v. With this solution, one can separate the contributions to the total vertical motion of the particles, W, into w and w_t.

How accurate is the recovery of u, v, and W using the triple Doppler solution? To estimate the accuracy, let us rewrite the geometric terms that multiply with the radial velocities in Eqs. (12.5)–(12.7) as

$$u = v_{r1}g_{u1} - v_{r2}g_{u2} + v_{r3}g_{u3} \tag{12.9}$$

$$v = v_{r1}g_{v1} - v_{r2}g_{v2} + v_{r3}g_{v3} \tag{12.10}$$

$$W = v_{r1}g_{W1} - v_{r2}g_{W2} + v_{r3}g_{W3} \tag{12.11}$$

One can then conceptually calculate the variance in u, v, W by imagining that we could take a number of realizations of the same volume of space with the particles in the pulse volumes shuffled about into different arrangements and/or could take averages over different PRFs. Each realization would give us a slightly different v_{r1}, v_{r2}, and v_{r3} and consequently a different u, v, W. The variance of u, v, W would be given by

$$\sigma^2(u) = g_{u1}^2 \sigma^2(v_{r1}) + g_{u2}^2 \sigma^2(v_{r2}) + g_{u3}^2 \sigma^2(v_{r3}) \tag{12.12}$$

$$\sigma^2(v) = g_{v1}^2 \sigma^2(v_{r1}) + g_{v2}^2 \sigma^2(v_{r2}) + g_{v3}^2 \sigma^2(v_{r3}) \tag{12.13}$$

$$\sigma^2(W) = g_{W1}^2 \sigma^2(v_{r1}) + g_{W2}^2 \sigma^2(v_{r2}) + g_{W3}^2 \sigma^2(v_{r3}) \tag{12.14}$$

It is reasonable to assume that the variances of v_{r1}, v_{r2}, and v_{r3} would, at least in a statistical sense, be identical. We therefore assume that

$$\sigma^2(v_r) = \sigma^2(v_{r1}) = \sigma^2(v_{r2}) = \sigma^2(v_{r3}) \tag{12.15}$$

We can then calculate the normalized variance of u, v, W as

$$\sigma_N^2(u) = \frac{\sigma^2(u)}{\sigma^2(v_r)} = g_{u1}^2 + g_{u2}^2 + g_{u3}^2 \tag{12.16}$$

$$\sigma_N^2(v) = \frac{\sigma^2(v)}{\sigma^2(v_r)} = g_{v1}^2 + g_{v2}^2 + g_{v3}^2 \tag{12.17}$$

$$\sigma_N^2(W) = \frac{\sigma^2(W)}{\sigma^2(v_r)} = g_{w1}^2 + g_{w2}^2 + g_{w3}^2 \tag{12.18}$$

Let us stop here for a second to remind ourselves that the g terms on the right side of Eqs. (12.16)–(12.18) are geometric terms consisting of sines and cosines of azimuth and elevation angles of the three radars. These equations then say that the variance in u, v, and W will increase over the variance in the measured radial velocities by a factor equal to the sum of the squares of the geometric terms in the solution for u, v, and W. To say another way, *the accuracy of the recovery of u, v and W depends critically on the geometry of the radar triplet relative to the storm.* This is easiest to understand if we examine the geometry for the **two-radar solution.**

The equations for just two radars are

$$ua_1 + vb_1 = v_{r1} - Wc_1 \tag{12.19}$$

$$ua_2 + vb_2 = v_{r2} - Wc_2 \tag{12.20}$$

where we are now treating W as unrecoverable as we have three unknowns (u, v, W) and only two equations. The two equation solution, with the unknown W, is then

$$u = \frac{b_2 v_{r1}}{a_1 b_2 - a_2 b_1} - \frac{b_1 v_{r2}}{a_1 b_2 - a_2 b_1} + W \frac{b_1 c_2 - b_2 c_1}{a_1 b_2 - a_2 b_1} \tag{12.21}$$

$$v = \frac{a_1 v_{r2}}{a_1 b_2 - a_2 b_1} - \frac{a_2 v_{r1}}{a_1 b_2 - a_2 b_1} + W \frac{a_2 c_1 - a_1 c_2}{a_1 b_2 - a_2 b_1} \tag{12.22}$$

We again can simplify the geometric terms for the two equation solution:

$$u = h_{u1}v_{r1} - h_{u2}v_{r2} + \varepsilon_u W \tag{12.23}$$

$$v = h_{v1}v_{r1} - h_{v2}v_{r2} + \varepsilon_v W \tag{12.24}$$

where the h and ε terms are geometric terms consisting of sines and cosines of azimuth and elevation angles. The variance of u and v is then given by

$$\sigma^2(u) = h_{u1}^2\sigma^2(v_{r1}) + h_{u2}^2\sigma^2(v_{r2}) + \varepsilon_u^2\sigma^2(W) \tag{12.25}$$

$$\sigma^2(v) = h_{v1}^2\sigma^2(v_{r1}) + h_{v2}^2\sigma^2(v_{r2}) + \varepsilon_v^2\sigma^2(W) \tag{12.26}$$

and the normalized variance is given by

$$\sigma_N^2(u) = \frac{\sigma^2(u)}{\sigma^2(v_r)} = h_{u1}^2 + h_{u2}^2 + \varepsilon_u^2\frac{\sigma^2(W)}{\sigma^2(v_r)} \tag{12.27}$$

$$\sigma_N^2(v) = \frac{\sigma^2(v)}{\sigma^2(v_r)} = h_{v1}^2 + h_{v2}^2 + \varepsilon_v^2\frac{\sigma^2(W)}{\sigma^2(v_r)} \tag{12.28}$$

As we do not know a priori the value of W, the terms containing ε_u and ε_v and W are unknown. In practice, these terms are ignored, degrading the solution for u and v. That is, it is implicitly assumed that either $W=0$, or $\varepsilon_u=\varepsilon_v=0$, or their product is small enough that the term can be ignored. This assumption is not safe at high elevation angles, particularly in strong storms where the vertical velocities can at times exceed the horizontal velocities. Although W cannot be determined, ε_u and ε_v are geometric factors that can be calculated and tested. We can think of ε_u and ε_v as the geometric factors that, when multiplied by the vertical component W, would provide the correction to the two equation solution that was obtained by ignoring the W term.

Let us now examine the behavior of the terms that contribute to the increase in the variance of u and v for the two equation solution. We can calculate the **normalized standard deviation** of the u and v components, σ_u and σ_v, by taking the square root of the geometric terms (excluding the term containing W). We can think of σ_u and σ_v as multiplicative factors to any original errors in the measured radial velocities. We have previously discussed many reasons why the original radial velocities may have errors. These include such things as contributions from noise due to low signal-to-noise power ratios, nonuniform distributions of reflectivity or radial velocity in the pulse volume, inadequate number of samples in the average, inadequate transmitter–receiver fidelity, velocity folding, contamination from sidelobes or second-trip echoes, and contamination from non-meteorological targets.

Figures 12.15 and 12.16 show plots of σ_u and σ_v for a situation where two radars are located 40 km apart with the **baseline** (the line connecting the two radars) rotated clockwise 20° with respect to true north. One can see that σ_u is greater than unity in all regions except the purple area, is minimum toward the west (and east), and increases to a factor > 5 as the baseline is approached. The standard deviation, σ_v, on the other hand, is less than unity in the purple areas north and south of the radar pair and in a small area to the east and west of the radars. The quantity σ_v is maximum along the baseline and increases with distance west and east of the radars. One can interpret these patterns by considering the degree to which each radar is actually sampling the u and v components when measuring the radial velocity. To determine

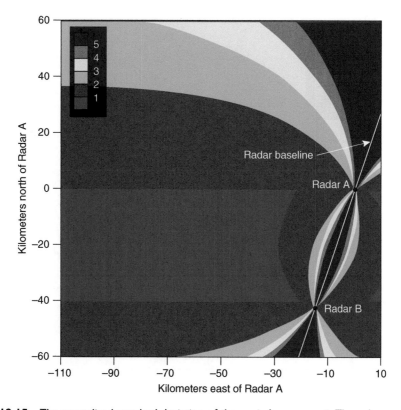

Figure 12.15 The normalized standard deviation of the u wind component. The colors on the diagram represent the increase in error in the u wind component relative to any error that may be present in the measured radial velocities (Courtesy L. Jay Miller, with changes)

the true horizontal wind vector at a point in space, measurements must be made from two different viewing directions to determine the individual wind components of the vector. The closer the angle between the beams from the two radars, δ, is to 90°, the more accurate the determination of both velocity components will be. Consider first measurements made along the baseline. When both radars point directly along the baseline, they measure the same component of the wind (ignoring for the moment the different contributions due to W at higher elevation angles). As $\delta = 0$ along the baseline, the u and v components cannot be determined. Doppler synthesis of wind fields cannot be done in this region.

Consider now regions away from the baseline. With the particular geometry of Figures 12.15 and 12.16, both radars measure the u component exactly when pointed at 270°, ignoring again the contribution of W at higher elevation angles. The radial velocity measured by each radar along this direction is the u component and therefore the multiplicative factor to the error, σ_u, is exactly unity. Between these two positions, in the purple area, the radars are pointed no more than 25° off the east–west axis and therefore primarily sample the u component. The u component is determined very well in this region, as the two radars are primarily sampling the u component from slightly different angles, leading to values of σ_u less than unity. In this same region, the v component is not well determined, as the rays are nearly normal to

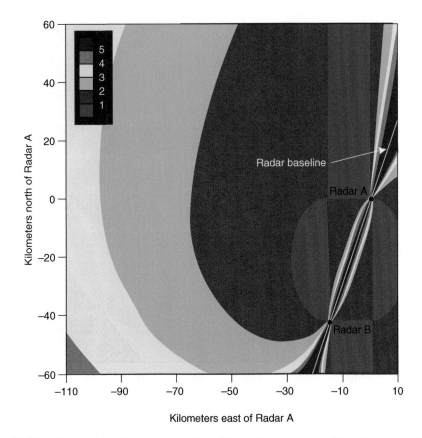

Figure 12.16 The normalized standard deviation of the v wind component. The colors on the diagram represent the increase in error in the v wind component relative to any error that may be present in the measured radial velocities (Courtesy L. Jay Miller, with changes)

the north–south direction. As the radar pair is approached from the west, the angle between the beams becomes greater and the determination of v becomes better. Correspondingly, σ_v approaches unity. The terms σ_u and σ_v are exactly unity when the beams are exactly 90° apart. Similar arguments can be made for the areas to the north and south of the radar pair. In this region, u is poorly determined and v is overdetermined. σ_v is exactly unity when either beam points directly north or south. The estimate of u becomes better as the radars are approached from the north or south because the angle between the beams becomes larger.

The normalized geometric error in the *total* horizontal wind vector (excluding the factors containing the term ε) can be obtained by calculating

$$\sigma_{\vec{V}} = (\sigma_u^2 + \sigma_v^2)^{1/2} \tag{12.29}$$

Figure 12.17 shows a plot of $\sigma_{\vec{V}}$ at 2 km elevation. Equal values of the function $\sigma_{\vec{V}}$ fall on arcs of circles whose centers lie on the perpendicular bisector of the baseline. All circles pass through both radars. The partial circles defined on Figure 12.17 by similar shades are called the **dual-Doppler lobes**. One can interpret the function $\sigma_{\vec{V}}$ again as a multiplicative factor to the root-mean-square error in the original measured radial

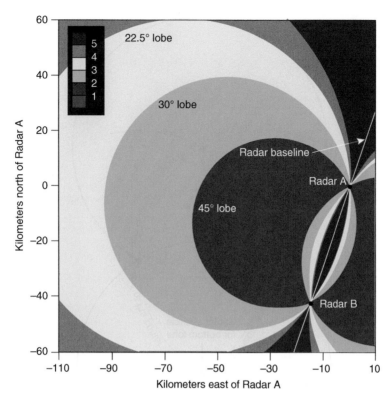

Figure 12.17 The normalized standard deviation of the horizontal wind vector from the two-radar solution. The colors on the diagram represent the increase in error in the wind relative to any error that may have been present in the measured radial velocities (Courtesy L. Jay Miller, with changes)

velocity estimates. As $\sigma_{\vec{V}}$ approaches unity, errors in the estimate of the u and v component of the wind will approach those of the original radial velocity components. This will occur in a geometric sense as the angle between the two beams approaches 90°. In fact, the beams are exactly 90° apart only along a single arc approximately in the middle of the innermost lobe. It is important to note that the *dual-Doppler lobe primary application is to outline the area where acceptable estimates of divergence, and therefore vertical motion, can be obtained*. The best estimates will be at the center of the lobes, with accuracy decreasing toward the edge of the lobes and beyond.

A simple interpretation of the geometry of $\sigma_{\vec{V}}$ is that a constant value of $\sigma_{\vec{V}}$ traces out the locus of points where the angle between the two beams, δ, is constant. The value of $\sigma_{\vec{V}}$ along the locus of points equals $\pi/(2\delta)$ with δ in radians. For example, along the innermost pair of arcs where $\delta = 45° = \pi/4$, $\sigma_{\vec{V}}$ has a value of 2. Along the next pair of arcs, $\delta = 30°$ and $\sigma_{\vec{V}} = 3$.

The characteristics of the meteorological situation one is studying, and the ultimate goals of the analysis will determine the required accuracy of the dual-Doppler wind synthesis. The more accuracy required, the tighter the requirements are on the choice of $\sigma_{\vec{V}}$ and the more limited is the area where dual-Doppler wind synthesis can be used. In general, the $\delta = 30°$ lobes are considered a good guide for most meteorological applications. These lobes can be drawn simply by drawing a circular arc through radar B with the center of the arc at radar A and vice versa, as shown in Figure 12.18.

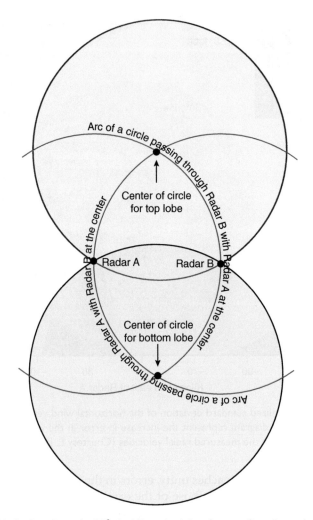

Figure 12.18 Method to draw the 30° dual-Doppler lobes for two Doppler radars

Then draw two circles centered on the arc intersections and passing through both radars.

The length of the baseline also determines the area where dual-Doppler analysis can be done with reasonable accuracy. The area comprising the dual-Doppler lobes is proportional to the square of the baseline length. The advantage of a longer baseline is that a broader area exists where reasonably accurate syntheses of the wind field can be done. The disadvantage is that the resolution of the analyses, which is determined by the longest distance from either radar, will be significantly degraded. The size and depth of the physical phenomenon to be studied ultimately determines the resolution required and, therefore, the desired size of the baseline. Typical baselines in most studies range in length from 20 to 50 km.

So far, we have ignored the terms containing ε_u and ε_v in Eqs. (12.23) and (12.24). Recall that these terms are also geometric factors contributing to the estimate of u and v. In Eqs. (12.23) and (12.24)), ε_u and ε_v are multiplied by the unknown vertical

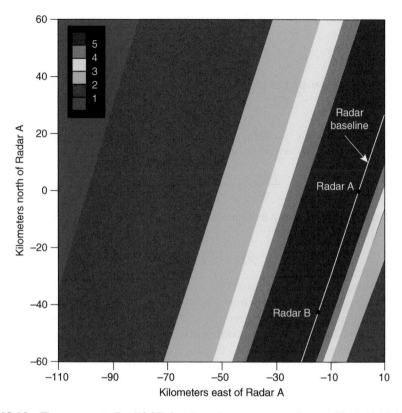

Figure 12.19   The term ε_u in Eq. (12.25) for the radar geometry in Figures 12.15–12.17 (Courtesy L. Jay Miller, with changes)

particle motion component W. For a given nonzero value of W, the estimates of u and v become poorer as ε_u and ε_v, respectively, increase closer to the baseline. Figure 12.19 shows the maximum value of ε_u at 10 km altitude. One can see that the maximum in ε_u increases toward the baseline and that the factor ε_u (and ε_v) are closely related to the elevation angle of the beam. As the elevation angle increases, the radar progressively samples more of the W component and less of u and v. The accuracy of the estimates of u and v therefore decreases at higher elevation angles. In practice, one can apply constraints on the region where dual-Doppler synthesis can be performed with required accuracy by estimating the likely maximum value of W to be encountered in the precipitation system under study from other information (such as previous studies, analysis of the stability on thermodynamic soundings, and considerations of the fall velocity of particles) and comparing the maximum value of $\varepsilon_u \times W_{max}$ or $\varepsilon_v \times W_{max}$ with typical magnitudes of u and v. One can then select to perform syntheses only in regions where the values of ε_u and ε_v are sufficiently small to insure that the accuracy required will be met in the most extreme situations. In meteorology, it is precisely these extreme cases that are usually the situations of most interest. Of course, for some applications, one need not choose the extreme value of W_{max}, as this extreme value is only likely to occur rarely and in a limited area. In general, Figure 12.19 shows that dual-Doppler analyses should not be done close to the baseline if one wants to avoid significant error associated with the fall speed of the particles.

12.4.3 Calculation of vertical motion from orthogonal wind components

To determine vertical air motion, w, (as distinct from vertical particle motion, $W = w + w_t$), the continuity equation in the form

$$\frac{\partial(\rho w)}{\partial z} + \frac{\partial(\rho u)}{\partial x} + \frac{\partial(\rho v)}{\partial y} = 0 \tag{12.30}$$

must be integrated. The term ρ in this equation is the air density, which varies exponentially with height such that

$$\rho(z) = \rho_0 \exp(-kz) \tag{12.31}$$

where k is a function of temperature and ρ_0 is a reference value at the lowest level, $z = z_0$. The integration procedure differs depending on whether one obtained the estimates of u and v from the two or three equation solution. We first consider the three equation solution. In this case, the continuity equation can be integrated from the level z_k to the level z_{k+1}:

$$\int_{z_k}^{z_{k+1}} \frac{\partial(\rho w)}{\partial z} dz = -\int_{z_k}^{z_{k+1}} \left(\frac{\partial(\rho u)}{\partial x} + \frac{\partial(\rho v)}{\partial y} \right) dz \tag{12.32}$$

Expansion of the right-hand side gives

$$\int_{z_k}^{z_{k+1}} \frac{\partial(\rho w)}{\partial z} dz = -\int_{z_k}^{z_{k+1}} \left(\rho \frac{\partial u}{\partial x} + \rho \frac{\partial v}{\partial y} + u \frac{\partial \rho}{\partial x} + v \frac{\partial \rho}{\partial y} \right) dz \tag{12.33}$$

As density is assumed to only vary in the vertical, this reduces to

$$\int_{z_k}^{z_{k+1}} \frac{\partial(\rho w)}{\partial z} dz = -\int_{z_k}^{z_{k+1}} \rho \left(\frac{\partial u}{\partial x} + \frac{\partial v}{\partial y} \right) dz \tag{12.34}$$

If we calculate the divergence and density at level k and level $k + 1$ and take the average value and apply it to the center of the layer, we can integrate the equation to give

$$(\rho w)_{k+1} = (\rho w)_k - \left\langle \rho \left(\frac{\partial u}{\partial x} + \frac{\partial v}{\partial y} \right) \right\rangle_{k+1/2} \Delta z \tag{12.35}$$

where the brackets denote an average value at the middle of the layer. Finally, solving for w

$$(w)_{k+1} = \frac{1}{\rho_{k+1}} \left[(\rho w)_k - \left\langle \rho \left(\frac{\partial u}{\partial x} + \frac{\partial v}{\partial y} \right) \right\rangle \Delta z \right] \tag{12.36}$$

In principle, one can integrate upward, downward, or in both directions level by level from the bottom to the top of the storm, provided a boundary condition is given for w at the starting point(s) of the integration. We will look at this issue after we examine the two equation solution.

When solving for u and v in the two equation solution, we explicitly ignored the terms containing the parameter ε. However, if one wishes to determine w correctly, the terms must be accounted for. The solution for u and v was given in Eqs. (12.23) and (12.24) as

$$u = h_{u1}v_{r1} - h_{u2}v_{r2} + \varepsilon_u W \tag{12.37}$$

$$v = h_{v1}v_{r1} - h_{v2}v_{r2} + \varepsilon_v W \tag{12.38}$$

Let us write that in a simpler form as

$$u = u' + \varepsilon_u W \tag{12.39}$$

$$v = v' + \varepsilon_v W \tag{12.40}$$

where we have substituted $u' = h_{u1}v_{r1} - h_{u2}v_{r2}$ and $v' = h_{v1}v_{r1} - h_{v2}v_{r2}$ for convenience.

The primed quantities denote the values of u and v estimated by assuming that the ε terms were zero. These are the values normally retained in the synthesis of the horizontal wind components. The first step to obtain the two equation solution for w is to write out u and v to include the precipitation fall speed and w such that

$$u = u' + \varepsilon_u \overline{w}_t + \varepsilon_u w \tag{12.41}$$

$$v = v' + \varepsilon_v \overline{w}_t + \varepsilon_v w \tag{12.42}$$

Although no direct measurement of the reflectivity-weighted average terminal velocity of the precipitation in the pulse volume is available, empirical relationships between Z and $\overline{w}_t$ have been derived by assuming that the drop size distribution follows a functional relationship, such as a Marshall–Palmer distribution. Such relationships can be used with the measured reflectivity factor to estimate $\overline{w}_t$ and then estimate u and v. The continuity equation

$$\frac{\partial(\rho w)}{\partial z} + \rho\left(\frac{\partial u}{\partial x} + \frac{\partial v}{\partial y}\right) = 0 \tag{12.43}$$

becomes

$$\frac{\partial(\rho w)}{\partial z} + \rho\left(\frac{\partial(u' + \varepsilon_u \overline{w}_t + \varepsilon_u w)}{\partial x} + \frac{\partial(v' + \varepsilon_v \overline{w}_t + \varepsilon_v w)}{\partial y}\right) = 0 \tag{12.44}$$

Integration of Eq. (12.44) between levels k and $k+1$ is not as simple and requires an iterative solution to obtain w. Nevertheless, it is computationally straightforward and is commonly done in dual-Doppler processing.

For an integration of Eq. (12.32) or Eq. (12.44) to proceed upward and/or downward, a boundary condition involving w must be given at every (x, y) point at, respectively, the bottom and/or top level. In principle, these boundary conditions could easily be estimated if the measurements extended from the ground to the top of the cloud. At the ground,

$$w = (u^2 + v^2)^{1/2}\frac{\Delta h_t}{\Delta n} \tag{12.45}$$

the product of the wind speed $(u^2 + v^2)^{1/2}$ and the slope of the terrain in the direction of the wind, $\frac{\Delta h_t}{\Delta n}$.

At the top of the cloud,

$$w = \frac{dh_{top}}{dt} \tag{12.46}$$

where dh_{top}/dt is the rate of change of the cloud top height from one observation time to the next. For stratiform clouds over flat terrain, $w_{top} = w_{bot} = 0$. Unfortunately, things are not so simple. Generally, both the top and bottom boundary conditions are unknown and are difficult to estimate. The problem arises because of the geometry of the radar measurement and the size distribution of particles in the cloud. Consider first downward integration from the cloud top. For downward integration to be applied, the radar must scan at least one level above the top of the cloud system to determine the position of the upper boundary. However, with a radar, the measured position of this boundary is not the true boundary of the cloud itself. Rather, it is the position where precipitation particles are of sufficiently large size and number concentration to scatter enough power so that the signal returned to the radar is above the noise level. This level of the cloud is not the top, and w is likely to have some nonzero value. For upward integration, the same problem applies in nonprecipitating regions of cloud systems (except those that contain other scatterers, such as insects). Even in precipitation areas, however, a more fundamental problem exists. Anywhere away from the radar site, the pulse volume will be a finite height above the surface of the earth. When the data are interpolated to Cartesian coordinates, the first analysis level will always be some finite height above the ground (e.g., 0.5 km). It is the vertical velocity at this level, and not the ground, that must be estimated. The difference can be appreciable in many cloud systems. In practice, the choice of boundary conditions is educated guesswork that depends on one's prior understanding of the storm system, additional measurements when available, and the physical sensibility and consistency of the final product.

Three approaches to integration are possible. As implied in previous discussions, integration of the continuity equation can be done from the top level of the domain to the bottom (**downward integration**), from the bottom to the top (**upward integration**), or simultaneously in both directions (**variational integration**). In the latter case, the two solutions must be averaged together using some prespecified weighting function.

12.4.4 Uncertainty in vertical motion retrievals

Let us now consider the behavior of the variance of w as we move away from the specified boundaries. One way to examine the behavior of $\sigma_{up}(w)$ or $\sigma_{down}(w)$ is to perform a large number of vertical integrations over a domain comprising divergence values generated using random velocities characteristic of a noise field and examine the statistics of these integrations. Figure 12.20 shows $\sigma_{up}(w)$ and $\sigma_{down}(w)$ for upward integration and downward integration. The mean w in each case, calculated from the integrations, as expected, is zero. However, the standard deviation *increases* upward away from the boundary for upward integration, whereas it *decreases* downward and is much smaller for a downward integration. In a noise field, one would expect any z level to statistically have the same characteristics as any other z level,

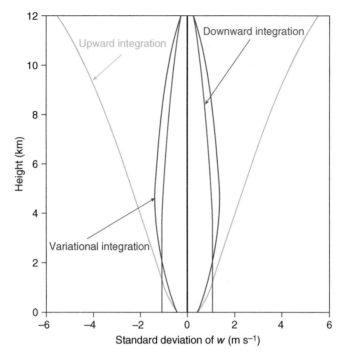

Figure 12.20 Standard deviation of w for a large number of vertical integrations of a field of random velocities characteristic of noise (Courtesy L. Jay Miller, with changes)

so the standard deviations $\sigma_{up}(w)$ or $\sigma_{down}(w)$ should be invariant with altitude. The reason that $\sigma_{up}(w)$ increases with altitude is because of the density term in the integral. As we move away from the boundary, density is decreasing, so mass moving upward from a lower level must accelerate. The variance of w therefore increases. What we are seeing in Figure 12.20 is the degradation in the solution for w due to the upward propagation and progressively greater influence of any error in the bottom boundary condition at higher altitudes. This upward propagation results because air density (the source of the exponent term) decreases with height.

Consider now the behavior of $\sigma_{down}(w)$ as we perform downward integration from the top boundary. The value of $\sigma_{down}(w)$ increases at a slower and slower rate as the integration continues downward. The influence of any error in the upper boundary condition becomes less and less as the integration proceeds. One can again understand this by considering the effect of density. As we move downward away from the upper boundary, density is increasing, so mass moving downward from a lower level must decelerate. The variance of w therefore decreases with decreasing altitude.

If one can obtain an estimate of the boundary condition for w at both boundaries, a way to minimize the errors in the solution for w is to perform a variational integration. The integration is done by integrating both upward and downward for w and then matching the solutions by using a weighting function that considers the variation in density with height. This weighting function should be chosen to optimize the solution for w based on statistics shown in Figure 12.20. Figure 12.20 shows the results of integrating a noise field using the variational integration scheme discussed above. Such a scheme will generally provide a considerable improvement in the results over the whole domain.

Let us think about these results in terms of how one should proceed with the integration to obtain the best estimate of w. Suppose that the system under investigation is deep, say 12 km, to conform to Figure 12.20. In such a case, because of the effect of density, a downward integration would statistically be likely to produce more accurate results than an upward integration over the domain from the top boundary down to a level of about 2 km. Below 2 km, the upward integration would be statistically superior. This assumes, of course, that the boundary conditions at the top and the bottom were equally well known. If, for one reason or another, one has more confidence in the boundary condition at one or the other boundaries, the height at which one integration scheme becomes superior over another will change. In general, however, one does not know either boundary condition adequately. Downward integration is generally superior to upward integration.

12.5 Summary

Multiple Doppler wind synthesis is a complicated but powerful process to recover winds in storms at very high resolution. The process begins with careful design of the observation network, with attention paid to the nature of the phenomenon to be studied and the limitations imposed by the design characteristics of each radar. In the design of the experiment, one must be careful to insure that the baseline is the correct length, and the scanning rates properly adjusted, to adequately resolve the phenomena of interest in both the horizontal and vertical direction. The data, once taken, must be carefully examined to determine if any fundamental errors exist, such as incorrect positioning of the antenna. Data must be edited to remove artifacts such as sidelobe and multiple-trip contamination and interpolated to Cartesian coordinates. After wind synthesis, the data must be carefully investigated to determine the cumulative effects of the many possible errors in analysis that were discussed in this chapter. With care, multiple Doppler analyses can allow a scientist to deduce the detailed kinematic structures of storm systems that would otherwise be impossible to attain with any other currently available instrumentation. Advanced techniques, beyond the scope of this book, are also available to perform wind syntheses. Procedures have also been developed with multiple Doppler syntheses to retrieve thermodynamic variables such as temperature and pressure perturbations. Because of this, multiple Doppler analysis has become one of the most important tools of mesoscale meteorology.

Important terms

baseline	downward integration
beam blockage	dual-Doppler lobes
Cartesian coordinate system	flare echo
Cressman weighting function	ground clutter

multiple Doppler wind syntheses
noise
normalized standard deviation
radius of influence
sea clutter
spherical coordinate system

three-body scatter spike
three-radar solution
two-radar solution
upward integration
variational integration
weighting function

Review questions

1. In summer, the sea breeze front regularly moves on shore in the late morning along the Florida coast and triggers thunderstorms inland by midafternoon. If you wished to use a pair of Doppler radars to study the sea breeze evolution, what might be some of the considerations you would have in siting the radars?

2. Based on both sampling considerations and wind retrieval techniques, is it most likely that dual-Doppler retrieval of 3-D wind fields will underestimate or overestimate updraft and downdraft intensity in thunderstorms? Explain your answer.

3. Why is it common to filter radial velocity data in range before interpolating the data to Cartesian coordinates and performing a multiple Doppler analysis?

4. What radar characteristics of ground clutter can be used to identify and remove ground clutter from a data set before dual-Doppler analysis?

5. What radar characteristic of noise can be used to identify and remove noise from a data set before dual-Doppler analysis?

6. Short-wavelength radars have the advantage that their antennas are small and can be easily mounted on trucks so that they can be used to study severe convection. From the perspective of multiple Doppler analysis, what is a disadvantage of short-wavelength radars?

7. One method to convert data in spherical coordinates to Cartesian coordinates involves use of range gates contained within a "radius of influence" surrounding a Cartesian point. What might be the advantages and disadvantages of choosing a small radius of influence? How about a large radius of influence?

8. What are the dual-Doppler lobes, and how are they related to dual-Doppler recovery of winds and vertical air motion?

9. Why is a downward integration of a radar-derived divergence field superior to an upward integration when calculating vertical air motion in a deep cloud system?

10. You are planning to perform a variational integration of a radar-derived divergence field to obtain vertical air velocity within a storm. What possible issues might you encounter when specifying the boundary conditions for vertical air velocity?

Challenge problems

11. Suppose you had a mobile Doppler radar and decided to deploy it 40 km from a WSR-88D radar and use the data from the two radars to study the circulations in a severe thunderstorm that, conveniently, is located in the center of the dual-Doppler lobe between the radars. The WSR-88D is using the VCP-121 scan. What problems might you encounter when you attempt to recover vertical motions in the storm using dual-Doppler processing (hint: look at Figure 10.4).

12. A radar has a beamwidth of 1°. Assume that the radar has collected data in a volume in which beams are spaced every 1° in elevation and azimuth, with a range gate spacing of 250 m. You wish to calculate the value of radial velocity at a Cartesian point in space and choose a spherical radius of influence of 1 km about the Cartesian point. Assume that a Cartesian point is located 30 km in range at an elevation angle of 5° from the radar. Estimate how many points in spherical space will be used to calculate the value of radial velocity at the Cartesian point by first calculating the density of points in a spherical shell in a volume centered at that location and then multiplying the result by the volume of a 1 km radius sphere.

13. Assume that two Doppler radars have a baseline of 50 km. What is the height and width (expressed in kilometers) of a 1° radar beam at the center of the 30° dual-Doppler lobe and at the intersection of the perpendicular bisector of the baseline with the dual-Doppler lobe? How wide would an updraft centered at each of these locations have to be for a dual-Doppler analysis to recover 90% of the updraft's amplitude?

14. Suppose you were a researcher that had a task of determining why some thunderstorms produce microbursts while others do not. To conduct the study, you will need to sample an entire thunderstorm, as well as any microburst it produces. You have two Doppler radars available to do the study. What dilemmas would you face in trying to determine the radar baseline to conduct this study? How might you at least partially resolve the dilemma if you had three radars instead of two?

15. When Hurricane Katrina came onshore in 2005, hurricane force winds extended 190 km from the center of the eye at landfall. For this problem, assume that hurricane force winds were present for 190 km along all radials extending from the center of the hurricane's eye. Suppose you wanted to study the region of hurricane force winds with two mobile Doppler radars by placing the radars along a coastline and have the entire region of hurricane force winds centered within the 30° dual-Doppler lobe as the hurricane approached from offshore.

    a. How long would the baseline have to be?

    b. How high would the 0.5° elevation radar beams be at the center of the eye (assume normal atmospheric refraction and that the center of the antenna is 10 m above sea level).

c. How many range gates from one radar would there be within the eye along an arc passing through the center of the eye? Assume that the eye is 10 km wide, the distribution of winds is the same along all radials outward from the eye, that the radars each have a 1° beamwidth, and that sequential beams are adjacent and do not overlap.

13

Precipitation Estimation with Radar

Objectives

By the end of this chapter, you should understand the following:

- The types of instrumentation used at the surface and on aircraft to measure precipitation rates, total precipitation, and particle size distributions.
- The various types of disdrometers and their operating principles.
- What an optical array probe is and how it is used.
- What is meant by an exponential and gamma particle size distribution.
- Why there is no exact mathematical relationship between precipitation rate and the reflectivity factor.
- What is meant by a Z–R relationship, and how these relationships are used to estimate rainfall.
- How Z–R relationships are determined.
- The many challenges in measuring precipitation with radar.
- How dual-polarization variables are used to improve radar-estimated rainfall.
- The challenges of measuring snowfall with radar.
- How spaceborne radars are being used to estimate global precipitation.

13.1 Introduction

The potential of radar to provide quantitative, region-wide estimates of both precipitation intensity and storm total precipitation has been recognized from the earliest days of radar meteorology and has been a driving force for decades of research related to precipitation estimation. The tremendous advantage of using radar is that a single radar can cover an area in excess of 30,000 km^2. Applications range from flash flood warning and water resource management to agricultural planning and insurance settlement and litigation. In this chapter, we examine how radar is used to estimate precipitation. Before we do, we will step back and examine how precipitation intensity and accumulated precipitation traditionally are measured with gauges, disdrometers, and other devices. We then examine the characteristics of rain, focusing particularly on raindrop size distributions and how they are parameterized. Armed with that information, we next examine how measurements of the radar reflectivity factor have

Radar Meteorology: A First Course, First Edition. Robert M. Rauber and Stephen W. Nesbitt.
© 2018 John Wiley & Sons Ltd. Published 2018 by John Wiley & Sons Ltd.
Companion website: www.wiley.com/go/Rauber/RadarMeteorology

been used to estimate precipitation and the caveats and complexities associated with quantitative interpretation of these measurements. We follow with a discussion of polarization radar measurements, and how these measurements are contributing to more accurate measurements of rainfall with radars. We also provide a brief discussion of winter precipitation and the challenges radar meteorologists face in deriving quantitative measurements of snowfall. The chapter closes by examining space-based radar estimation of global precipitation.

13.2 Measurement of precipitation rate, total precipitation, and particle size distributions

Precipitation intensity, also called **precipitation rate** (or **rainfall rate** if the precipitation is liquid), is defined as the volume of water passing downward through a horizontal surface, per unit area, per unit time. This definition implies that precipitation rate can be measured at any altitude in the atmosphere, not just at the surface. Based on this definition, the precipitation rate will have units of volume/area/time, or $m^3/(m^2 s) = m s^{-1}$, and therefore precipitation rate has units of *velocity*. More commonly, precipitation rate is expressed in units of $mm\,h^{-1}$ or $in.\,h^{-1}$.

Practically, the only level people care about is the precipitation rate *at the Earth's surface*. There, as precipitation does not necessarily pass into the ground, precipitation rate is defined as the depth to which water would accumulate on a runoff-free surface per unit time. **Total accumulated precipitation**, the integral of the precipitation rate over time, can be measured in any arbitrary time period, such as hourly or daily. The **storm total precipitation** is the depth of water accumulating on a runoff-free surface during a complete storm event.

13.2.1 Precipitation gauges

A precipitation gauge is a device that gathers and measures precipitation. Several types of gauges exist, some in which the accumulated precipitation is read manually from time to time, and others that record precipitation accumulation either electronically or on chart paper. Some gauges are designed to capture precipitation reasonably well in both rain and snow, whereas others function well only in rain. Some gauges fail to register light precipitation, while some have difficulty measuring very heavy precipitation. Most gauges do not perform well in high winds because turbulent flow over and around the gauge can deflect rain or snow away from the catchment area. For this reason, gauges are normally sited in protected areas, such as a clearing in a stand of trees, or a location protected by buildings or other structures.

The US National Weather Service uses both non-automated, 8 in. diameter **standard rain gauges** that are read manually, as well as **tipping bucket rain gauges** in their Automated Surface Observing System (ASOS) network. Non-automated gauges, such as the standard 4 in. diameter gauge used in the Community Collaborative Rain, Hail and Snow Network (CoCoRaHS, Figure 13.1), are used for measuring precipitation accumulation between successive visits by an observer, normally every few hours to a day. The tipping bucket rain gauge (Figure 13.2) has a collection area that funnels water into a small seesaw device that tips and dumps

Figure 13.1   A standard 4-in. orifice manual rain gauge used by the Community Collaborative Rain, Hail and Snow Network (CoCoRaHS) (Photo: David Bodine)

Figure 13.2   A tipping bucket rain gauge. The inset shows the seesaw tipping mechanism (Photo: Ali Tokay)

accumulated water every time a specific volume of water fills a small collection area at one end of the seesaw (see Figure 13.2 inset). The precipitation rate is determined automatically by the time between successive tips and the total precipitation by the number of tips in a specific time interval. Tipping bucket gauges work well in rain, but do not work well in snow unless the tipping device is heated. **Weighing bucket precipitation gauges** are designed for operation in both warm and cold conditions. In these gauges, precipitation falls through the orifice at the top of the gauge and into a container within the gauge. The weight of the container is recorded either electronically or on a chart. If the gauge is intended for use in winter, the container is normally filled partially with antifreeze, so any snow falling into the container will melt. The change in the weight of the container in a specific time period is a measure of the precipitation rate.

13.2.2 Disdrometers

The size distribution of raindrops can be measured by various methods. In the past, this involved counting and sizing the raindrops manually from impact impressions on filter paper or by examining photographs taken by cameras. Today, these measurements are made electronically either as drops fall to the ground or from aircraft using optical spectrometers.

A **disdrometer** is either an electromechanical or optical device used at the surface to measure rainfall rate, raindrop size distributions, and in some devices, raindrop shape. There are three common types, termed impact, video, and optical disdrometers. A disdrometer has to average over a sampling time (i.e., 1–5 min) to allow a sufficient statistical sample of drops to accurately characterize the size distribution of hydrometeors.

An **impact disdrometer** measures the momentum of falling raindrops as they hit a sensor. The sensor transforms the mechanical momentum of an impacting drop into an electrical pulse, where the pulse amplitude is approximately proportional to the drop's momentum. The momentum of a drop is related to its mass and fall velocity, which are both functions of it **equivalent spherical diameter** (ESD), the diameter that the drop would have if its volume was contained in a sphere. Drops with diameters greater than about a millimeter distort into "hamburger" shape as they fall because of frictional drag (see Chapter 7), necessitating the use of the ESD to characterize their size. From the measurement of momentum, the drop's mass and size can be derived from the amplitude of the electrical pulse. Figure 13.3 shows a *Joss–Waldvogel disdrometer*, a common impact disdrometer used in radar and precipitation research applications.

A **Video disdrometer** has very different operating principles. One type of video disdrometer, called the *two-dimensional video disdrometer* (2DVD, Figure 13.4), records orthogonal image projections of raindrops as they cross its sensing area. The instrument is designed to measure drop size, shape, and fall velocity simultaneously. Within the instrument, a white light source generates a sheet of light that is projected onto a line-scan camera containing a photodiode array. A second sheet of light is generated on a second photodiode array that is vertical and normal to the first array. When a drop passes through the array, it shadows the linear array of photodiodes, which are sampled to determine the drop dimensions. The two orthogonal projections provide

Figure 13.3   A Joss–Waldvogel impact disdrometer (Photo: Ali Tokay)

Figure 13.4   A two-dimensional video disdrometer (Photo: Ali Tokay)

three-dimensional raindrop shape information because it directly measures the distortion that occurs in the raindrop shape due to frictional drag. Rapid sampling allows determination of the fall velocity as the drop moves across the arrays. Shape information permits the drop volume to be computed and from that, the ESD, as well as the drop oblateness. By sampling many drops, the drop size distribution is measured.

An **optical disdrometer** employs laser technology. For example, in the *Parsivel Disdrometer*, a light sheet produced by a laser is focused onto a single photodiode in the receiver. The transmitter and receiver are mounted in housings for protection from the weather (Figure 13.5). The receiver produces a 5-V signal at the sensor output when no drops intercept the light sheet. Particles passing through the light sheet decrease the signal by extinction producing a voltage reduction, which depends linearly on the fraction of the light sheet blocked. The amplitude of the signal drop is a measure of particle size, and the duration of the signal allows an estimate of particle fall velocity.

13.2.3   Optical array probes

An alternative to measuring raindrop size distributions and precipitation rate at the ground is to measure these quantities from aircraft. The rational for this from the perspective of radar measurements is that the radar beam intercepts rain above the ground, with the radar beam progressively higher in altitude the further the sample is from the radar. Sampling precipitation aloft at the beam altitude can provide direct comparisons between radar and rainfall. An optical array probe on an aircraft can provide a particle size distribution much faster (5–10 s) than a surface disdrometer due to the relatively large sampling volume due to the motion of the aircraft.

Droplet spectrometers used to measure drop size distributions from aircraft are called **optical array probes**. Figure 13.6 shows two such probes mounted under the wing of the Hercules C-130 aircraft owned by the National Science Foundation and

Figure 13.5   A Parsivel optical disdrometer

Figure 13.6   Two-dimensional cloud (right) and precipitation (left) optical array probes mounted under the wing of the National Science Foundation/National Center for Atmospheric Research C-130 Hercules Aircraft

operated by the National Center for Atmospheric Research. The design of individual probes vary somewhat, but they all work by projecting a laser through a gap between two arms of the probe. The laser then impinges on a linear photodiode array, often containing 32 or 64 diodes, although this number varies with probe design. As the aircraft moves forward through a cloud, raindrops and/or ice particles pass through the laser beam and create a shadow on all or part of the photodiode array. Every time the aircraft moves forward a very small distance (typically 25, 50, or 200 μm, depending on the probe), the photodiode array is sampled and the shadow pattern recorded. The shadow patterns from many successive samples allow researchers to observe an image of the particle and reconstruct its two-dimensional shape (Figure 13.7). Knowledge of the sample area of the laser, the true airspeed of the aircraft, and the number and size of the particles passing through the laser allows reconstruction of the raindrop or ice particle size distribution, from which the rainfall rate can be estimated.

All of the devices described above, even simple rain gauges, have inherent sources of error and measurement uncertainties that must be understood and, to the degree possible, corrected to obtain accurate measurements of precipitation rate. Measurements from these devices provide "ground truth" against which radar measurements must be compared to establish relationships between radar measurements and precipitation rate.

13.3   Nature of particle size distributions

Raindrop size distributions and **ice particle size distributions** have been measured in many types of weather systems across many regions of the earth. Figure 13.8, for

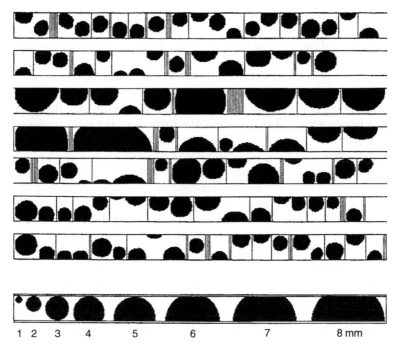

Figure 13.7   Raindrop images from a two-dimensional cloud optical array probe (From Rauber, R.M., Beard, K.V., and Andrews, B.M. (1991) A mechanism for giant raindrop formation in warm, shallow convective clouds. *J. Atmos. Sci.*, **48**, 1791–1797. © American Meteorological Society, used with permission)

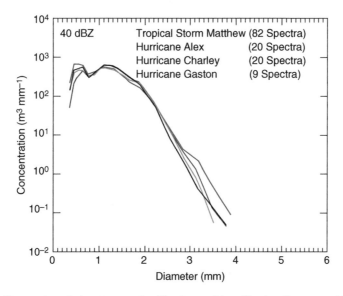

Figure 13.8   Composite raindrop spectra for Hurricanes Alex, Charley, Gaston, and Tropical Storm Matthew, all which occurred in 2004. The number of 1 min spectra comprising the composite are listed. All spectra correspond to a radar reflectivity factor of 40 dBZ (Adapted From Tokay, A., Bashor, P.G., Habib, E., and Kasparis, T. (2008) Raindrop size distribution measurements in tropical cyclones. *Mon. Weather Rev.*, **136**, 1679–1685, © American Meteorological Society, used with permission)

example, shows composite raindrop size distributions measured using an impact disdrometer in four land-falling tropical cyclones, three of which were hurricanes. These raindrop size distributions in Figure 13.8 show typical characteristics of sampled particle size distributions in many weather systems—a peak concentration in the smaller size range, a rapid decrease in the concentration of particles with size, a smaller decline in the concentration of the smallest particles, and a large size cutoff beyond which no particles are sampled. Note that in Figure 13.8, there are minima in the spectra near 0.8 mm. These minima are artifacts of the sampling with the disdrometer and are not real features. Such artifacts occur often in real data and must be recognized and accounted for to interpret data properly.

Aside from the minima at 0.8 mm, the general features of the spectra shown in Figure 13.8 are quite common. The total concentration, shape of the distribution, and diameter at which the peak concentration occurs will vary considerably from sample to sample, but the basic characteristics of particle size distributions illustrated in Figure 13.8 are nearly universal.

Particle size distributions are characterized by the size distribution function $n(D)$, which has standard units of m^{-4}, although $n(D)$ is more often expressed as m^{-3} mm^{-1} as the volume of air containing the raindrops is given in meters and the size of raindrops is expressed in millimeters. The quantity $n(D)\Delta D$ represents the number of drops in a size interval ΔD. By adding up the number of drops in each size interval, ΔD_j, encompassing the smallest to the largest drop in a sample volume of air,

$$N = \sum_j n(D)\Delta D_j \tag{13.1}$$

one obtains the total concentration, N, which is the total concentration of drops in a volume. The total concentration, N, has units of m^{-3}. As we have seen in Chapter 5, the radar reflectivity factor for Rayleigh scattering is dependent on the sixth power of the diameter of the particles in a sample volume and is therefore very sensitive to the size distribution of the raindrops in the volume. As particle size distributions display common characteristics, and quantitative expressions are needed to characterize particle size distributions in models, researchers have develop mathematical equations that fit observed particle size distributions. The two common equations used are termed the **exponential size distribution** and the **gamma size distribution**.

13.3.1   The exponential size distribution

When measured particle size distributions are plotted on a semi-logarithmic scale, as in Figure 13.8, the data often fit a straight line over a large part of the size range, implying that $n(D)$ is approximately exponential within this range. In 1948, J.S. Marshall and W. Mc K. Palmer first introduced the exponential size distribution function

$$n(D) = n_0 \exp(-\Lambda D) \tag{13.2}$$

to describe the measurements.[1] Their data suggested that n_0 was a constant with a value of 8×10^6 m^{-4} and that the slope parameter, Λ, was a function of the rainfall

---

[1]Marshall, J.S. and Palmer, W. Mc K. (1948) The distribution of raindrops with size. *J. Meteor.*, **5**, 165–166.

rate R given by

$$\Lambda = \Lambda_R \left(\frac{R}{R_0} \right)^c \tag{13.3}$$

where R_0 is a reference value taken as $1\,\text{mm}\,\text{h}^{-1}$, R is the rainfall rate in $\text{mm}\,\text{h}^{-1}$, and c is a constant equal to -0.21. Their data suggested values of $\Lambda_R = 4.1\,\text{mm}^{-1}$ when the drop diameter is expressed in millimeters. Subsequent to the Marshall–Palmer study, a large number of measurements were obtained and compared to the Marshall–Palmer distribution. When these data were fit to exponential size distribution functions, it became clear that the parameters n_0, Λ, and c varied with both the type of weather system (e.g., tropical cyclone, frontal, and thunderstorm) and the nature of the clouds (e.g., convective or stratiform) producing the rainfall. Furthermore, the small size end of the spectrum, where a drop off in concentration occurred, was not adequately parameterized, as the exponential size distribution predicts the greatest concentration of raindrops at near-zero diameter sizes. These deficiencies led to the introduction of the gamma size distribution.

13.3.2 The gamma size distribution

The inability of the exponential function to properly parameterize the small drop end of observed particle size distributions has led to adoption of the gamma distribution in many applications. The gamma distribution can be written as

$$n(D) = n_1 D^\mu \exp(-\Lambda D) \tag{13.4}$$

where n_1, μ, and Λ are termed the concentration, shape, and size parameters, respectively. It can be easily seen that when $\mu = 0$, the gamma distribution reduces to the exponential distribution. Figure 13.9 shows examples of a gamma drop size distribution and an exponential drop size distribution for two simulated rain shafts that have identical liquid water content and median volume diameters of the drop size distribution (the median volume diameter is the diameter at which half the volume of rain is contained in larger drops and half in smaller drops). The gamma drop size distribution appears very similar to those in Figure 13.8. However, the calculated radar reflectivity factor (assuming Rayleigh scattering) and rainfall rate are almost identical for both distributions in Figure 13.9. This happens because the radar reflectivity factor is sensitive to the sixth power of the drop diameters and the rainfall rate to approximately the fourth power, so most of the contributions to both these quantities come from the large end of the drop size distributions.

13.4 Radar remote sensing of precipitation

Because of the importance of precipitation, radar meteorologists have long conducted research to establish relationships between the radar reflectivity factor and precipitation rate. The fundamental conundrum faced by radar meteorologists attempting to measure precipitation with radar is that a radar, assuming Rayleigh

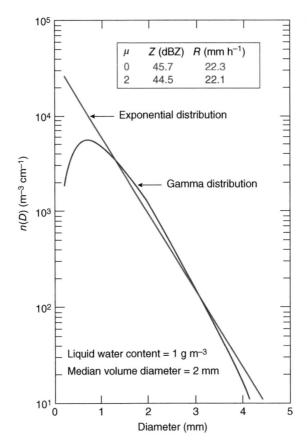

Figure 13.9 Comparison of an exponential raindrop size distribution with a gamma distribution where both correspond approximately to the same radar reflectivity factor and rainfall rate (Adapted from Ulbrich, C.W. (1983) Natural variations in the analytical form of the raindrop size distribution. *J. Appl. Meteor. Climatology*, **22**, 1764–1775. © American Meteorological Society, used with permission)

scattering, measures the radar reflectivity factor, defined from Eq. (5.40), as

$$Z = \frac{\sum_j D_j^6}{V_c} \tag{13.5}$$

whereas the total precipitation rate for spherical raindrops is the sum of the contributions to the precipitation rate of each drop, divided by the contributing volume. This can be expressed as

$$R = \frac{\sum_j R_j}{V_c} = \frac{\pi \sum_j D_j^3 W_j}{6 V_c} \tag{13.6}$$

where W_j is the total fall speed of the jth drop in the contributing volume V_c. The total fall speed is $W_j = w + w_{t,j}$, where w is the vertical air velocity and $w_{t,j}$ is the terminal velocity of the jth drop. For smaller raindrops $(0 < D < 2\,\mathrm{mm})$, the terminal velocity can be approximated as

$$w_t = aD \tag{13.7}$$

where a is a constant. In a situation where there is no vertical air motion, we can state approximately that

$$R \, \alpha \sum D^4 \tag{13.8}$$

while

$$Z \, \alpha \sum D^6 \tag{13.9}$$

The conundrum is, although it is true that

$$D_j^6 = (D_j^4)^{1.5} \tag{13.10}$$

it is also true that

$$\sum_j D_j^6 \neq \left(\sum_j D_j^4 \right)^{1.5} \tag{13.11}$$

The unfortunate conclusion we are forced to draw from this last equation is that there is *no mathematical relationship between the radar reflectivity factor and the precipitation rate.* The situation is even gloomier when we consider the case where vertical air motion does exist so that raindrops might be suspended in a convective updraft or driven downward in a convective downdraft. In the case where the drops are suspended, the reflectivity factor could have a large value, whereas the precipitation rate would be zero!

Despite all these caveats, empirical studies comparing measured radar reflectivity factors and precipitation rates in different types of weather systems suggest that relationships of the form,

$$Z = Z_R \left(\frac{R}{R_0} \right)^b \tag{13.12}$$

where Z_R, and b are empirically determined constants, can be used to obtain reasonable estimates of precipitation rate and total precipitation. The value of R_0 is normally $1 \, \text{mm} \, \text{h}^{-1}$ and is placed in Eq. (13.8) to avoid raising a quantity with units to a non-integer power. In general, in terms of percent error, estimates turn out to be better for events of longer duration and poorer for short duration.

In weather radar applications, the precipitation rate R is normally estimated by measuring the radar reflectivity factor Z. In this case, an inverse form of the Z–R relationship is used:

$$R = R_Z \left(\frac{Z}{Z_0} \right)^{\frac{1}{b}} \tag{13.13}$$

here, R_Z is the value of R when $Z = Z_0$ and Z_0 is a reference value of Z, nominally $1 \, \text{mm}^6 \, \text{m}^{-3}$. The parameters of Eqs. (13.12) and (13.13) are related by

$$R_Z = R_0 \left(\frac{Z_0}{Z_R} \right)^{\frac{1}{b}} \tag{13.14}$$

13.4.1   *Determining Z–R relationships*

Hundreds of studies have been conducted over the decades since radar was introduced into meteorology to determine values of Z_R and b for specific regions, seasons, and different types of rainfall, so as to estimate precipitation rates from radar. These studies follow two general approaches, which we refer to here as the direct and indirect methods. The *direct method* involves making measurements of the radar reflectivity factor using a radar and comparing those measurements with other measurements, such as those from a precipitation gauge or disdrometer located below the radar volume, or with the precipitation rate derived from near-collocated raindrop size spectra measured by an aircraft. The *indirect method* involves measuring the raindrop size distribution with either a disdrometer or airborne spectrometers and calculating both the rainfall rate and the reflectivity factor from the raindrop spectra. In the indirect method, actual radar measurements are not involved at all, hence the term "indirect." In both cases, a large number of measurements are collected that span the range of rainfall rates characterizing the region, weather system, or type of precipitation that is of interest. Once these data are collected, a regression is done to determine the appropriate value of Z_R, and b that best fits the collected data. For example, Figure 13.10 shows rainfall rates and radar reflectivity values calculated from raindrop spectra that were measured by an aircraft penetrating rain shafts emerging from trade wind clouds over the Atlantic Ocean just east of the island of Barbuda near the Caribbean Sea. The $Z–R$ relationship on Figure 13.10 was derived from these measurements using the indirect method and applied to

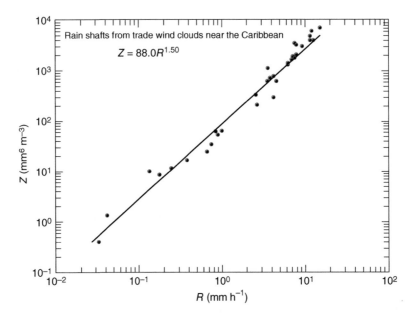

Figure 13.10   A $Z–R$ relationship calculated using the indirect method from measurements of raindrop size distributions measured over the western Atlantic Ocean beneath precipitating trade wind clouds (From Snodgrass, E.R., Di Girolamo, L., and Rauber, R.M. (2009) Precipitation characteristics of trade wind clouds during RICO derived from radar, satellite, and aircraft measurements. *J. Appl. Meteor. Climatology*, **48**, 464–483. © American Meteorological Society, used with permission)

interpret rainfall rates from an S-band radar that viewed precipitation over a much longer period than the flight. In general, Z–R relationships derived from the direct or indirect method can be applied to radar data collected during rainfall events that have a similar character.

In practice, both approaches have difficulties. In the case of the direct method, the contributing volume for the radar measurements is normally located some distance above the rain gauge or disdrometer with which it is being compared. Precipitation in the volume is dispersed by vertical wind shear and the differential fall velocity of the various sized raindrops, so the rain does not have the same drop size distribution at the ground as it did at the altitude where the radar sampled it. In convection, where there are strong horizontal gradients of both reflectivity and rainfall rate, the transport and dispersion of the rain can be significant so that the character of the rain arriving at the gauge may be nothing like that observed by the radar aloft. Even when the rain is directly sampled by aircraft, the sample volumes of the radar and aircraft are so different (cubic kilometers vs cubic meters) that the sampled rain spectra by the aircraft may not be representative of that sampled by the radar. With the indirect method, the primary problem is obtaining a statistically robust sample in the time of an aircraft pass. As the spectra in Figures 13.8 and 13.9 show, there are few drops occupying the large end of the droplet spectra. The odds of encountering a sufficient sample of large drops during an aircraft pass is low, but it is precisely these drops that contribute significantly to the reflectivity factor and the rainfall rate.

The WSR-88D radars now have complicated algorithms for rainfall estimation using polarization technology. However, before the introduction of polarization, the radars used five different selectable Z–R relationships. The specific relationship in any event was chosen by the operator based on the meteorology. These Z–R relationships are now called "legacy" relationships because they have been superseded by algorithms that use polarization information. The WSR-88D legacy Z–R relationships are listed in Table 13.1.

Figure 13.11 shows these five relationships graphically. The shaded area beneath the five lines is an envelope that encompasses 69 different Z–R relationships summarized in Battan's 1973 text, *Radar in Meteorology*. It should be clear from this figure that rainfall estimates will vary substantially depending on the specific Z–R relationship chosen to characterize the rainfall. Choice of a Z–R relationship is one of several challenges in radar precipitation estimation.

13.4.2   Challenges in precipitation estimation with radar

Factors that complicate precipitation estimation with radar include meteorological phenomena such as the radar bright band; ice versus water clouds; anomalous

Table 13.1   Legacy Z–R relationships used by the WSR-88D radars

Default	$Z = 300R^{1.4}$
Rosenfeld tropical	$Z = 250R^{1.2}$
Marshall–Palmer	$Z = 200R^{1.6}$
East Cool Season	$Z = 130R^{2.0}$
West Cool Season	$Z = 75R^{2.0}$

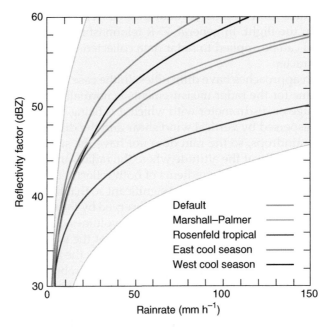

Figure 13.11   The relationship between the reflectivity factor and rainfall rate for the five legacy WSR-88D Z–R relationships (solid lines). The shading denotes the range of relationships for 69 other Z–R relationships in the published literature

propagation; attenuation; geometric issues such as beam blocking, beam height, and contributing volume resolution; and radar problems such as errors in absolute calibration.

The radar bright band, which develops at and below the melting level, is a common feature of stratiform rain. The high reflectivity factors characterizing the bright band appear on a plan position indicator (PPI) scan as a ring around the radar site (see Figure 15.16). The rapidity at which ice particles melt depends on the thermal structure of the atmosphere below the level at which particles first start to melt. In the cool season, a near 0 °C deep isothermal layer can develop in which melting occurs slowly causing the bright band to have some depth. In other cases, the bright band occupies a more narrow range of altitudes. In any case, algorithms to calculate rainfall will overestimate the rainfall within the bright band ring. A dramatic example of this effect appears in Figure 13.12, where rainfall estimates exceeded 6 in. in the northern part of the bright band ring during a winter rain event in Indiana, that actually produced about an inch of rain. The bright band ring is obvious in the rainfall pattern.

At altitudes above the bright band, the opposite effect occurs. As radar algorithms normally are coded to assume dielectric properties of water (the variable K in the weather radar equation, see Eq. (5.43)), the radar reflectivity factor in regions of ice particles will have lower values than water regions with the same water content. As a result, precipitation rates will appear to have lower values when the beam is above the bright band. This effect is also obvious in Figure 13.12.

Anomalous propagation is another problem that can lead to errors in estimating precipitation with radar. Anomalous propagation can lead to extensive ground clutter echo, which is typically characterized by high reflectivity values. Furthermore,

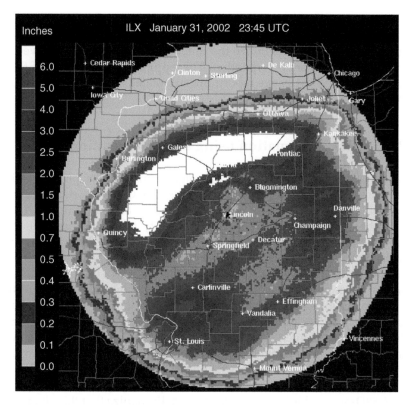

Figure 13.12 A example of effects of the radar bright band on estimates of precipitation using Z–R relationships. The actual total storm precipitation was closer to 1 in. across the domain of the radar

ground targets are stationary, so the "accumulated rainfall" in these high reflectivity regions will be large. A solution is to ignore the lowest beam, with the trade-off being that the next higher beam will be at a higher altitude, and therefore less representative of the surface rainfall. In general, the higher the beam, the poorer the estimate of the surface rainfall, so estimates can be expected to be less accurate at locations more distant from the radar. Attenuation will also influence precipitation estimates, particularly at wavelengths shorter than S-band. Attenuation will always reduce the estimated precipitation, with the estimates progressively worse farther from the radar when the beam is passing through regions of precipitation. Partial or total beam blocking obviously will bias precipitation estimates within the blocked sectors.

Another factor affecting the accuracy of measurements at greater distances from the radar is the increasing dimensions of the contributing volume. The contributing volume becomes quite wide and covers a larger depth of atmosphere at farther distances, so the precipitation estimates represent averages of progressively larger volumes. In inhomogeneous echo as occurs with convection, this can lead to underestimates of localized heavy precipitation. Finally, any error in absolute calibration of a radar will translate into error in estimated precipitation rate. Avoiding this problem requires frequent calibration to insure proper measurement of power by the radar system.

One technique to mitigate biases in estimated precipitation caused by these various problems is to employ an algorithm that does gauge data adjustment. This involves adjusting radar estimates to minimize the difference between radar-derived precipitation and gauge-measured precipitation. This approach has potential if the gauge density is sufficient and the gauge reports are automated so the data can be accessed in real time.

13.5   Precipitation estimation using dual polarization

Incorporation of polarization variables into algorithms for precipitation estimation remains a primary goal of polarization radar research. One important motivation is that the variables Z_{DR} and K_{DP} are immune to radar calibration uncertainties and partial beam blocking and thus provide opportunity for rainfall estimation under a broader range of conditions compared to using the radar reflectivity factor alone. However, each of these variables have their own inherent uncertainties, as none are mathematically related to precipitation rate.

Substantial research has been conducted to develop optimization algorithms to estimate precipitation from dual-polarization radars that address the challenges described in the previous sections. New algorithms are continually being tested and are likely to be implemented if the estimates they produce outperform existing algorithms. These efforts are ongoing, and without question, the algorithms in place at the time of this writing will be replaced by the time you are reading this book. As an example of an algorithm, we focus here on one specific algorithm that was in place for the WSR-88Ds when this book was developed. Recognize that this algorithm may now be superseded and no longer in use.

One might assume that data from the lowest elevation sweep would provide the best information to derive rainfall rates at the ground. Unfortunately, using the lowest sweep can also result in contamination by ground clutter, beam blocking, and anomalous propagation. An approach employed by the WSR-88Ds to minimize these and other issues is to construct a "hybrid scan" from data collected from several elevation cuts. The algorithm creating the hybrid scan employs both terrain data and the hydrometeor classification algorithm. Together, this information limits the radar data selected for precipitation estimation to those data not influenced by blocking, ground clutter, and anomalous propagation.

Before estimating precipitation, the melting layer detection algorithm is first used to determine the top and bottom of the melting layer. This is done by identifying, along higher elevation scans (4.5–10°), the altitudes of radar sample bins where Z_{HH}, Z_{DR}, and CC (or ρ_{HV}) meet specific thresholds characteristic of wet snow. The bright band appears more prominently on the higher elevation scans. In the event the data are ambiguous, the top is assumed to be the altitude of the 0° isotherm and the bottom 500 m below that altitude. Bins within the melting layer are classified as wet snow. A fuzzy-logic hydrometeor classification scheme, similar to that described in Chapter 7, is then used to assign 11 other classifications, as shown in Table 13.2.

Specific Z_H–R, Z_H–Z_{DR}–R, or K_{DP}–R equations are then used to calculate precipitation rates for the precipitation echoes. The choice of equation depends on the hydrometeor class assigned to a bin, which is itself based on the values of the polarization parameters fed into the fuzzy logic algorithm. The polarization variables used in

Table 13.2 WSR-88D hydrometeor classification at time of writing

Symbol	Description
GC	Ground clutter/anomalous propagation
BI	Biological
DS	Dry snow
WS	Wet snow
IC	Ice crystals
BD	Big drops
RA	Light to moderate rain
HR	Heavy rain
GR	Graupel
HA	Hail (mixed with rain)
UK	Unknown
ND	No data (less than threshold)

Table 13.3 WSR-88D equations used in precipitation algorithm as a function of hydrometeor class

Hydrometeor class	Variables in precipitation rate equation
Non-meteorological echo (BI, ND)	None, no precipitation
Light to moderate rain (RA)	$R(Z_H, Z_{DR})$
Heavy rain or big drops (HR, BD)	$R(Z_H, Z_{DR})$
Hail mixed with rain below melting layer top (HA)	$R(K_{DP})$
Wet snow (WS)	$0.6 \times R(Z_H)$
Graupel (GR)	$0.8 \times R(Z_H)$
Hail, mixed with rain above top of melting layer (HA)	$0.8 \times R(Z_H)$
Dry snow (DS) below the top of melting layer	$R(Z_H)$
Dry snow (DS) above the top of melting layer	$2.8 \times R(Z_H)$
Ice crystals	$2.8 \times R(Z_H)$

the equations to calculate precipitation rate for the various classifications appear in Table 13.3.

The equations used in this algorithm were developed based on data collected during the Joint Polarization Experiment in Oklahoma. A variety of weather systems were sampled between 2002 and 2005, including mesoscale convective systems, stratiform rains, isolated convection, and remnants of tropical convection. The equations in the algorithm are

$$R(Z_H) = (1.7 \times 10^{-2}) Z_H^{0.714} \tag{13.15}$$

$$R(K_{DP}) = 44.0 |K_{DP}|^{0.822} \, \text{sign}(K_{DP}) \tag{13.16}$$

$$R(Z_H, Z_{DR}) = (1.42 \times 10^{-2}) Z_H^{0.770} Z_{dr}^{-1.67} \tag{13.17}$$

Eq. (13.15) is simply the inversion of the default WSR-88D relation $Z_{HH} = 300R^{1.4}$, where Z_H is expressed in mm^6 m^{-3} and R in mm h^{-1}. In Eq. (13.16), K_{DP} is in deg km^{-1} and the sign(K_{DP}) term permits negative values of R. The lowercase subscript in Z_{DR} in (13.17) means that linear units should be used rather than logarithmic units.

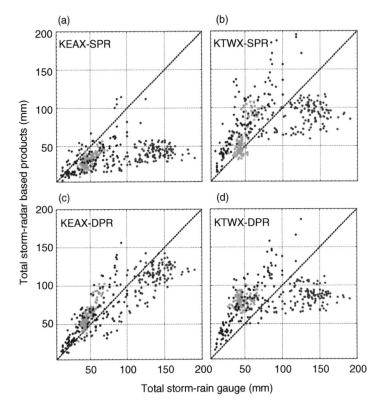

Figure 13.13 Scatterplot of gauge vs radar rainfall accumulation for three rainfall events (denoted by red, green, and blue dots) for single polarization retrieval (SPR) using the WSR-88D default Z–R relationship (a,b), the dual-polarization retrieval (DPR, c,d) for the KEAX (a,c), and KTWX (b,d) WSR-88D radars (From Cunha, L.K., Smth, J.A., Baeck, M.L., and Krajewski, W.F. (2013) An early performance evaluation of the NEXRAD dual-polarization radar rainfall estimates for urban flood applications. *Weather Forecasting*, **28**, 1478–1497. © American Meteorological Society, used with permission)

This algorithm is not unique. Other algorithms have been proposed in the scientific literature, and without doubt, continued refinements will appear as more research becomes available.

Once precipitation rates are determined, it is a simple matter to integrate them over time to obtain the total precipitation in specific time periods. For example, standard products from the WSR-88Ds include 1-h accumulation and storm total precipitation, where the start time of the "storm" is operator selected.

Studies continue to test various rainfall estimation algorithms for dual-polarization radar systems. Results generally show improvement over single polarization estimations, but biases remain that depend both on the type of precipitation particles sampled and distance the sampled precipitation is located from the radar. As an example,[2] consider the data in Figure 13.13. These data were obtained by two

---

[2] The full study can be found in Cunha, L.K., Smth, J.A., Baeck, M.L., and Krajewski, W.F. (2013) An early performance evaluation of the NEXRAD Dual-Polarization Radar Rainfall Estimates for Urban Flood Applications. *Weather Forecasting*, **28**, 1478–1497.

overlapping radars, one located at Kansas City, MO (KEAX), and the other at Topeka, KS (KTWX). A rain gauge network consisting of 136 tipping bucket rain gauges was located between the radars in Overland Park and Johnson County, KS, with the gauge network 23–72 km from KEAX and 104–157 km from KTWX. Figure 13.13 shows a comparison for three weather systems that produced various amounts of rainfall, each denoted by different colors. Panels (a) and (b) show rainfall estimates using the single polarization default WSR-88D Z–R relationship, whereas panels (c) and (d) show comparisons using the dual-polarization precipitation retrieval algorithm described earlier in this section. Note first that the two radars produce different estimates of precipitation. This can largely be attributed to the geometry and altitude of the respective beams—the KTWX beam is higher than the KEAX beam above the gauge network, its beam width is larger, and, because of its altitude, at least part of the KTWX beam either intersects the melting level or extends upward into subfreezing parts of the cloud. Note also that there is a distinct improvement with dual-polarization for KEAX, but a much less noticeable improvement with KTWX. These data provide evidence that better estimates are obtained with dual polarization within close range of the radars. The evidence is less certain or inconclusive at far ranges. Even with polarization, the errors in total storm precipitation estimated from this study remained as large as 25%. This, of course, is one study, and results may differ in another geographic region or storm type. Much research remains to be done to fully understand the advantages and limitations of dual-polarization technology in precipitation estimation in a wide variety of storms and environments.

13.6   Winter precipitation

Measuring snowfall with radar poses additional complexities when compared to rain. Rayleigh scattering theory for spheres, on which the weather radar equation is based, does not apply because ice crystals and snowflakes have complicated shapes, necessitating the use of the "equivalent reflectivity factor." The equations used in radar-processing algorithms to convert received power to the reflectivity factor assume a dielectric factor appropriate for water and are rarely if ever designed to switch in winter conditions to the appropriate dielectric factor for ice. As ice particles fall slowly compared to water drops, the distance ice particles drift downwind between the altitude of a radar sample and the ground is significantly greater than for raindrops. This complicates the development of Z_e–S relationships using comparison with gauge measurements. Finally, snow is easily lifted over a gauge orifice by turbulence induced by flow around a gauge, making ground-based measurement more complicated. Even a simple measurement of snow depth with a ruler is complicated because snow drifts and the compacting of snow with time. All of these complexities make radar measurement of **snow accumulation**, or alternatively, the **snow water equivalent precipitation** (the accumulated water that would exist if all the snow was melted on a runoff free surface), one of the most difficult, and largely unsolved problems in radar meteorology.

Despite these complexities, there have been attempts to establish Z_e–S relationships for specific regions and applications. These generally show a larger degree of scatter in the measurements compared to data used to establish Z–R relationships. An example appears in Figure 13.14. The snow water equivalent precipitation at the

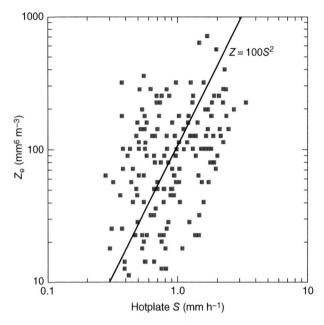

Figure 13.14 Z_e–S scatterplot with the equivalent radar reflectivity factor measured using a 0.5° ray path and the precipitation measured with a hotplate (see Figure 13.15) (Adapted from Wolfe, J.P. and Snider, J.R. (2012) A relationship between reflectivity and snow rate for a high-altitude S-band radar. *J. Appl. Meteor. Climatology*, **51**, 1111–1128. © American Meteorological Society, used with permission)

ground in this study was measured with a device called a **Hotplate**.[3] The Hotplate has two vertically stacked 13-cm diameter plates mounted on a pedestal (Figure 13.15). The temperature of the Hotplate's top and bottom surfaces are maintained at 75 °C. For the top plate, a portion of the electrical heating goes into melting and evaporation of snow impacting the plate. The difference between the electrical power provided to the plates provides a measure of the snow water equivalent precipitation rate, S. Hotplate measurements of S during snowfall events were made at a site 25 km northwest of the Cheyenne WSR-88D radar. One can see from Figure 13.14 that for any given Z_e value, the Hotplate-measured liquid water equivalent snowfall rate varied over an order of magnitude (e.g., 0.2–2.0 mm h^{-1} at $Z_e = 100$ mm^6 m^{-3}). This type of variability is common in snow, making quantitative estimates of snowfall less certain than for rain.

13.7   Measuring precipitation from space

Vast reaches of the earth, including its oceans, deserts, steppes, mountain ranges, polar regions, and unpopulated areas, are not instrumented with precipitation gauges or within ground-based radar coverage. The only way precipitation can

---

[3]*Hotplate* is a registered trademark of the University Corporation for Atmospheric Research Foundation.

Figure 13.15   A hotplate precipitation gauge for measurement of snow water equivalent precipitation rate (From Rasmussen, R.M., Hallett, J., Purcell, R., Landolt, S.D., and Cole, J. (2011) The hotplate precipitation gauge. *J. Atmos. Oceanic Technol.*, **28**, 148–164. © American Meteorological Society, used with permission)

be routinely monitored in these regions is from space. Spaceborne estimates of precipitation are currently accomplished through analysis of data collected using an array of passive and active remote sensing platforms including microwave radiometers that operate on a number of satellites and use a range frequencies, infrared radiometers that measure cloud-top radiances, and short-wavelength (K_u- and K_a-band) radars. The data from one or more of these sensors is processed using increasingly sophisticated algorithms that are first tested and validated by comparison with traditional ground-based measurements. Precipitation estimates derived from a specific algorithm is referred to as a **product**. Before the launch of the Tropical Rainfall Measuring Mission (TRMM) in 1997, which carried the first **precipitation radar** in space, all precipitation estimates were derived from passive sensors. Since the launch of TRMM in 1997, and the Global Precipitation Mission core satellite in 2014, precipitation radars in space have been used to refine the algorithms used by satellites carrying only passive sensors. The process of developing a precipitation product is complicated. NASA carries out extensive field campaigns in conjunction with the launch of each of the satellite missions to calibrate and validate the measurements made from space. These field campaigns include vertically pointing and/or scanning ground-based radar systems. They also include research aircraft, typically carrying radars similar to those in space and a suite of probes to measure cloud properties. All field studies also include ground-truth measurements of precipitation from a network of gauges and disdrometers. Teams of scientists analyze these data in conjunction with the space-based measurements with the purpose of constructing algorithms that can be applied to the space-based data to measure precipitation in different environments around the globe.

13.7.1 Tropical Rainfall Measuring Mission

The Tropical Rainfall Measuring Mission was a joint space mission between NASA and the Japan Aerospace Exploration Agency. The TRMM satellite platform flew in space between 1997 and 2015, at which time it reentered the atmosphere and crashed into the Pacific. The TRMM satellite carried three rain-measuring instruments, the TRMM Microwave Imager (TMI), the Visible Infrared Scanner (VIRS), and the Precipitation Radar (PR). The TRMM satellite had a circular, non-sun-synchronous orbit at an altitude of 350 km and an inclination of 35° to the equator, limiting its coverage to the tropics. The non-sun-synchronous orbit allowed locations on the earth to be covered at a different local time each day, so that the satellite could observe the diurnal cycle of precipitation. The TMI was a passive, multichannel microwave radiometer. The device operated at four frequencies at dual-polarization and a single additional frequency at a single polarization. The TMI provided information on the integrated column precipitation content, cloud liquid water, cloud ice, rain intensity, and rainfall types (e.g., stratiform or convective). The VIRS was a five-channel, cross-track scanning radiometer operating at five frequencies that provided high-resolution observations on cloud coverage, cloud type, and cloud top temperatures. The **precipitation radar** was an electronically scanning radar, operating at 13.8 GHz frequency. Its primary mission was to map the 3-D rainfall distribution over both tropical lands and oceans and define the layer depth of the precipitation. The precipitation radar had a minimum detectable signal of 17 dBZ, a horizontal resolution of 4.3 km at nadir, a vertical resolution of 0.25 km, and a cross-track scan range of 17° corresponding to a 215 km wide swath. Figure 13.16 shows a map of globally averaged precipitation based on 16 years of TRMM data collection. TRMM was also used to study high-impact events such as hurricanes during overpasses.

13.7.2 Global Precipitation Mission

The Global Precipitation Measurement (GPM) mission is comprised of a network of satellites that together provide global observations of both rain and snow. The GPM core satellite (Figure 13.17) carries a dual-wavelength radar system as well as a radiometer. The core satellite is intended to act as the reference standard for precipitation measurements for the other research and operational satellites in the constellation that measure precipitation using passive techniques. The GPM mission involves international space agencies.

The GPM Core Observatory launched from Japan in February 2014. The GPM Core Observatory uses a non-Sun-synchronous orbit that extends to high latitudes, extending the coverage of TRMM. A key advancement of GPM over TRMM is its

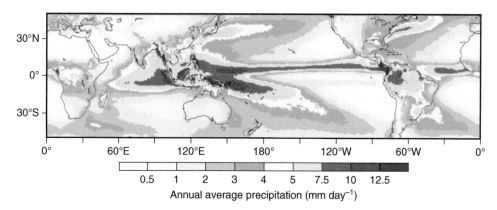

Figure 13.16   Annual average precipitation across the tropics between 1998 and 2013 as measured by the Tropical Rainfall Measuring Mission (courtesy NASA)

Figure 13.17   Image of the Global Precipitation Mission Core Observatory (Courtesy NASA)

capability to measure light rain and snow and obtain information about the microphysical properties of hydrometeors. The GPM Core Observatory carries the first spaceborne dual-frequency precipitation radar, as well as a multichannel Microwave Imager. The dual-frequency radar consists of a K_a-band precipitation radar operating at 35.5 GHz and a K_u-band precipitation radar operating at 13.6 GHz. The swaths of the two radars overlap, but are not the same size, the K_a-band swath being 125 km and the K_u-band swath 245 km. The simultaneous measurements in the overlapping region provide information on particle drop size distributions. The radar measurements are complimented by measurements from a 13-frequency, conical-scanning multichannel microwave radiometer that covers a swath of 885 km. The radiometer frequencies are optimized to retrieve precipitation at various intensities using the polarization difference from each channel.

At the time of this writing, the GPM core satellite has been in space for just over 3 years. Field studies are still being conducted worldwide to develop high-quality algorithms and products to measure precipitation, particularly in winter when snow dominates the precipitation in the middle and high latitudes. GPM is expected to contribute to improvement of weather and precipitation forecasts through data assimilation, as well as contribute new understanding of global precipitation processes across the world, particularly in remote and underdeveloped locations where precipitation measurements are rare or nonexistent.

Important terms

disdrometer
equivalent spherical diameter
exponential size distribution
gamma size distribution
Global Precipitation Measurement
    Mission
Hotplate
ice particle size distribution
impact disdrometer
optical array probes
optical disdrometer
precipitation intensity
precipitation radar
precipitation rate

product
raindrop size distribution
rainfall rate
standard rain gauge
snow accumulation
snow water equivalent precipitation
storm total precipitation
tipping bucket rain gauge
total accumulated precipitation
Tropical Rainfall Measuring Mission
video disdrometer
weighing bucket precipitation gauge
Ze–S relationship
Z–R relationship

Review questions

1. What are the three common disdrometer types and how do their operating principles differ?
2. Explain the difference between an exponential drop size distribution and a gamma size distribution. Which more accurately describes natural drop size distributions?
3. Explain one direct and one indirect method for determining Z–R relationships.
4. Can the precipitation rate be negative? If not, why not? If so, where might this occur in the atmosphere?

5. There are well over a hundred Z–R relationships that have been proposed for use. Why is not there just one Z–R relationship that applies universally to storms?

6. Write a bullet list summary of the challenges affecting the use and application of Z–R relationships to estimate rainfall.

7. What is a major advantage of using the variables K_{DP} and Z_{DR} in estimates of precipitation rate?

8. Explain how hydrometeor classification is being used as a component of precipitation estimation with polarization radars.

9. Why is estimation of storm total snowfall much more difficult with radar compared to estimation of storm total rainfall?

10. What are the primary differences between the TRMM and GPM procedures for precipitation estimation from space?

Challenge problems

11. Assume in a particular storm that raindrops fall at a speed of $4\,\mathrm{m\ s^{-1}}$. A radar observes these raindrops 3 km above the surface, directly above a rain gauge. The horizontal wind increases with height at that location such that the wind speed is given by $V(z) = az$, where z is height above the ground in kilometers and the constant $a = 10\,\mathrm{m\ s^{-1}\ km^{-1}}$.

    a. How far downwind of the rain gauge will a drop observed at 3 km elevation be located by the time it reaches the ground?

    b. Perform this same calculation again, but this time for a snowflake falling to the ground at $1\,\mathrm{m\ s^{-1}}$.

12. It is late October and it is raining. The clouds are stratiform and the rain is steady, so steady in fact that all of the rain gauges in the area report exactly 0.3 in. of rain during the last 5 h. While working at an NWS office, you examine the WSR88D radar data. You are unaware of the gauge reports because they only come into the office once a day. You decide instead to use the radar to estimate the amount of rainfall that fell over the state during the last 5 h. You determine that the radar reflectivity factor in dBZ has the following relationships as a function of altitude:

$$\mathrm{dBZ} = 25 + 10H \quad (0.50\,\mathrm{km} > H \geq 0.00\,\mathrm{km})$$

$$\mathrm{dBZ} = 60H \quad (0.75\,\mathrm{km} > H \geq 0.50\,\mathrm{km})$$

$$\mathrm{dBZ} = 120 - 100H \quad (1.00\,\mathrm{km} > H \geq 0.75\,\mathrm{km})$$

$$\mathrm{dBZ} = 25 - 5H \quad (3.00\,\mathrm{km} \geq H \geq 1.00\,\mathrm{km})$$

(continued)

You switch among the five approved Z/R legacy relationships that can be used with WSR-88D radars (where Z is in units of $mm^6 \, m^{-3}$ and R is in units of $mm \, hr.^{-1}$) to examine the differences in predicted total rainfall over the 5 h.

a. Plot a vertical profile of the radar reflectivity factor and provide a detailed physical explanation for the form of the profile (from top to bottom).

b. Assume that you are using the 0.5° scan, and the radar beam crosses the maximum unambiguous range when the beam rises above 3 km altitude. Determine the radar-estimated total rainfall (in inches) that fell during the 5 h period as a function of distance from the radar site for each of the assumed Z–R relationships. To illustrate your calculations, plot a graph that shows radar estimated 5-h rainfall total (in inches) as a function of distance, S, from the radar site (kilometers) for each of the five Z/R relationships. (Put all five plots on one graph to make it easy to compare them). You may assume that the index of refraction decreases linearly with height at $1.25 \times 10^{-7} \, m^{-1}$, that the earth radius is 6250 km, and that the radar antenna is on the ground ($z = 0$). (Hint: you will need Eq. (4.15)).

c. Explain the results in (b). Qualitatively, which Z/R relationship worked best for the situation described? Discuss problems that hamper the estimation of precipitation with radar in the "cool" season based on your results.

13. During a rainfall event, you use a disdrometer to measure the raindrop size distribution. The data you obtain are below:

Diameter range (mm)	Mid-range diameter (mm)	Drop concentration (m^{-3})
0.1–0.3	0.2	1
0.3–0.5	0.4	3
0.5–0.7	0.6	8
0.7–0.9	0.8	90
0.9–1.1	1.0	247
1.1–1.3	1.2	499
1.3–1.5	1.4	520
1.5–1.7	1.6	360
1.7–1.9	1.8	210
1.9–2.1	2.0	145
2.1–2.3	2.2	126
2.3–2.5	2.4	74
2.5–2.7	2.6	51
2.7–2.9	2.8	29
2.9–3.1	3.0	16
3.1–3.3	3.2	14

Diameter range (mm)	Mid-range diameter (mm)	Drop concentration (m^{-3})
3.3–3.5	3.4	9
3.5–3.7	3.6	6
3.7–3.9	3.8	6
3.9–4.1	4.0	3
4.1–4.3	4.2	1
4.3–4.5	4.4	1

a. Calculate the reflectivity factor for this raindrop size distribution.

b. Which size drops contribute the most toward the reflectivity factor?

c. Calculate the rainfall rate, using each of the five legacy Z–R relationships used with the WSR-88D radars. Which produces the highest and lowest rainfall rate?

14. Suppose you measured the reflectivity factor as 40 dBZ, with $Z_{DR} = 1$ dB and $K_{DP} = 2.0$ deg km^{-1}, but you had no idea what the particle classification was. You decide to calculate the precipitation rate for each hydrometeor classification in Table 13.3. What is the range of precipitation rates that would be possible? Which hydrometeor classification produces the highest and the lowest precipitation rate?

15. Suppose you lead a team of scientists who charged with developing an algorithm that uses data from the dual-frequency radar on the GPM core satellite to measure high-latitude wintertime precipitation from space. List the complexities you might face in developing such an algorithm.

14

Warm Season Convection

Objectives

By the end of this chapter, you should understand the following:

- The radar evolution of a mesoscale convective system as it relates to its lifecycle.
- The structural features of a mesoscale convective system as they appear on radar.
- How to use radar to identify strong straight-line winds and tornadoes associated with squall lines.
- The key structural features of a supercell thunderstorm and the relation of these features to radar echoes.
- How to use radar to identify tornadoes and hail within supercell thunderstorms.
- The characteristic radar signatures of downbursts.

14.1 Introduction

Thunderstorms produce a range of severe weather phenomena detectable with meteorological radars. Severe thunderstorms normally require four elements for formation: a source of moisture, a conditionally unstable atmosphere, a lifting mechanism to initiate the thunderstorm's updraft, and vertical wind shear—a rapid change in wind speed and/or wind direction with altitude. Destructive thunderstorms develop most often in environments characterized by large conditional instability and strong vertical wind shear. The US National Weather Service considers a thunderstorm severe if it produces (1) a tornado, (2) has hail with diameter of 1 in or larger at the ground, and (3) winds with gusts of 50 knots or greater near the earth's surface.

Thunderstorms organize differently depending on many factors including the lifting mechanism that initiates the storms, the vertical wind shear, and the degree of instability in the atmosphere. Lifting occurs along boundaries between airmasses called fronts, or other less-distinct boundaries such as the leading edge of a cool air outflow that is remnant air of a past thunderstorm. These more subtle boundaries become increasingly important in the warm season when larger scale fronts become less distinct or nonexistent. Thunderstorms forming along subtle boundaries in the warm season often undergo self-organization that leads to the formation of a **mesoscale convective system**, or **MCS**. In the cool season, when fronts are more

Radar Meteorology: A First Course, First Edition. Robert M. Rauber and Stephen W. Nesbitt.
© 2018 John Wiley & Sons Ltd. Published 2018 by John Wiley & Sons Ltd.
Companion website: www.wiley.com/go/Rauber/RadarMeteorology

distinct and lifting is focused along fronts, very long lines of thunderstorms called **frontal squall lines** develop along frontal boundaries.

If sufficient vertical wind shear is present in the atmosphere, thunderstorms may develop into violent, rotating storms called **supercell thunderstorms**. Supercells can occur along a line or individually, but each has its own individual circulation that is related to the storm's rotation and the vertical wind shear in the atmosphere. Supercells most often develop when strong winds are present in the upper troposphere and winds in the lowest kilometer or two of the atmosphere increase speed rapidly and change direction with altitude. Most hail that is golf ball-sized and larger and most large, strong **tornadoes** are produced by supercells.

Downbursts, another hazard of convective storms, are localized downdrafts that create strong straight-line winds and horizontal wind shear near the ground, an extreme hazard to aviation. Downbursts originate within the lower part of towering cumulus clouds or thunderstorms, descend to the ground, and diverge outward at the surface as a burst of strong winds. Downbursts were the primary motivation for the installation of Terminal Doppler Weather Radars (TDWRs) near major airports.

In this chapter, we examine radar signatures associated with important modes of severe thunderstorm organization—mesoscale convective systems, frontal squall lines, and supercells. We will also consider radar signatures associated with downbursts. Our focus will be on radar signatures associated with severe weather hazards detectable with operational radars (WSR-88D and TDWRs). We will also examine features of these storms deduced from research radars that have the capability and advantage of rapid scanning and/or scanning in range-height indicator (RHI) or other modes not associated with the standard Volume Coverage Patterns of the WSR-88D and TDWR radars.

14.2 Mesoscale convective systems

Mesoscale convective systems produce much of the summertime rainfall in North America, as well as flash floods, destructive straight-line winds, and sometimes hail and tornadoes. The storms typically move across a large geographic region during their life cycle, and their cloud shields can cover thousands of square kilometers.

14.2.1 Radar-observed life cycle of an MCS

The radar-observed life cycle of a typical MCS is illustrated in Figure 14.1. This storm system passed over St. Louis, Missouri, and produced severe straight-line winds and weak tornadoes in Missouri and Illinois.[1] The event occurred within the range of the KEAX and KSLX radars.

The structural features of this MCS are common to many MCSs. The sequence of radar images cover a period of 7 h, starting at 18:00 UTC (1 p.m. LT). An MCS typically initiates as thunderstorms develop along a weak airmass boundary,

---

[1] A summary of the severe weather associated with this MCS can be found in Atkins, N.T. *et al.* (2005) Damaging surface wind mechanisms within the 10 June 2003 Saint Louis Bow Echo during BAMEX. *Mon. Weather Rev.*, **133**, 2275–2296.

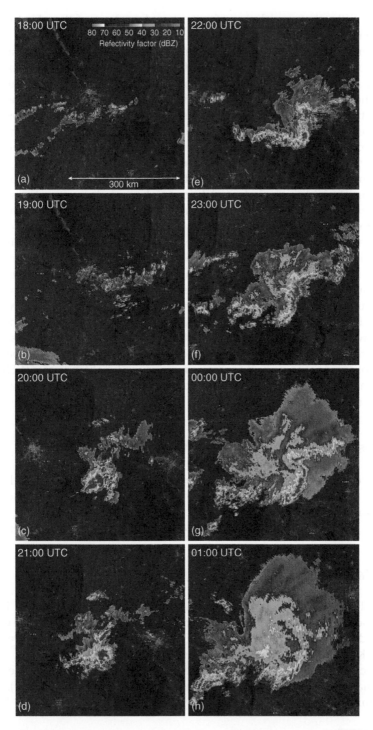

Figure 14.1   The life cycle of a mesoscale convective system. Over a period of 7 h, initial convection intensifies, grows upscale as a cold pool develops, forms a convective line with a bowing segment, and a trailing stratiform region

particularly in late afternoon as air on the warm side of the boundary reaches its warmest temperatures of the day. In this event, the thunderstorms erupted somewhat earlier (Figure 14.1a). Sometimes, the initial thunderstorms align along the position of the boundary, whereas at other times, they appear as a disorganized cluster (Figure 14.1b). The storms may be ordinary, nonrotating thunderstorms or possess supercellular characteristics.

Water-phase changes (sublimation of ice and evaporation of rain) associated with thunderstorm precipitation lead to the development of a **cold pool** of air at the surface. The cooler air spreads outward toward the warm, moist inflow air feeding the thunderstorms, as well as rearward away from the storms. New updrafts form as conditionally unstable air is lifted along and over the advancing cold pool. The reorganization of updrafts along the leading edge of the developing cold pool leads to the formation of a **squall line**. From a radar perspective, as time proceeds, the storms typically appear to organize first as a cluster (Figure 14.1c) and then become more intense and align into an arc-shaped squall line as the cold pool develops (Figure 14.1d). One or more segments along a squall line may bow outward, producing a radar feature called a **bow echo** (Figure 14.1e). As the line of thunderstorms is typically not straight, radar meteorologists have adopted the term **quasi-linear convective system** (QLCS) as a general description of this type of thunderstorm organization. During this time, a region of widespread less-intense precipitation develops to the rear (typically, west or northwest) of the squall line (Figure 14.1f,g). This **trailing stratiform region** becomes progressively larger and more widespread as the thunderstorm cells spread along a progressively widening arc (Figure 14.1g). Eventually, the thunderstorms decay, leaving in their wake a wide stratiform region (Figure 14.1h) that may produce rainfall for several more hours. During this time, new thunderstorm cells often develop ahead of, on the periphery, or even to the rear of the original MCS (Figure 14.1g,h).

Near the end of the MCS life cycle, the trailing stratiform region decays, leaving a broad area of clouds. Latent heat released during the condensation process will sometimes cause an area of low pressure to develop within the trailing stratiform precipitation that may persist long enough to induce a broad cyclonic rotation. This rotation, called a **mesoscale convective vortex**, can sometimes be detected with radar (i.e., a mesoscale area of rotation is evident in reflectivity factor and Doppler velocity fields) and can be the focal point for thunderstorm development the following day and beyond.

14.2.2 Conceptual model of an MCS as observed with a research radar

Figure 14.2 summarizes the key structural features of a mature MCS that contains a trailing stratiform region, and how these features appear on radar. The radar data were collected using the tail radar system of the NOAA P-3 aircraft during the Bow Echo and Mesoscale Convective Vortex Experiment (BAMEX) on June 29, 2003. The width of the heavy precipitation region of an MCS squall line typically ranges between 5 and 15 km, whereas the entire precipitation region of a mature MCS typically ranges between 100 and 200 km but can be much larger on occasion. Heavy rain falls from the convective region of the storm just to the rear (west) of the updraft. Lighter rain falls from the stratiform region farther to the west. Air

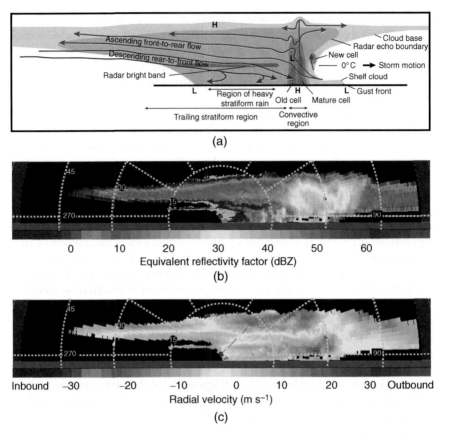

Figure 14.2 (a) Schematic cross section of the structure and storm-relative flow within a mature MCS with a trailing stratiform precipitation region. (b) Radar reflectivity factor and (c) radial velocity measured by the NOAA P-3 aircraft within a mature MCS (Panel (a) adapted from Houze, R.A., Jr., Rutledge, S.A., Biggerstaff, M.I., and Smull, B.F. (1989) Interpretation of Doppler weather radar displays in midlatitude mesoscale convective systems. *Bull. Am. Meteorol. Soc.*, **70**, 608–619. ©American Meteorological Society, used with permission)

ascending in the updraft exhausts both forward, creating the storm's forward anvil, and rearward, creating the stratiform region to the west. At higher altitudes in the stratiform region, air rises slowly as it flows westward (**front-to-rear flow**) relative to the convective line. On radar cross sections of the stratiform region, a **bright band** of radar reflectivity will appear at the melting level.

Low pressure develops in the middle troposphere in the vicinity of the convective line as a result of the vertical distribution of latent heating/cooling. This forces development of **rear-to-front flow**, a feature called the **rear-inflow jet**. The rear-inflow jet is evident in the radial velocity field in the radar cross section in Figure 14.2c. Air within the rear-inflow jet often reaches the ground and spreads out behind the advancing gust front, which can lead to the appearance of strong radial velocities to the rear of the convective line (Figure 14.3). The rear-inflow jet, front to rear flow, and the convective updraft are more easily visualized in Figure 14.4, which shows a dual-Doppler

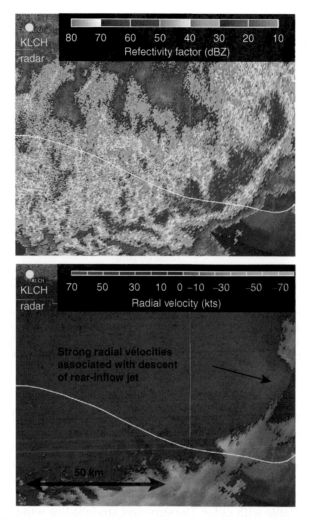

Figure 14.3 Reflectivity factor and radial velocity from the KLCH WSR-88D showing an MCS convective line with strong outbound radial velocities associated with a descending rear-inflow jet. The white line is the Gulf Coastline

synthesis of the wind field in a cross section parallel to the rear-inflow jet in the same MCS as in Figure 14.2b,c.[2]

14.2.3   Radar signatures of hazardous weather in MCSs

At some critical point in the early evolution of an MCS, the downdrafts forming immediately behind the convective line cause the cold pool to consolidate and rush

[2]This event is described by Grim, J.A. *et al.* (2009) Development and forcing of the rear inflow jet in a rapidly developing and decaying squall line during BAMEX. *Mon. Weather Rev.*, **137**, 1206–1229.

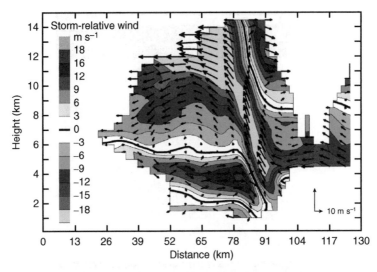

Figure 14.4 Quad-Doppler analysis normal to the convective line of a mature MCS showing the cross section parallel storm-relative wind component (colors) and two-dimensional flow vectors in the plane of the cross section. Hot (cold) colors denote flow to the right (left). The descending rear-inflow jet, divergent flow at the top of the updraft, and front-to-rear flow above the rear-inflow jet are all evident (From Grim, J.A., Rauber, R.M., McFarquhar, G.M., Jewett, B.F., Jorgensen, D.P. (2009) Development and forcing of the rear inflow jet in a rapidly developing and decaying squall line during BAMEX. *Mon. Weather Rev.*, **137**, 1206–1229. ©American Meteorological Society, used with permission)

outward toward the warm air feeding the thunderstorms. The leading edge of this out-rushing air is the **gust front** and is sometimes marked by a **radar fine line**. Figure 14.5 shows an example of a radar fine line associated with an advancing cold pool. The fine line is often associated with a shelf cloud that forms directly over the gust front, as well as lofted dust and insects. Although the shelf cloud does not produce much precipitation, droplets composing the cloud are sometimes large enough to be detectable with radar. The strongest straight-line winds associated with outrushing air typically occur during the early stages of the formation of the cold pool and bow echo, although in the strongest cases, the outrushing air may continue to produce severe winds for many hours, especially if thunderstorms regenerate frequently to replenish the cold pool.

The ability of a radar to observe strong winds depends on the orientation of the wind vector relative to the beam. For example, Figure 14.6 shows two points along the line of thunderstorms associated with the MCS in Figure 14.1. A bow echo is just developing along the northern end of the line, where the measured radial velocity exceeded severe criteria, 63 kts (32 m s^{-1}). At the southern point, behind a mature bow echo, the radial velocity was well below severe criteria, 17 kts (8.7 m s^{-1}). The difference between the radial velocity at these two points may be due to a difference in the magnitude of the wind, the orientation of the wind vector relative to the beam, or both. Beam orientation is an important consideration in determining whether the true wind exceeds severe criteria.

Surface winds behind an MCS-generated gust front can be severe, sometimes reaching 80–100 kts (~40–50 m s^{-1}). In many cases, the winds remain below the 50 kts

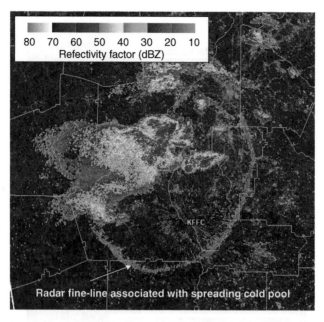

Figure 14.5 Radar fine line associated with a spreading cold pool generated by convection near the KFFC WSR-88D radar

($\sim$25 m s^{-1}) severe threshold. The character of the outward rush of air depends on many factors including the availability of hydrometeors to change phase as they descend to the ground, the existence of subsaturated and high-temperature lapse-rate air to maintain negative buoyancy within the downdrafts, and the low-level wind shear in the environment ahead of the squall line. The outrushing air sometimes creates smaller bow echoes 10–20 km long, whereas at other times, a large bow echo 150–200 km long may develop.

When convection is occurring over a gust front marked by strong horizontal wind shear, small tornadoes can form along "cusps" that develop as a result of shearing instability along the gust front. Typically, these tornadoes are short-lived, small, and weak. Figure 14.7 shows an example of four cusps in the radial velocity field at two times separated by 10 min for the MCS shown in Figure 14.1. Each cusp marked the location of a weak tornado, as confirmed by damage surveys. Cusps such as these are sometimes the only radar indication of the possible presence of tornadoes in these systems. The cusps are sometimes associated with small radial velocity couplets in the storm-relative radial velocity field if the storm motion vector is oriented with a significant component along the direction of the radar beam.

14.2.4 Frontal squall lines

Thunderstorms are often aligned with either surface or upper-level cold fronts associated with extratropical cyclones, particularly in the cool season. Lines of thunderstorms can sometimes stretch over 1000 km (Figure 14.8). In the case of forcing by an upper-level front, the storms often generate a cold pool that appears much like a

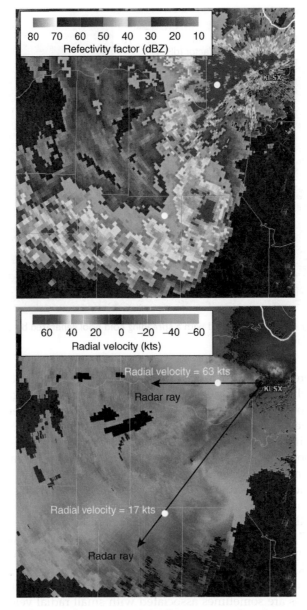

Figure 14.6 Radar reflectivity factor and radial velocity measured in the vicinity of a severe wind producing bow echo squall line by the KSLX WSR-88D

synoptic-scale surface cold front (the true surface cold front can be tens to hundreds of kilometers west of the line of storms). In the case of a surface-based cold front, evaporation of rain from the stratiform region to the rear of the convection can further cool the cold air behind the front.

Storms forming along a synoptic-scale frontal boundary may or may not be super-cellular depending on the ambient vertical shear and conditional instability that characterizes the warm air ahead of the front. If the horizontal shear across the front is

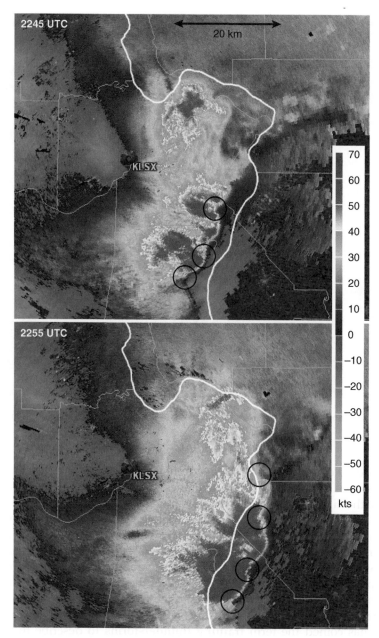

Figure 14.7   Radial velocity fields measured by the KSLX radar at two times 10 min apart during which weak tornadoes were reported along the line. The circles denote regions of local rotation and are marked by cusps in the line

sufficient, the front can undergo shearing instability and produce cusps with localized rotation, similar to those shown in Figure 14.7. These can produce tornadoes in a manner similar to that described for warm season squall lines. Because of these many possible outcomes, no one conceptual model describes all frontal squall lines.

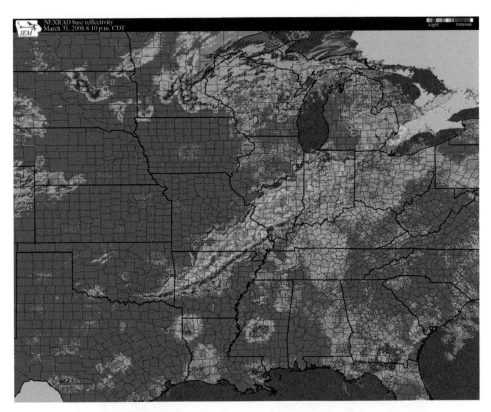

Figure 14.8 A frontal squall line stretching across several states (Courtesy Iowa Environmental Mesonet)

In the absence of supercellular storms, frontal squall lines, from a structural and radar perspective, appear similar to squall lines associated with warm season MCSs, except that the line of storms is typically much longer. Typically, frontal squall lines have a leading line of convection, a trailing stratiform region, front-to-rear storm-relative flow aloft above the frontal interface, and rear-to-front flow below the frontal interface, similar to their warm season counterparts. There are cases, however, where the vertical wind shear vector is oriented parallel to the front. In these cases, stratiform precipitation can straddle both sides of the convective line or develop on the north end of the line. These events are often associated with flash flooding, as thunderstorm cells propagate along the line, whereas the line itself is near-stationary. The media has coined the term **training thunderstorms** to describe these events, the analogy that the thunderstorms are like boxcars of a train moving along a track. MCSs with **leading stratiform** precipitation have also been observed[3] (Figure 14.9). Convection with leading stratiform precipitation that reaches the ground generally can be expected to have a short survival time because surface air ahead of the front feeding the storms is no longer insolated and is cooled by evaporating precipitation

[3]This event is described by Storm, B.A. *et al.* (2007) A convective line with leading stratiform precipitation from BAMEX. *Mon. Weather Rev.*, **135**, 1769–1785.

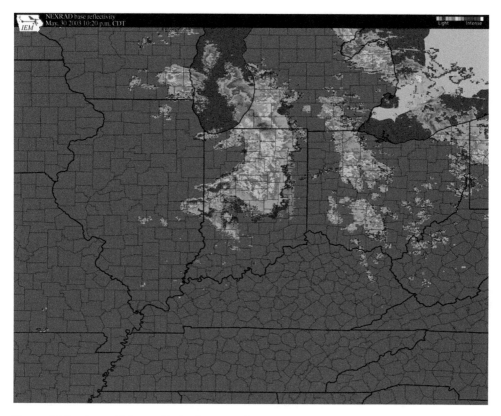

Figure 14.9   A squall line with leading stratiform precipitation (Courtesy Iowa Environmental Mesonet)

(although in some cases when the precipitation evaporates before reaching the ground, the conditional instability ahead of the MCS can be increased).

14.3   Supercell thunderstorms

Supercells (Figure 14.10) are the world's most violent thunderstorms. Their defining property is rotation of the storm updraft. The source of rotation is ultimately tilting of vorticity associated with environmental vertical wind shear. A supercell's circulations dominate its immediate area so that the entire storm behaves as a single entity, rather than as a group of individual cells, like an MCS. Supercells produce most tornadoes, virtually all strong tornadoes, most large hail, and damaging straight-line winds. Updrafts, determined using multiple Doppler radar measurements, are typically 20–40 m s^{-1} but have been estimated to approach 50 m s^{-1}. The strongest of supercell updrafts can support hailstones as large as grapefruits.

Supercells generally move to the right of the mid-tropospheric winds, which are typically from the west. Airflow within a supercell is closely related to precipitation processes and the environmental winds. Figure 14.11 shows the time evolution of the flow in a supercell at various levels, including updrafts and downdrafts. Figure 14.12

Figure 14.10   Structure of a supercell thunderstorm as viewed from southwest (Courtesy NOAA)

shows the radar reflectivity in a supercell from a 0.5° scan of a WSR-88D, with the location of the updraft, downdrafts, gust fronts, and low-level flow features superimposed.

The northeast side of a supercell is called the **forward flank** of the storm, whereas the southwest side is called the **rear flank**. The directions describing the storm flanks are relative to the storm updraft. The same terminology is used to describe the downdrafts and gust fronts in these sectors of a supercell. The inflow to the initial updraft approaches the core of the updraft from all directions. Precipitation formed within the storm's updraft is carried downwind of the updraft core by the southwesterly and westerly middle and upper-level winds, respectively, shown by the long arrows at 4 and 9 km altitude in Figure 14.11a. As precipitation falls to the north, northeast, and east, evaporative cooling of precipitation and precipitation drag leads to the formation of downdrafts. The **forward flank downdraft** (FFD) forms to the northeast of the updraft. Most of the air in the FFD originates in the lower to middle levels from altitudes of about 2 to 4 km. The downdraft air reaches the surface and spreads outward in all directions, creating at its leading edge the **forward flank gust front** (Figures 14.11 and 14.12). The precipitation in the FFD creates the main part of the radar echo.

As mid-level air approaches the storm from the west and encounters the updraft, cloud and precipitation particles along the rear flank of the storm mix with the dry air and evaporate. Air cooled in this region descends to the surface creating the **rear flank downdraft** (RFD, Figure 14.11b). Precipitation within the RFD wraps around the updraft due to the storm rotation, creating the **hook echo** seen on radar in the southwest sector of the storm (Figure 14.12). As air from the RFD reaches the ground and spreads out, it creates the **rear flank gust front** (Figures 14.11c and 14.12). New cells

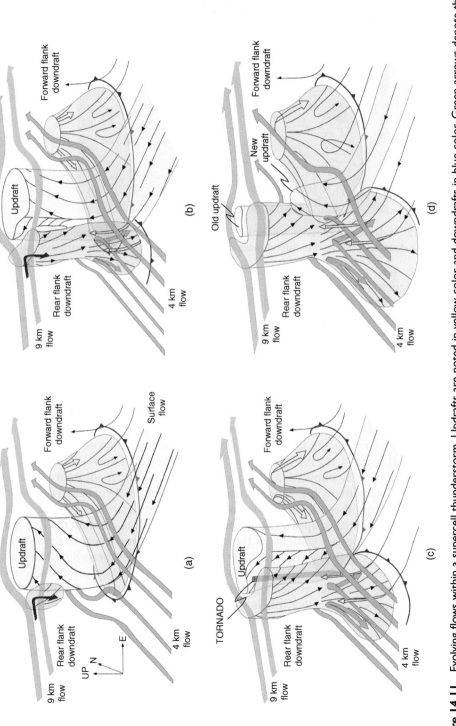

Figure 14.11 Evolving flows within a supercell thunderstorm. Updrafts are noted in yellow color and downdrafts in blue color. Green arrows denote the mid-level flow and tan arrows denote the upper-level flow. Gust fronts are noted by the blue lines with barbs. Time progresses from (a) to (d) (Adapted from Lemon, L.R. and Doswell, C.A. (1979) Severe thunderstorm evolution and mesocyclone structure as related to tornadogenesis. *Mon. Weather Rev.*, **107**, 1184–1197. ©American Meteorological Society, used with permission)

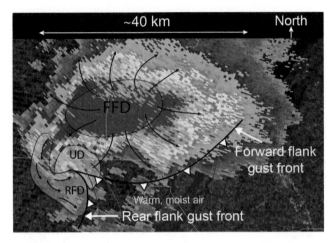

Figure 14.12 Relationship of radar echoes in a supercell to key structural features including the forward flank downdraft (FFD), rear flank downdraft (RFD), updraft (UD), and forward and rear flank gust fronts

are often triggered along the rear flank gust front as it advances, creating a **flanking line** of convection that extends southwest of the updraft region (Figure 14.10).

Tornado formation within a supercell is believed to occur in three steps. The first step is the tilting of the vertical wind shear that causes the storm's updraft to rotate. This rotation occurs well above the surface and leads to the formation of the **mid-level mesocyclone**. The second step occurs when strong rotation (not yet tornadic) develops very close to the ground underneath the mid-level mesocyclone. The exact processes responsible for the development of this near-ground rotation are still not fully understood, but a horizontal circulation that forms along the forward-flank gust front through a process called **baroclinic generation of vorticity** is thought to play an important role. Air typically flows along this gust front and then enters the updraft from the east (Figure 14.11). Air behind, and typically north, of the forward flank gust front is cool, negatively buoyant, and descending. The warm air adjacent to the gust front is typically positively buoyant and ascending. Horizontal vorticity is generated along the interface of these flows. As the horizontally rotating air along this boundary advances under the strong updraft, it is tilted into the vertical, creating the **low-level mesocyclone** that appears visually as the rotating **wall cloud** beneath the rain-free base of the storm's updraft. The final step occurs when the supercell's RFD moves under the low-level mesocyclone. As with the processes that develop the low-level mesocyclone, the processes associated with the RFD that lead to a tornado are still not fully understood. But a key role is likely played by horizontal rotation near the ground that is generated by the buoyancy gradient at the boundary of the descending RFD. This horizontal rotation is then tilted upward creating a horizontally rotating "tube" of air. This tube is then stretched by the updraft to become the tornado.

14.3.1 Tornado detection

Tornadoes loft debris into the air. Debris provides unique radar signatures that can pinpoint the location of a tornado. There are four key signatures that identify the

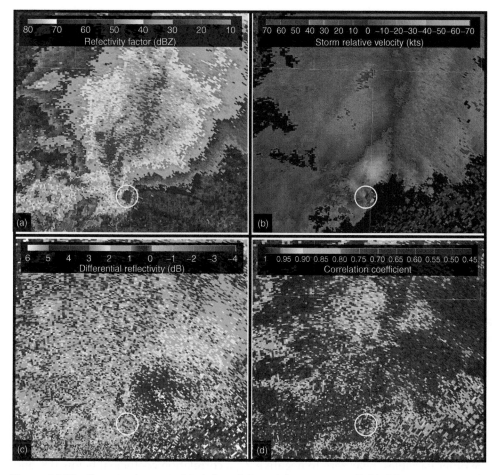

Figure 14.13 Tornado signatures (circled area) from Arkansas supercell: (a) debris signature in the reflectivity factor field; (b) velocity couplet in the radial velocity field; (c) low or negative values of differential reflectivity; (d) low values of the correlation coefficient

position of a tornado on radar, three associated with debris, and one associated with rotation. Figure 14.13 shows a tornadic supercell thunderstorm, one of many that occurred during the super outbreak in late April 2011 over the southeast USA (Figure 14.14). The white circle denotes the location of the tornado. Note in the reflectivity field (Figure 14.13a) the intense echo at the tip of the hook on the southwest side of the storm. This feature is called a **tornado debris signature**, colloquially called a **debris ball** by tornado enthusiasts. In Figure 14.13b, at the same location, a velocity couplet, signifying strong rotation, is present in the storm-relative radial velocity field. The differential reflectivity field in Figure 14.13c shows Z_{DR} values that are near zero or negative. These values are expected as, unlike raindrops, debris has random orientation. Finally, in Figure 14.13d, the correlation coefficient shows values well below 0.8. The echoes from debris are decorrelated compared to precipitation.

As a second example, consider Figure 14.15a–d. These data, from the University of Oklahoma RAXPOL research radar, are from the El Reno, Oklahoma tornado that,

Figure 14.14 A large outbreak of tornadic supercell thunderstorms over northern Alabama and nearby states on April 27, 2011

at the time of this writing, was the largest tornado observed. The same variables are shown as in Figure 14.13, but the radar was much closer to the tornado. The tornado signatures in the four variables again are easy to identify. In addition to these signatures, the close proximity of the RAXPOL radar to the tornado reveals a weak "hole" in the radar reflectivity field that marks the center of the tornado (Figure 14.15a). This feature is the result of centrifuging of debris away from the center of the tornado. The other variables, near zero to negative Z_{DR}, correlation coefficient below 0.8, and a radial velocity couplet, are obvious. These examples demonstrate how these four variables together can provide an unambiguous signature of a tornado.

Unfortunately, not all tornadoes produce as clear of a signature as these examples. If a tornado is far from a radar, the beam may be high enough to not intercept debris. Rotation may not be obvious at higher altitudes. Even in cases where the tornado is close to the radar, the ground over which a tornado passes may be dry, hard, and compacted, so that little debris is lofted. Nevertheless, for most situations, the debris signatures, combined with rotation in the storm-relative velocity field, provide sufficient evidence that a tornado is on the ground, and a warning can be issued. It is, of course, highly desirable to issue tornado warnings before tornado formation. Before tornado formation debris signatures will not be present, and the nowcaster must rely on the Doppler radial velocity signature of strong rotation, particularly at the location near the tip of the hook echo.

14.3.2   Radar signatures of supercells

The appearance of a supercell thunderstorm on radar depends on the size of the storm, the current phase of its lifecycle, and its position relative to the radar.

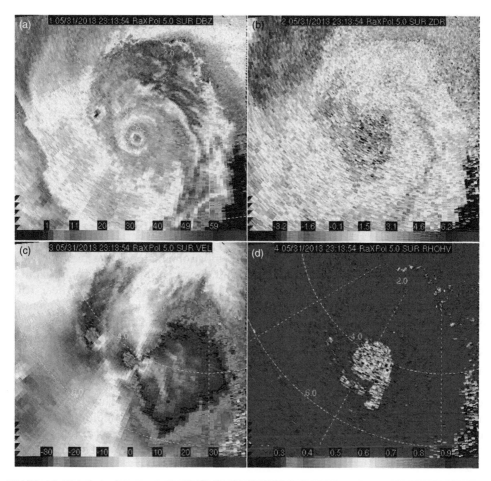

Figure 14.15 A tornado viewed by the Rapid Scan X-Band dual-Polarization Radar (RAXPOL). The panels show (a) the radar reflectivity factor, (b) the differential reflectivity, (c) radial velocity, and (d) correlation coefficient (Courtesy of Howard Bluestein)

Figure 14.14 shows an example image from the April 2011 super outbreak. Nearly every thunderstorm cell on this image is supercellular. The WSR-88D radar that obtained the data, KHTX, is located in northeast Alabama. Storms appear in various stages of their individual lifecycles. The radar echoes from some storms are merging with others. Storms A–C have clear debris signatures at the tips of their respective hook echoes. The shapes of the echoes correspond to the classic supercell depicted in Figure 14.12. The echo from the forward flank of Storm C has merged with the hook region of Storm D, making it more difficult to discern a debris signature within Storm D. Echoes from Storms E and F are also merging. The debris signature from Storm G is hard to identify—it appears separated from the body of the storm echo. Storms H and I are 218 and 254 km from the radar, respectively. The beam height at the location of these storms is 4600 and 5900 m. At this distance, the storms show no evidence of a hook echo, debris signature, or rotation signature, even though all might be present if the storms were closer to the radar. While nowcasting severe weather with radar, it is important to keep in mind the geometry of the beam and the complexities of storm evolution.

14.3.3 Hail detection

When large raindrops fall, aerodynamic forces acting on the drops cause them to become squeezed into a shape resembling a hamburger bun. The distortion begins when a drop's diameter grows to about 1 mm. The distortion becomes quite large as the horizontal diameter of a drop grows to sizes exceeding 3 mm. The electric field of a radar beam that is transmitted at horizontal polarization, when passing through a distorted drop, will sense a drop that appears to have a large diameter, whereas at vertical polarization, the drop appears smaller, leading to a Z_{DR} value in the range of 2–3 dB. When a hailstone grows, it typically tumbles. Some hailstones are spherical, and others have different shapes, but overall, hailstones fall randomly with no preferred orientation. When radar views a large number of hailstones, the average shape of the stones appears nearly spherical. As a result, the radar reflectivities at horizontal and vertical polarizations are nearly equal, and the differential reflectivity is near 0 dB. The difference in differential reflectivity between rain and hail allows the differential reflectivity to be a good discriminator between these two precipitation types.

Figure 14.16 shows the reflectivity and differential reflectivity measured in a supercell thunderstorm that produced softball-sized hail near Dallas, TX. Two areas, indicated by the circle and the oval, show regions where the radar reflectivity exceeds 60 dBZ. The Z_{DR} values coinciding with the >60 dBZ pixels within the circle are all near 0 dB, indicating that this is a region where hail is present. The Z_{DR} values in the region with the oval vary from 1 to 3 dB, suggesting that this region is probably either rain or more likely a mixture of rain and hail. There are no exact cutoffs for discriminating hail from rain, but generally, the higher the reflectivity and lower the Z_{DR}

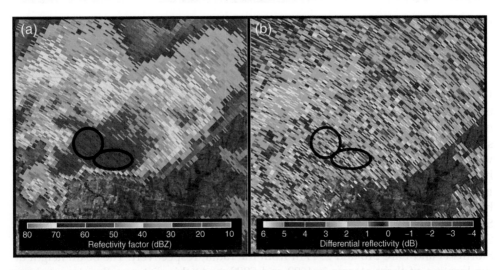

Figure 14.16   Reflectivity factor (a) and differential reflectivity (b) within a severe hailstorm over the Dallas–Fort Worth area as observed by the KFWS WSR-88D. The circle and oval areas are described in the text

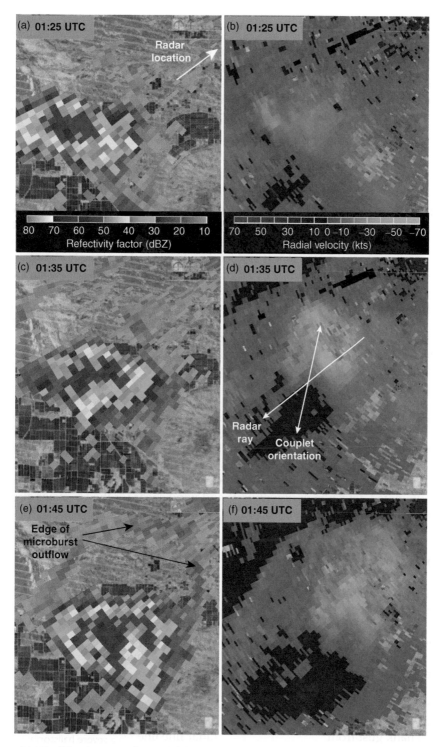

Figure 14.17 Microburst outflow as observed by the Phoenix, Arizona WSR-88D (KIWA) in the reflectivity factor (a,c,e) and radial velocity (b,d,f) fields at three times, each separated by 10 min. The maximum wind gust was 67 miles h⁻¹

values, the higher the probability of hail. Another feature that sometimes occurs with hail is a spike-shaped feature that extends outward radially behind a high-reflectivity region (>55 dBZ). This feature, the three-body scatter spike, results from a three-way scattering process, where hail scatters energy toward the ground, part of the signal is then reflected back to the hail, and then scattered back to the radar. An example is shown in Figure 12.11. Typically, the strong negative velocities that characterize the feature are due to the large fall velocity of hail.

14.4   Downbursts and wind shear

A downburst is a strong downdraft that originates within the lower part of a cumulus cloud or thunderstorm and descends to the ground. As air within a downburst reaches the surface, it spreads rapidly outward creating strong straight-line winds. Winds from downbursts can exceed 30–40 m s^{-1}. Downbursts are extremely hazardous to aircraft during takeoff and landing because the strong horizontal wind shear across a downburst can lead to a rapid reduction in true air speed as an aircraft traverses across it. Reduction of true air speed reduces lift, which can be catastrophic near landing and takeoff when an aircraft is flying near its stall speed. As a cumulus cloud can evolve from the updraft stage to a downburst in a matter of minutes, special monitoring and warning systems, particularly the TDWR radars, have been deployed to alert pilots to the presence of downbursts and associated wind shear.

The radar signature of a downburst is a radial velocity couplet, where the axis connecting the maximum and minimum radial velocity is oriented along the radar beam. Figure 14.17 shows a developing downburst from a storm near Phoenix, Arizona, that produced maximum outflow winds of ~30 m s^{-1}. The radial velocity couplet is not as distinct as those with tornadoes because the flow extends outward with time. A radar fine line associated with lofted dust is evident at the edge of the microburst outflow. The radial velocity couplet in this case is oriented slightly off axis, suggesting some rotation to the flow as well as divergence. Currently, the TDWR radar network provides automated detection of downbursts and wind shear for the US major airports. Details about the operation of these radars are discussed in Chapter 10.

Important terms

baroclinic generation of vorticity	gust front
bow echo	hook echo
bright band	leading stratiform
cold pool	low-level mesocyclone
debris ball	mesoscale convective system
downburst	MCS
flanking line	mid-level mesocyclone
forward flank	mesoscale convective vortex
forward flank downdraft	quasi-linear convective system
forward flank gust front	radar fine line
frontal squall line	rear flank
front-to-rear flow	rear flank downdraft

rear flank gust front

rear-inflow jet

rear-to-front flow

squall line

supercell thunderstorm

tornado

tornado debris signature

trailing stratiform region

training thunderstorms

wall cloud

Review questions

1. What radar features on a 0.5° surveillance scan from a WSR-88D radar would alert you to the presence of strong straight-line winds associated with an MCS?

2. What radar features on a 0.5° surveillance scan from a WSR-88D radar would alert you to the possible presence of a weak tornado along a squall line?

3. Do the strongest straight-line winds typically occur when a bow echo first forms, or when the bow is well developed?

4. Suppose the gust front associated with an MCS squall line was approaching an S-band Doppler radar. What key structural features of the MCS could be discerned from the reflectivity and radial velocity fields from an RHI scan oriented perpendicular to the convective line through the trailing stratiform region?

5. Are the strongest radar echoes in a supercell thunderstorm associated with the forward flank, or rear flank downdraft? Why?

6. What four radar signatures help pinpoint the location of a tornado in a supercell thunderstorm?

7. What radar variables can be used to distinguish hail from rain in a supercell thunderstorm?

8. Is the updraft of a supercell a region of weak or strong radar echo? Why? (Justify your answer by considering hydrometeor growth).

9. Suppose regions northeast of the hook echoes of three supercell storms had the following radar characteristics: Storm 1 : $Z = 55$ dBZ, $Z_{DR} = 4$ dB, Storm 2 : $Z = 58$ dBZ, $Z_{DR} = 0.1$ dB, Storm 3 : $Z = 59$ dBZ, $Z_{DR} = 2$ dB. Which storm is most likely to be producing hail? Which storm is most likely to have mixed hail and rain? Which storm is most likely to be producing heavy rain only?

10. How does a microburst radial velocity couplet differ from the radial velocity couplet associated with a tornado?

Challenge problems

11. If your institution has software such as GRAnalyst® available to view Level-2 radar data from WSR-88Ds, download radar data from a recent

(continued)

tornado event and attempt to track the tornado by identifying the debris and rotation signatures.

12. You are driving across a deadly supercell thunderstorm that is producing baseball-sized hail and an EF-5 Tornado. Luckily, your armored vehicle can withstand these conditions. During your drive, you decide to compare the radar data you retrieved on your cell phone with the conditions outside at points A, B, C, and D below. You check the reflectivity, Z_{DR}, and correlation coefficient. What values of these variables might you expect at points A, B, C, and D, and what are the likely scatterers (what do you see out the window) that are causing the radar returns at each of the four points?

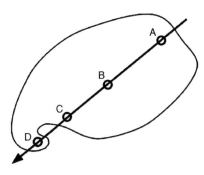

13. You work at the National Weather Service and would like to have an automatic detection algorithm to detect tornadoes. You decide to develop a "fuzzy logic" system with the full radar suite of variables available on a WSR-88D to pinpoint tornadoes on the radar display. Based on your knowledge of tornado characteristics on radar, develop a series of membership functions (you can draw graphs) that together would have a high probability of pinpointing a tornado.

14. Suppose a WSR-88D radar with polarization capability is observing three supercell thunderstorms simultaneously at different locations on a particular day. At the tip of the hook echo region, the radar signature had the following radar characteristics:

Storm 1: $Z = 43$ dBZ, CC $= 0.90$, radial velocity couplet present

Storm 2: $Z = 55$ dBZ, CC $= 0.99$, radial velocity couplet absent

Storm 3: $Z = 59$ dBZ, CC $= 0.70$, radial velocity couplet present

Below are three statements. Which best goes with which storm?

a. A dangerous tornado is on the ground.

b. A tornado may be present but cannot be confirmed with radar.

c. A tornado is unlikely at this time.

15. Some supercell thunderstorms are called low-precipitation or LP supercells. Search the internet to identify the time and location that a low-precipitation supercell was observed. Now obtain the Level-2 WSR-88D radar data for the WSR-88D closest to the storm. Examine the radar echo pattern. How does the radar echo pattern differ from a classic supercell echo?

15

Extratropical Cyclones

Objectives

By the end of this chapter, you should understand the following:

- The radar techniques that have been used to observe mesoscale features in extratropical cyclones.
- The difference between an anafront and katafront and how the rainbands associated with these fronts appear on radar.
- How gap and core structures in the narrow cold-frontal rainband relate to the position of weak tornadoes.
- Where elevated convection and thundersnow typically occurs within the comma head of winter cyclones.
- The nature of precipitation banding within the comma head and the various mechanisms that can lead to banding.
- What generating cells are, where they occur, and why they are important for precipitation production.
- The relationship between radar variables and the rain–snow line at the surface.

15.1   Introduction

Extratropical cyclones, the large comma-shaped storms of earth's middle-latitudes, create a wide variety of weather over the areas they influence, weather that ranges from blizzards to supercell thunderstorms. A single cyclone can influence weather over a vast geographic area covering several million square kilometers, an area much larger than the ten-thousand square kilometer domain of any single radar. Individual radars at any given time observe **mesoscale** weather features within small regions of a cyclone, although radar reflectivity composites created from many radars can depict the precipitation structure of an entire cyclone (Figure 15.1).

Figure 15.2 shows archetypical tracks of cyclones affecting the North American Continent. Cyclones influencing the west coast of North America typically originate over the Pacific and arrive at the coast in an occluded state. As a cyclone moves over the coast, the mesoscale radar features, typically bands of heavier rain oriented near-parallel to the frontal boundaries, separated by regions of lighter rain, are

Radar Meteorology: A First Course, First Edition. Robert M. Rauber and Stephen W. Nesbitt.
© 2018 John Wiley & Sons Ltd. Published 2018 by John Wiley & Sons Ltd.
Companion website: www.wiley.com/go/Rauber/RadarMeteorology

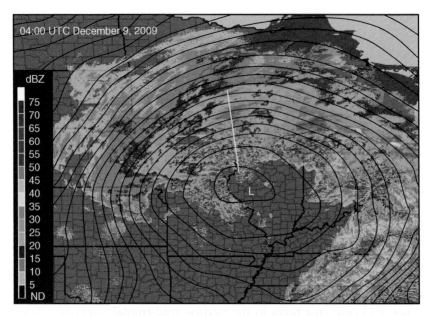

Figure 15.1   Composite radar image of an extratropical cyclone over the central USA. The yellow line refers to the location of cross sections in Figure 15.4 (Image courtesy of the Iowa Environmental Mesonet)

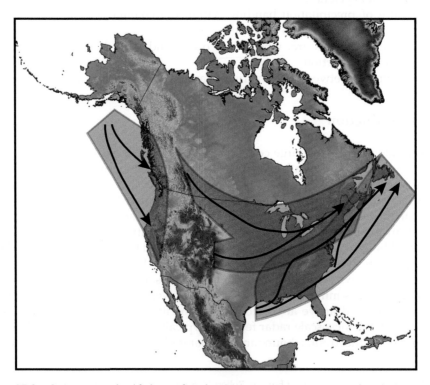

Figure 15.2   Common tracks of the surface low-pressure centers associated with major winter cyclones impacting the North American Continent

disrupted as they encounter the steep topography of the Coastal, Cascade, and Sierra Nevada ranges.

Cyclones affecting interior North America typically originate east of the Rockies, particularly at the latitude of Colorado, and over the plains near Alberta, Canada. The mesoscale weather produced by these cyclones varies considerably depending on the ambient environment in which the cyclone forms, location within the cyclone, and even time of day. Clouds and storms that form the tail of the **comma cloud** normally align with either the upper-level or surface-based cold front. Depending on the meteorological conditions, clouds along the comma tail may produce showers or more violent weather such as squall lines or supercells. The **comma head** consists of clouds overriding the warm and/or occluded fronts. These clouds can produce rain, ice pellets, freezing rain, and/or snow. Much of the hazardous weather across central North America is produced by extratropical cyclones. Tornado outbreaks, hail, and strong straight-line winds develop from squall lines and supercell thunderstorms that form along the **comma tail**. Heavy rain and flooding can occur locally along the comma tail or comma head when fronts remain stationary and storms pass repeatedly over the same location. Ice storms, blizzards, and heavy snowstorms commonly occur beneath the comma head in winter.

East Coast cyclones preferentially form in two regions along the US coastline: the first just off the East Coast, particularly around the latitude of North Carolina, and the second just off the Gulf Coast centered near Texas–Louisiana. East Coast cyclones typically track northeast along the coast and out to sea northeast of Canada's Maritime Provinces. Gulf cyclones that affect the East Coast commonly track either eastward along the Gulf Coast and then northward along the Atlantic coastline, whereas those affecting the eastern half of the interior continent typically track inland along the Mississippi and Ohio River Valleys. Mesoscale weather features observable with radar associated with cyclones tracking along the East Coast are normally associated with the comma head of these storms, as the comma tail is located over the Atlantic Ocean. As East Coast cyclones predominately occur in the cold season, the primary concern with these storms is winter weather.

Although the discussion in this chapter focuses on radar features within cyclones affecting North America, the information applies well to cyclones over Europe and along Asia's East Coast. Cyclones moving off the Atlantic over Great Britain and the European Continent have similar mesoscale structures to those observed moving into the North American Continent from the Pacific. In Europe, the cyclones are not as strongly influenced by topography, so mesoscale features maintain their structure as they move deeper into the continent. Cyclones along the East Coast of Asia evolve similarly to those along the East Coast of North America.

15.2   Radar approaches to monitor cyclone mesostructure

Three basic complementary approaches have been employed using radar to study and monitor mesoscale weather features within extratropical cyclones: (1) 360° scanning at a series of progressively higher elevation angles, (2) pointing a radar vertically and monitoring the weather as it passes across a site, and (3) flying across a storm with airborne radar that can either be scanning, vertically pointing (up and down), or pointing at an angle to either nadir or the zenith. The first approach is used operationally by the S-band WSR-88D radars. The latter approaches typically use shorter wavelengths, X-band, K-band, or W-band radars, that are more sensitive to smaller cloud particles. Use of these radars is largely confined to short-duration field experiments, although vertically pointing radars are coming into more common use, particularly at universities.

Scanning radars have the advantage of monitoring the time evolution of mesoscale features as they propagate across the radar domain, allowing a forecaster to estimate the start and stop time of precipitation, precipitation rate, and integrated precipitation, at least in rain. For example, Figure 15.3 shows the radar reflectivity measured on the 0.5° scan of the KARX WSR-88D radar at La Crosse, Wisconsin, during a snowstorm associated with the December 9, 2009 cyclone appearing in Figure 15.1. The northern section of the precipitation region can be seen to be organized into two primary (A and C) and several somewhat narrower (e.g., B, D) near-parallel, linear bands. The southern section of the precipitation contained discrete cells (e.g., cells a–g) with no obvious organization. This type of organization has been noted in other strong winter cyclones.

Figure 15.4 shows vertical cross sections along the white line in Figure 15.3 (see also yellow line in Figure 15.1). Figure 15.4a shows the WSR-88D reflectivity data from the volume scanned between 03:22 and 03:31 UTC, interpolated to the plane

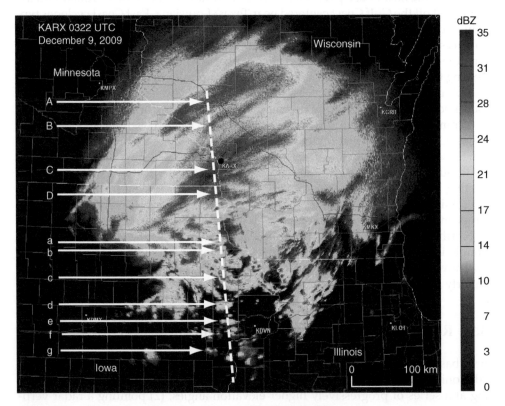

Figure 15.3 Radar reflectivity factor from the KARX WSR-88D in the same storm as shown in Figure 15.1. The white dashed line corresponds to the yellow line in Figure 15.1. The features labeled with letters correspond to the features in the cross sections shown in Figure 15.4

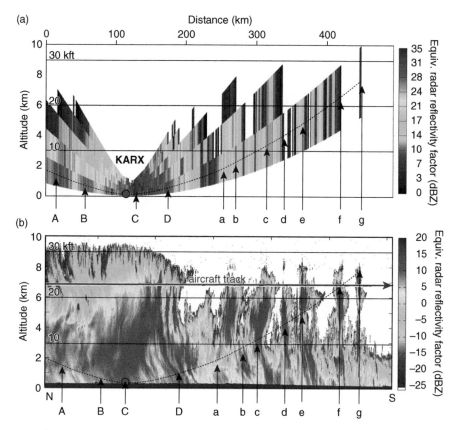

Figure 15.4 Radar reflectivity factor measured by (a) the KARX WSR-88D and (b) by the W-band Wyoming Cloud Radar flown aboard the NSF/NCAR C-130 aircraft at approximately the same time and in the same location. The features labeled with letters correspond to the features in Figure 15.3

of the cross section. Figure 15.4b shows data along the same cross section collected by the University of Wyoming Cloud Radar (WCR), a vertically pointing W-band radar that was flown on the NSF/NCAR C-130 research aircraft during this storm. Comparing Figure 15.4a,b, one can see that the WSR-88D poorly resolved many of the key features responsible for snowfall generation. The elevated convective cells rising to the tropopause on the right side of Figure 15.4b (cells a–f on Figure 15.3) were poorly resolved and their vertical depth is undeterminable. The cells at cloud top on the left side of Figure 15.4b went undetected by the WSR-88D. The poor sensitivity of the S-band WSR-88D to small ice particles and the low resolution of the data projected into a vertical cross section from its standard plan position indicator (PPI) mode account for the differences between Figure 15.4a,b. The airborne data, on the other hand, provide no information about the horizontal organization or time evolution of the precipitation, a distinct disadvantage of the airborne radar.

Vertically pointing radars present a time–height depiction of cyclone mesostructure as the storm passes over a site. For example, Figure 15.5 shows a WSR-88D composite of a weak cyclone that moved across the Ohio Valley on February 14–15, 2010. The dot in Indiana shows the location of the University of Alabama-Huntsville's vertically pointing X-band Polarization Radar (XPR). A 1 h segment of reflectivity

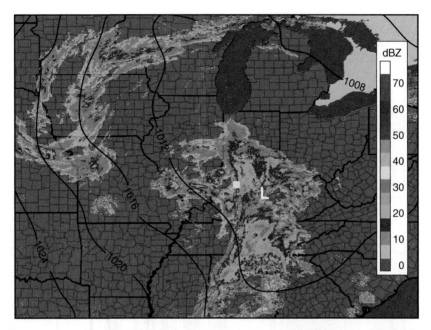

Figure 15.5   WSR-88D composite of a weak cyclone over the Ohio Valley. The yellow square in Indiana shows the location of the University of Alabama-Huntsville's vertically pointing X-band Polarization Radar (XPR) that collected the data shown in Figure 15.6 (Image courtesy of the Iowa Environmental Mesonet)

and vertical radial velocity from the XPR radar are shown in Figure 15.6. The high-resolution view of the clouds within the comma head shows the presence of cloud-top generating cells with upward and downward velocities exceeding $1–2\,\mathrm{m\,s^{-1}}$ and the passage of an upper-level front near 07:45 UTC that was accompanied by a rapid reduction in cloud depth.

The observation techniques described above illustrate advantages and disadvantages of different techniques used to probe cyclone mesostructure with radar. It is important to recognize the strengths and limitations of various radar systems when interpreting data and features they show.

15.3   Mesoscale structures observable with radar

The broad-scale vertical motions that produce stratiform clouds and precipitation within an extratropical cyclone are forced primarily by secondary circulations associated with frontogenesis. Superimposed within the broad-scale stratiform radar echoes are often narrow, long lines of higher reflectivity termed **precipitation bands** (see Figure 15.3). At the top of the stratiform cloud echo in cyclones, shallow convective circulations are very common. These circulations are called **generating cells** because they generate plumes of precipitation that descend through the stratiform clouds as **fall streaks** observable in the reflectivity field. In some cyclones, potentially unstable layers develop aloft in the comma head region as dry air from the cyclone's dry slot intrudes over a lower level moist airmass. This can lead to **elevated convection**—convection occurring in the middle troposphere above a

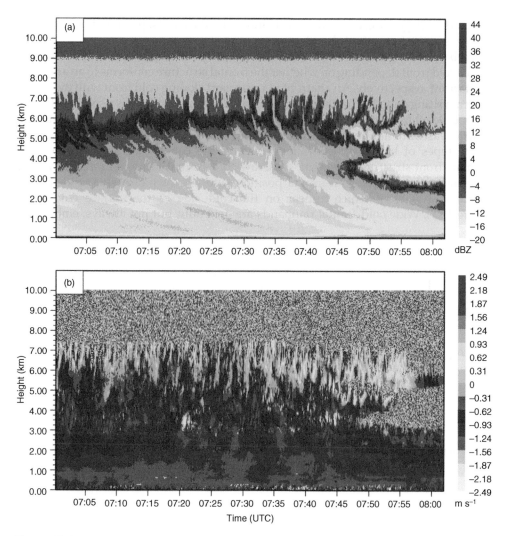

Figure 15.6 (a) Radar reflectivity factor and (b) vertical radial velocity measured by the University of Alabama-Huntsville's X-band vertically pointing radar (XPR) in the cyclone depicted in Figure 15.5

stable layer—and occasional lightning in winter snowstorms. Surface-based **deep convection** is also common, particularly along the comma tail.

15.3.1 The comma-cloud tail

The organization of the precipitation field appearing on radar along the tail of an extratropical cyclone comma cloud depends on the type of front forcing the circulation, the atmospheric stability, vertical and horizontal wind shear, and moisture availability. Over land during the transitional and warm seasons, frontal squall lines and/or lines of supercell thunderstorms often develop along the comma-cloud tail. Deep convective phenomena, including frontal squall lines, will not be considered in this chapter, as they were already discussed in Chapter 14. During the cool season over land, and more generally over the ocean and along the coasts, precipitation normally organizes in the form of **rainbands**.

Rainbands along the tail of the comma cloud may develop ahead of or behind the surface cold front depending on whether the frontal structure conforms to an **anafront** or **katafront**, respectively. An anafront is a rearward sloping cold front with the warm airstream rising over the frontal surface (Figure 15.7a). A katafront is a split front, with an upper-level front leading the surface cold front. With a katafront, descending dry air flows eastward aloft over the surface-based cold front (Figure 15.7b).

Two types of bands typically appear along the comma tail within a cyclone when anafront structure is present, a **narrow cold-frontal rainband** (NCFR), which conforms to the surface cold-frontal position, and one or more **wide cold-frontal rainbands** (WCFRs), which appear on the cold side of the surface cold front (Figure 15.8). Wide cold-frontal rainbands are generally, but not always, embedded

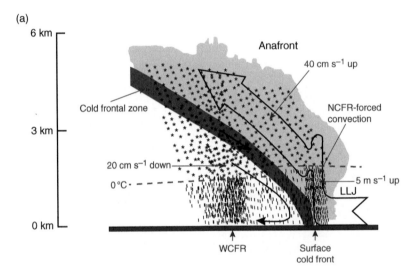

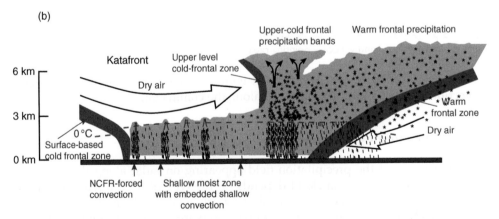

Figure 15.7 Schematic cross sections through an anafront (a) and katafront (b). Frontal zones are indicated in blue, clouds in gray, circulations with arrows, and precipitation as rain (short lines) and snow (stars) (Adapted from Browning, K.A. (1986) Conceptual models of precipitation systems. *Weather Forecasting*, 1, 23–41. © American Meteorological Society, used with permission)

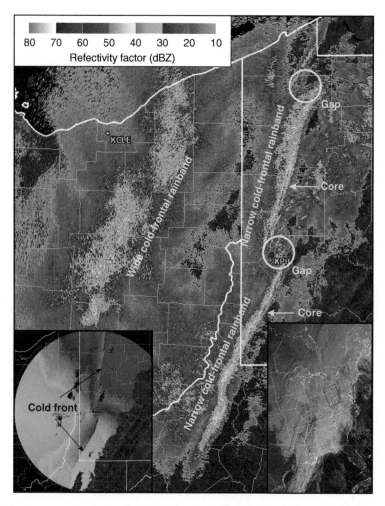

Figure 15.8 An example of a narrow cold-frontal rainband and a wide cold-frontal rainband along the tail of a cyclone that passed over eastern Ohio and Pennsylvania. The larger panel shows the radar reflectivity factor, whereas the left inset shows the radial velocity and the right inset the cloud cover from a visible satellite image

in weaker precipitation echo on radar displays (e.g., Figure 15.8). The narrow cold-frontal rainband is only about 3–5 km wide but can have vertical velocities exceeding 5 m s^{-1}. Measurements suggest that these bands often occur in a near-zero convective available potential energy environment. Forced lifting at the cold-frontal interface triggers the NCFR. NCFRs often have a core and gap structure that develops from horizontal shearing instability along the front. In some cases, weak tornadoes, formed from stretching of vortices associated with shearing instability, have been observed to form in the vicinity of the gaps.

WCFRs have widths in the range 60–100 km. The dynamical/microphysical mechanisms responsible for WCFRs are not well established. Hypotheses have been put forward relating the bands to the release of conditional symmetric instability. The bands may also be related to the merging of fall streaks from cloud-top convective generating cells. There is evidence that at least some bands are associated with a change in frontal slope. In any case, vertical velocities in the bands are generally weak, characteristic of stratiform rather than convective vertical motions.

A katafront is a split front in which the upper-level cold front leads the surface cold front (Figure 15.7b). With a katafront, dry air descends over the cold-frontal surface. The leading edge of the dry air aloft corresponds to the location of the upper-level front. On satellite images, the descending dry air aloft appears as the cyclone's dry slot (Figure 15.9 inset). In some cyclones, a NCFR may be present along the surface cold front (Figure 15.9). Shallow clouds, typically stratocumulus or stratus topped by convective generating cells, are typically located between the surface cold front and the upper-level front. Deep rainbands typically form along and ahead of the upper-level front. These rainbands can be isolated (Figure 15.9) or embedded in weaker echo. There is evidence indicating that in some cases, the rainbands develop as a result of the release of elevated potential instability. As noted earlier, particularly in the warm and transition seasons, a squall line, or even supercell thunderstorms, may develop ahead of the upper-level front. These deep convective storms are often surface based.

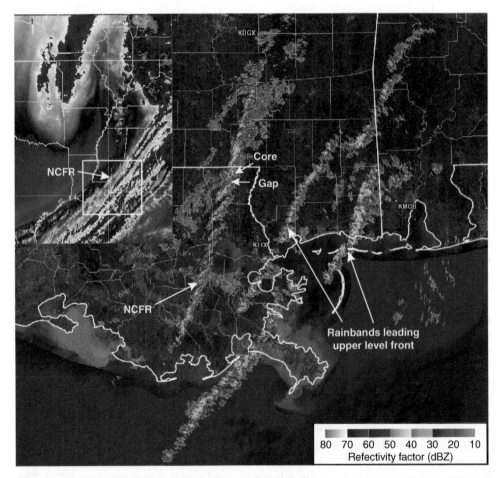

Figure 15.9 Rainbands along an upper level front and a surface cold front. The narrow cold-frontal rainband along the surface front exhibits core and gap structure

15.3.2   *The comma-cloud head*

The comma-cloud head in the cool season is the locus for winter weather. Blizzards, ice storms, heavy snow, sleet, mixed precipitation, and even winter thunder snow-storms all can occur. From a scanning radar perspective, such as with the WSR-88Ds, the comma head typically appears as a broad region of relatively uniform echo with embedded stronger echoes associated with precipitation bands, cellular elevated con-vection, and local reflectivity enhancements where the beam intersects the melting layer. From a vertically pointing radar perspective, the comma head appears as a stratiform layer, sometimes with a bright band if the surface temperature is above 0 °C, and typically topped with convective cloud-top generating cells that produce fall streaks of precipitation. Elevated deeper convective cells also are observed with vertically pointing radar.

Elevated convection

Upper tropospheric dry air associated with a cyclone's dry slot can sometimes intrude over the lower level moist air located above the warm front, particularly in strong cyclones. This creates two zones of precipitation within the comma head, a northern zone characterized by deep stratiform clouds and often topped by cloud-top generating cells, and a southern zone with dry air aloft that is marked by elevated convection. Figures 15.1, 15.3, and 15.4 show an example of the comma head of a cyclone exhibiting this structure.[1] Winter lightning (so-called **thundersnow**), when it occurs, appears to originate from the elevated convection within the southern zone. Figure 15.10 shows two elevated convective cells observed along the south side of the comma head of the storm in Figure 15.1. These cells are similar to those in Figure 15.4, but shown at a 1:1 projection, rather than the vertically stretched view in Figure 15.4. Radar measurements of vertical radial velocity from airborne radar and ground-based vertically pointing radars show that elevated convection in winter storms is associated with updrafts that can approach 6–8 m s^{-1} (Figure 15.11). The precipitation plumes falling from these cells typically produce cellular or short linear feature echoes on radar (Figure 15.3). If the plumes reach the ground as snow, the reflectivity may not differ substantially from that of the nearby stratiform echo, unlike radar signatures of rain from convection in the warm season.

Precipitation bands

Variability in the location, type, and intensity of precipitation within the comma head is often associated with precipitation banding—either rainbands or **snowbands**—on scales of approximately 5–100 km. Banding is particularly common in narrow regions in the northwest quadrant in cyclones where frontal structures and associated frontal

---

[1]This cyclone is described in Rosenow, A.A., Plummer, D.M., Rauber, R.M., McFarquhar, G.M., Jewett, B.F., and Leon, D. (2014) Vertical velocity and physical structure of generating cells and convection in the comma head region of continental winter cyclones. *J. Atmos. Sci.*, **71**, 1538–1558, and Rauber, R.M., Wegman, J., Plummer, D.M., Rosenow, A.A., Peterson, M., McFarquhar, G.M., Jewett, B.F., Leon, D., Market, P.S., Knupp, K.R., Keeler, J.M., and Battaglia, S. (2014) Stability and charging characteristics of the comma head region of continental winter cyclones. *J. Atmos. Sci.*, **71**, 1559–1582.

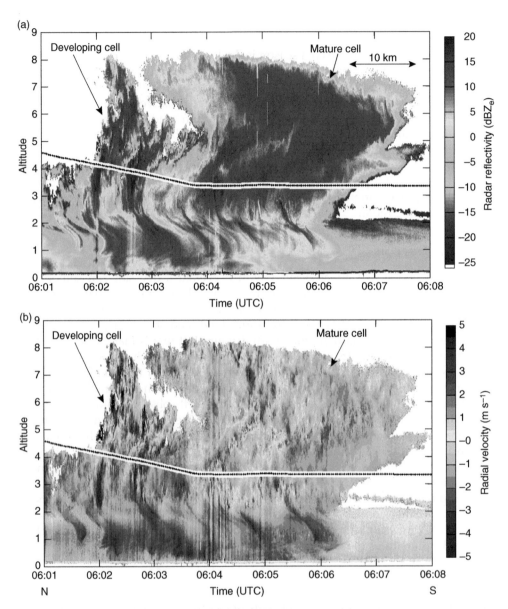

Figure 15.10   Cross section from the Wyoming Cloud Radar W-band radar showing the radar reflectivity factor (a) and vertical radial velocity (b) in an elevated convective cell within the southern comma head of the cyclone depicted in Figure 15.1. The dashed line is the aircraft flight track (From Rauber, R.M., Wegman, J., Plummer, D.M., Rosenow, A.A., Peterson, M., McFarquhar, G.M., Jewett, B.F., Leon, D., Market, P.S., Knupp, K.R., Keeler, J.M., and Battaglia, S.M. (2014) Stability and charging characteristics of the comma head region of continental winter cyclones. *J. Atmos. Sci.*, **71**, 1559–1582. © American Meteorological Society, used with permission)

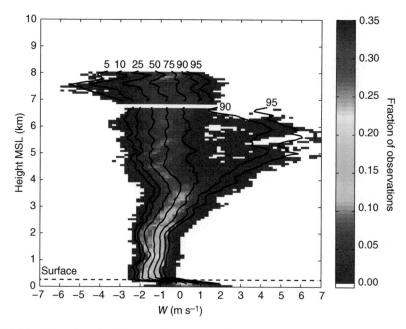

Figure 15.11 Wyoming Cloud Radar observations of vertical radial velocity depicted in a contour frequency by altitude diagram. The diagram shows the percentage of observations at each altitude falling into each 0.2 m s⁻¹ velocity bin. The break in the diagram below 7 km is the aircraft altitude. The numbers on the contours denote the percentage of observations with values to the left of the contour (From Rosenow, A.A., Plummer, D.M., Rauber, R.M., McFarquhar, G.M., Jewett, B.F., and Leon, D. (2014) Vertical velocity and physical structure of generating cells and convection in the comma head region of continental winter cyclones. *J. Atmos. Sci.*, **71**, 1538–1558. © American Meteorological Society, used with permission)

circulations are modified by deformation flow. As with the cold-frontal rainbands along the comma tail, a number of dynamic and microphysical mechanisms have been hypothesized to explain the existence of bands including ageostrophic circulations associated with frontogenetical forcing, the release of conditional instability, potential instability, or a form of moist symmetric instability, gravity wave forcing, organization of fall streaks by deformation flow, and others. Reviewing the underlying dynamics of all these possible forcing mechanisms is beyond the scope of this text.

Unfortunately, radar measurements of the fine-scale circulations within banded substructures and direct microphysical observations that can help interpret the enhanced radar reflectivity and polarization signatures are currently limited to a few field studies. With a single scanning Doppler radar, it is generally not possible to identify which of the possible mechanisms might be acting in a particular event. For example, Figure 15.12 shows an image of precipitation bands during a winter cyclone from the KOKX WSR-88D radar near New York City.[2] Three types of bands are clearly evident. The largest band inland of the coast is along a deformation zone

---

[2]This event is described in Novak, D., Colle, B.A., and Aiyyer, A.R. (2010) Evolution of mesoscale precipitation band environments within the comma head of northeast U.S. cyclones. *Mon. Weather Rev.*, **138**, 2354–2374.

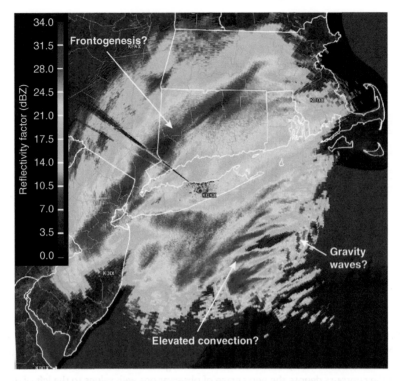

Figure 15.12   Precipitation bands within a winter cyclone over the US East Coast

and is likely forced by the ageostrophic circulation associated with frontogenesis. The forcing for the larger bands over the ocean is less obvious. Their position close to the cyclone's dry slot suggests that they might be associated with elevated convection (e.g., compare Figure 15.1 with Figure 15.12), but without high-resolution data, no definitive statement can be made. The small bands nearly normal to the other bands are an isolated feature. Their short wavelength suggests that they may be associated with gravity waves, but again, this is conjecture. The multiple-banded structure displayed here is common within the comma head of winter cyclones.

Cloud-top generating cells

Cloud-top generating cells were one of the early mesoscale structures observed with meteorological radars. Since the measurements in the early 1950s, many studies have shown them to be ubiquitous at the top of otherwise stratiform clouds in extratropical cyclones. Examples of generating cells can be seen at the top of the deep stratiform clouds in Figure 15.4 and at cloud top in Figure 15.6. A close-up of the radar reflectivity field within generating cells in a winter cyclone, along with the associated Doppler vertical radial velocity field, is shown in Figure 15.13 using a one-to-one aspect ratio. This view clearly differentiates the stable lower portion of the cloud from the generating cells. This perspective also shows that the cloud-top convection has similar horizontal and vertical scales. Although irregularities appear in the generating cell structure and spacing, the cells are approximately 1–2 km deep and 0.5–1.5 km wide. Figure 15.14 shows a statistical analysis of the vertical radial velocity measured

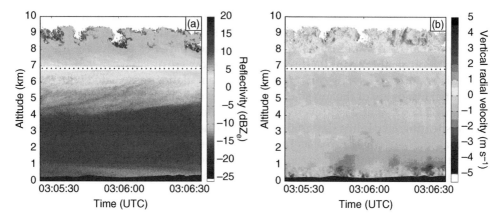

Figure 15.13   Wyoming Cloud Radar (a) reflectivity factor and (b) vertical radial velocity shown at a one-to-one aspect ratio depicting cloud-top generating cells at the top of a stratiform cloud layer. The aircraft track is the dotted line (From Rosenow, A.A., Plummer, D.M., Rauber, R.M., McFarquhar, G.M., Jewett, B.F., and Leon, D. (2014) Vertical velocity and physical structure of generating cells and convection in the comma head region of continental winter cyclones. *J. Atmos. Sci.*, **71**, 1538–1558. © American Meteorological Society, used with permission)

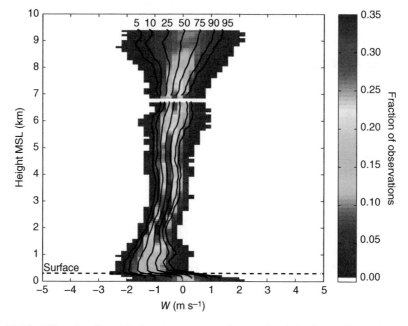

Figure 15.14   Wyoming Cloud Radar observations of vertical radial velocity depicted in a contour frequency by altitude diagram. The diagram shows the percentage of observations at each altitude falling into each $0.2\ \mathrm{m\,s^{-1}}$ velocity bin. The break in the diagram below 7 km is the aircraft altitude. The numbers on the contours denote the percentage of observations with values to the left of the contour (From Rosenow, A.A., Plummer, D.M., Rauber, R.M., McFarquhar, G.M., Jewett, B.F., and Leon, D. (2014) Vertical velocity and physical structure of generating cells and convection in the comma head region of continental winter cyclones. *J. Atmos. Sci.*, **71**, 1538–1558. © American Meteorological Society, used with permission)

through the depth of a 9.5 km deep stratiform area topped by generating cells. The cloud top exhibits both updrafts and downdrafts of 1–2 m s^{-1}, after accounting for the approximate 1 m s^{-1} terminal velocity of the ice particles producing the radar echo. The vertical motions in the stratiform area below are much weaker. Evidence to date indicates that generating cells are convective circulations associated with release of potential instability at the moist–dry interface at cloud top. These circulations can be enhanced or, in some cases, even be forced by cloud-top radiative cooling and/or vertical shearing instability. Generating cells produce fall streaks of precipitation on radar that may appear sheared as the precipitation encounters changing winds at different levels and is mixed into the stratiform echo (Figure 15.15a). Sometimes, the fall streaks retain their identity as they fall from the generating level to the ground (Figure 15.15b).

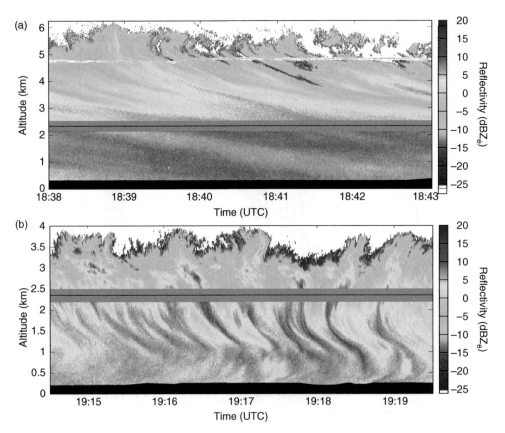

Figure 15.15 Precipitation fall streaks emanating from generating cells atop the comma-head clouds within an extratropical cyclone (From Rauber, R.M., Plummer, D.M., Macomber, M.K., Rosenow, A.A., McFarquhar, G.M., Jewett, B.F., Leon, D., Owens, N., and Keeler, J.M. (2015) The role of cloud-top generating cells and boundary layer circulations in the finescale radar structure of a winter cyclone over the great lakes. *Mon. Weather Rev.*, **143**, 2291–2318. © American Meteorological Society, used with permission)

Bright-band effects and the rain–snow line

Reflectivity enhancements associated with melting ice particles are common within the comma head of cyclones both near the rain–snow line, and where snow is melting above the surface and the radar beam intersects the melting layer producing the radar **bright band**. It is often difficult to distinguish melting effects in the reflectivity field from true bands associated with dynamic forcing. Figures 15.16 and 15.17 show two examples of reflectivity enhancements in the comma heads of two different cyclones. In Figure 15.16, the melting layer appears an annulus of stronger echo concentric with the radar (Figure 15.16). If the guide circles on Figure 15.16 were not there, the northern section could easily be mistaken for a band of heavier precipitation.

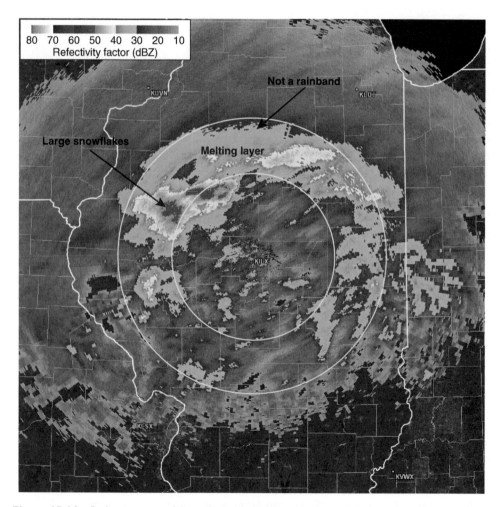

Figure 15.16 Radar signature of the radar bright band, with strong echoes associated with very large wet snowflakes. The inner and outer yellow circles denote the base and top of the bright band

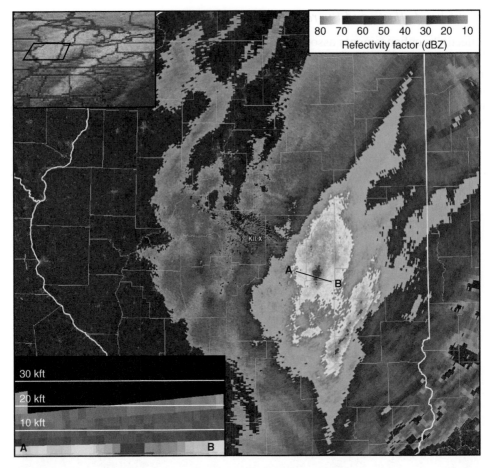

Figure 15.17 Radar signature of "silver dollar"-sized snowflakes (Giant snowflakes were observed at the ground at the location of the highest reflectivity at the time of this image.) The insets show a reconstructed RHI from points A to B and a satellite image showing the location of the radar image

In some cases, it is hard to discern whether what is observed is a bright band effect. For example, Figure 15.17 shows a radar reflectivity image of a brief episode of "silver dollar snowflakes" that occurred in central Illinois. Wet snowflakes with diameters ranging from 4 to 7 cm were observed at the surface beneath the region of highest reflectivity. The linear feature of enhanced echo to the east of the radar is entirely due to wet snowflakes. The radar reflectivity reached values between 50 and 55 dBZ. In fact, the media at the time reported the phenomenon as a thunderstorm. The cross section on Figure 15.17 shows that the high reflectivity was confined to the lowest beam, characteristic of a low-elevation melting layer. The surface temperature at the time the snowflakes fell was 0 °C.

Determining the location where rain transitions to snow in a winter cyclone is an important challenge in winter forecasting. An important question is whether polarization radar can be useful in identifying the transition. Melting of snow, in addition to enhancing the radar reflectivity, is also associated with an increase in

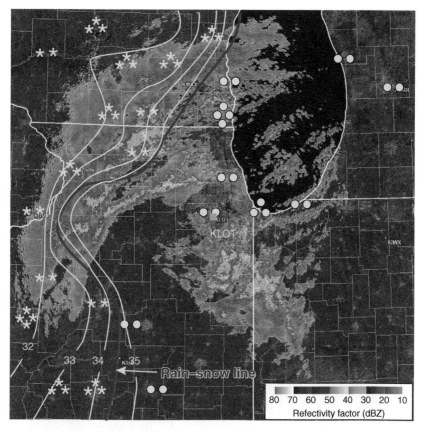

Figure 15.18 The radar reflectivity factor measured by the KLOT WSR-88D within the comma-head of a winter cyclone. The temperature field (°F) and the best estimate of the position of the rain–snow line based on surface observations are overlaid

the differential reflectivity and a decrease in the correlation coefficient. Figure 15.18 shows the reflectivity across the comma head of a winter cyclone as observed by the Chicago, IL (KLOT) radar. The **rain–snow line** is estimated from surface observations and the surface temperature distribution. Figure 15.19 shows the corresponding Z_{DR} and ρ_{HV} fields. The zone of high Z_{DR} and low ρ_{HV} is close to, but not align well with, the rain–snow line. This example points to a complexity of using radar to determine the rain snow line. The rain–snow line is a surface-based line (where rain changes to snow at the surface, not aloft). The radar bright band, which marks the transition, can occupy a considerable depth if melting proceeds slowly or a narrow altitude range if melting occurs rapidly. The bright band also slopes upward toward the warm air from its position at the ground at the rain snow line. Complicating the geometry, radar beams slope upward away from the radar because of earth curvature. Ultimately, the 3-D geometry of the intersection of the radar lowest scan with the sloping bright band determines where in space the band of low ρ_{HV} and high Z and Z_{DR} will appear.

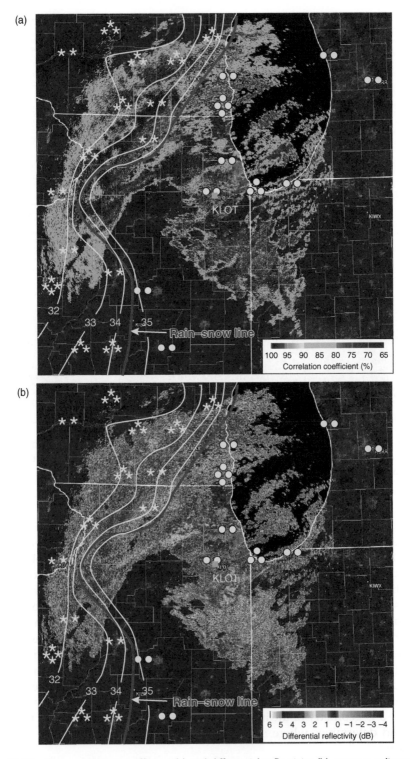

Figure 15.19 The correlation coefficient (a) and differential reflectivity (b) corresponding to Figure 15.18. The rain–snow line, the temperature field (°F), and the best estimate of the position of the rain–snow line based on surface observations are overlaid

Important terms

Anafront

Bright band

Comma cloud

Comma head

Comma tail

deep convection

Elevated convection

Extratropical cyclone

Fall streak

Generating cell

Katafront

Mesoscale

Narrow cold-frontal rainband

Precipitation band

Rainband

Rain–snow line

Snowband

Thundersnow

Wide cold-frontal rainband

Review questions

1. What type of hazardous weather is most commonly associated with cyclones impacting North America's West Coast, Central Plains, and East Coast, and how can radar be used to identify and warn against these hazards?

2. What are the advantages and disadvantages of using a scanning S-band radar to study winter precipitation?

3. What are the advantages and disadvantages of using a vertically pointing X or K-band radar to study winter precipitation?

4. Explain how precipitation bands would appear on a radar display for an anafront versus a katafront as the front moved across a radar domain.

5. How is the character of the precipitation different for a narrow cold-frontal rainband and a wide cold-frontal rainband?

6. A narrow cold-frontal rainband is approaching a city for which you have weather warning responsibility. A local police officer calls into say that she thinks she can see a tornado. Where along the rainband would you look for a signature of rotation?

7. You observe what appears to be a precipitation band on radar. What approaches can you take to be sure it is a true enhancement of precipitation and not just a bright band effect?

8. Where in the comma head region of a winter cyclone is a likely location for elevated convection? How might it appear on radar?

9. What is a generating cell? What are they found, what are their dimensions, and what typical vertical velocities characterize their updrafts and downdrafts?

10. What polarization signatures can you use to help identify the bright band and the rain–snow line in winter cyclones?

Challenge problems

11. Assume that ice particles formed in a long-lived cloud-top generating cell at 10 km altitude fall to the ground at a terminal velocity given by $V_t = 1$ m s^{-1} through an atmosphere with no background vertical air motion, but horizontal winds from the west that vary with height according to $V = 10$ m s$^{-1} + 5$ m s^{-1} km$^{-1} \times H$, where H is the height above the ground in kilometers. Suppose a scanning radar is located due east of the generating cell. Draw a diagram showing how the fall streak might appear on a range-height indicator (RHI) scan along the 270° azimuth. Suppose instead that the cell and fall streak passed over a vertically pointing radar. Draw a second diagram showing how the fall streak might appear on a time–height diagram. Assume both radars have the capability to resolve the fall streak as the ice particles descend to the ground.

12. Assume that a surface-based cold front is approaching a WSR-88D radar from the west. The synoptic scale winds are southerly ahead of the front and westerly behind the front. A narrow cold-frontal rainband exhibiting core and gap structure is along the front. Using red and green pencils, sketch how the Doppler radial velocity field might appear on a WSR-88D display in the vicinity of two core regions separated by a gap. Assume that the radar beam is pointing directly west when pointed at the center of the gap.

13. You are awarded a grant to investigate precipitation formation and organization across the comma head of winter cyclones and are allocated the use of a single mobile X-band Doppler polarization radar that has full 360° (PPI) scanning and 90° elevation (RHI) scanning capability. How would you set up your scanning strategy to optimize your opportunities to make scientific breakthroughs regarding our understanding of winter storms?

14. While sitting in the radar trailer during the study described above, you see on the 0.5° PPI scan a small region where the radar reflectivity has a maximum value of 54 dBZ embedded in a band of strong echoes organized in a curved line. Design a scanning strategy to determine if the phenomenon producing the reflectivity is a thunderstorm along a line of convection or melting snowflakes within the bright band.

15. If you have software capable of displaying Level-2 radar data, identify a WSR-88D radar that was beneath the comma head of a recent strong cyclone. Retrieve from the National Climatic Data Center (NCDC) Level-2 radar data for an hour in which precipitation occurred at the radar site. Examine the data and identify all the precipitation bands and other mesoscale features on the radar screen. Use any ancillary data available (surface maps, upper air maps, forecast model data, etc.) to try to associate the mesoscale features with specific forcing mechanisms that may be responsible for their existence. The NCDC site to download the data (as of this writing) can be found at http://has.ncdc.noaa.gov/pls/plhas/HAS .FileAppSelect?datasetname=6500.

16

Tropical Cyclones

Objectives

By the end of this chapter, you should understand the following:

- Why radar is an important tool to study tropical cyclone structure.
- The advantages and disadvantages of various radar platforms for studying tropical cyclones.
- The physical structure of tropical cyclones and how that structure appears on radar displays.
- How the radar reflectivity factor differs in tropical cyclones compared to mid-latitude thunderstorms.
- How radar depictions of primary band features in tropical cyclones relate to their evolution.
- How storm winds are estimated from single Doppler radars during landfalls.
- How tropical cyclone hazards (extreme winds, surge, heavy rainfall, and tornadoes) are detected using radar.

16.1 Introduction

Radar is one of many important tools scientists and forecasters have to understand, monitor, and predict tropical cyclones. Much of the technological development in airborne radar has indeed been motivated by the scientific and operational needs to observe tropical cyclones over the remote oceans, as well as to transmit radar data in real time to operational forecast centers for data assimilation into numerical weather prediction models. The first observation of a tropical cyclone on radar was captured during the Great Atlantic Hurricane of 1944 as it passed nearby a US Navy Research radar at Lakewood Naval Air Station in New Jersey. This devastating storm (which incidentally was also the first storm observed by a research reconnaissance aircraft) did tremendous damage to the major population centers in the Mid-Atlantic. Soon after that, radar imagery was collected in many other storms that affected the USA (e.g., Figure 16.1). The installation of a network of radars along the East and Gulf Coast of the USA then allowed more consistent observations of tropical cyclones making landfall in the USA.

Radar Meteorology: A First Course, First Edition. Robert M. Rauber and Stephen W. Nesbitt.
© 2018 John Wiley & Sons Ltd. Published 2018 by John Wiley & Sons Ltd.
Companion website: www.wiley.com/go/Rauber/RadarMeteorology

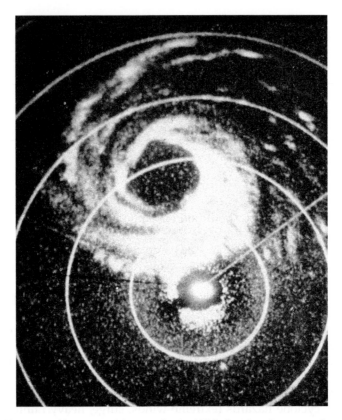

Figure 16.1   Hurricane Alice imaged by a radar in 1954 showing the eye and eyewall structure (Courtesy NOAA)

Figure 16.2 shows a climatology of tropical cyclone tracks in the Atlantic and Eastern Pacific Basins along with the location and low-level coverage areas of operational ground-based coastal radar stations throughout North America and the Caribbean. It should be apparent from this figure that the main development and tracking regions of tropical cyclones lie well outside the coverage area of ground-based radars. In addition, a single hurricane occupies an area much larger than the coverage of an individual radar (compare Figures 16.2 and 16.3). Most societal impacts of tropical cyclones occur during and after the storms make landfall; however, it should be apparent from Figure 16.2 that tropical cyclones form and intensify over oceans, often far beyond the range of ground-based radar stations.

Because of the geographically remote nature of tropical cyclones, other types of radars are required to observe tropical cyclones during their life cycle from their genesis and intensification, to their weakening, landfall, or extra-tropical transition. Airborne radars have been used over the oceans for many years to study and monitor tropical cyclones. These platforms offer coverage over a wide geographic area and can study hurricanes at times when they undergo rapid changes in intensity; however, airborne radar systems have limitations compared with ground-based radar systems. Because of the trade-offs of antenna size and power on radar capabilities, airborne radars may have wider beamwidths and lower effective resolution, lower transmit powers, non-Gaussian antenna patterns, and/or attenuating wavelengths.

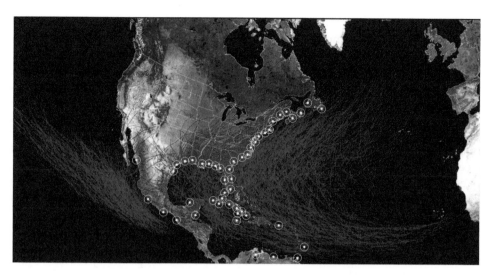

Figure 16.2 Hurricane tracks in the Atlantic and East Pacific basins with coastal and island radar locations superimposed (Track image courtesy NOAA.) The circles denote the radar coverage area below 1.2 km for a 0.5° elevation scan

Figure 16.3 Hurricane Katrina (2005) (Courtesy NASA)

In addition, their on-station durations are limited by the amount of fuel the aircraft can carry and the transit time from the airport to the storm. However, the advantages of aircraft system deployability and unique sampling geometries have made these systems useful for both operational monitoring of tropical cyclones and for collecting research-quality radar data on tropical cyclone structure, flow, and dynamics. Satellite precipitation radar data, first from the **Tropical Rainfall Measuring Mission** (TRMM, 1997–2015) and now from the **Global Precipitation Mission** (GPM, launched in 2014), have also been used since the late 1990s to operationally monitor tropical cyclone structure and post-landfall impacts, as well as to understand internal processes within tropical cyclones. Satellite-based radars are the only radar systems currently available that monitor tropical cyclones over all tropical oceans. Both the TRMM and GPM radars were discussed in Chapter 13, so we limit our discussion here to the uses of these radars in tropical cyclone research and operations.

16.2   Airborne and satellite radar systems for tropical cyclone research and operations

Tropical cyclones are often within reach of research aircraft operated by agencies of the USA while over the Atlantic and East Pacific ocean basins. Two NOAA Lockheed W3-PD propeller-driven research aircraft probe these storms regularly and carry onboard sophisticated radar systems. Occasionally, research aircraft supported by other federal agencies such as NASA, the National Science Foundation (NSF), and the Department of Defense (DOD) also carry out radar studies of tropical cyclones. Before examining data from these airborne radar systems, we will briefly review their characteristics.

16.2.1   NOAA WP-3D radar systems

Each NOAA aircraft has three radars on board, a nose C-band radar that is used for flight safety (the data are not recorded), a C-band (5.59 cm) wavelength lower fuselage radar (LFR) used for large-scale surveillance and research, and an X-band (3.22 cm) tail Doppler radar (TDR) system used for research and dual-Doppler recovery of winds. The LFR performs surveillance scans and uses a low-pulse repetition frequency (200 Hz) to achieve a large unambiguous range. The C-band LFR wavelength suffers from relatively little attenuation, allowing the aircraft to observe the full storm structure and design optimal flight tracks for storm penetrations well before encountering the storm's core (Figure 16.4). Because of the physical limitation that the antenna must be mounted on the base of the aircraft, the beam is oblong rather than circular, with a 1.1° beamwidth in the horizontal and a 4.1° beamwidth in the vertical. A major drawback of the LFR for research applications is that the large vertical beamwidth leads to inadequate illumination of the targets in the beam. More generally, inadequate beam filling becomes a severe problem in the estimation of the reflectivity of a storm at ranges exceeding 60–90 km.

The tail radar system is designed to perform successive conical scans to the side of the aircraft, first forward and then aft, at an angle of 19° to a direction normal to the direction of flight (Figure 16.5). A display of a single conical scan appears much

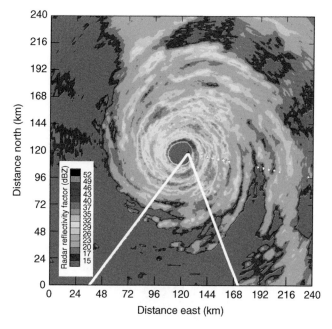

Figure 16.4 Hurricane Olivia (1994) as imaged by the lower fuselage radar of a NOAA WP-3D aircraft. The white lines indicate flight paths (Courtesy NOAA)

Figure 16.5 Scanning procedure used with the tail Doppler radar on the NOAA WP-3D aircraft (Image courtesy NOAA)

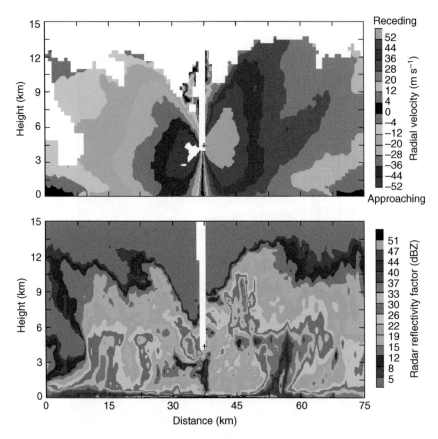

Figure 16.6 (a) Radial velocity (corrected for aircraft motion and unfolded) and (b) radar reflectivity factor from a conical sweep of the tail Doppler radar in Hurricane Olivia (1994) (Courtesy NOAA)

like a range-height indicator (RHI), with the caveat that the data are collected on a cone rather than a circle (Figure 16.6). The advantage of this scanning procedure can be understood by examining Figure 16.7. As the aircraft moves forward, successive forward and aft scans overlap at a sufficient angle that dual-Doppler recovery of the wind field can be achieved. As the radar is X-band, rain attenuation limits the maximum range at which Doppler estimates can be obtained. The radar uses a high-pulse repetition frequency, which, coupled with the short wavelength results in a low Nyquist interval. At the time of this writing, the radars were undergoing an upgrade that will employ a dual-PRF approach to extend the Nyquist interval from $\pm 50\,\text{m s}^{-1}$.

16.2.2 Other airborne radars used in hurricane research

Several other airborne radars have been deployed on aircraft flying over and into hurricanes for specialized field campaigns. The NSF/NCAR ELDORA radar, which is no longer available, flew into hurricanes during the 2005 active season while deployed on a U.S. Navy WP-3D aircraft. ELDORA was a dual-radar system that employed a sophisticated multiple frequency waveform. ELDORA operated with

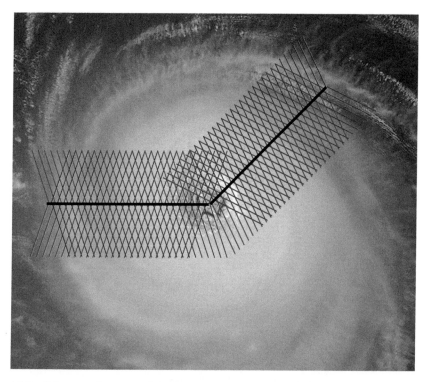

Figure 16.7 Schematic depiction of the beam locations for fore (blue) and aft (red) scanning of the tail Doppler radar on the NOAA WP-3D aircraft, as well as the ELDORA radar formerly deployed on the U.S. Navy W3-PD aircraft (Image courtesy NASA, NOAA)

a similar scanning strategy to the NOAA TDR but had a fast antenna rotation rate providing along-track sampling of 0.40 km with 0.15-km gate spacing and large Nyquist interval. A National Aeronautics and Space Administration (NASA) X-band radar system called the ER-2 Doppler radar (EDOP) flies on NASA's ER-2 aircraft. This radar has two fixed antennae, one pointing 33° forward and the other pointing at nadir, providing 0.10-km along-track sampling and 0.04-km gate spacing. NASA has conducted several experiments where a dual-frequency radar identical to that on the GPM satellite was flown over the top of hurricanes to obtain data to develop precipitation and microphysical retrieval algorithms for the GPM mission. One NOAA WP-3D aircraft flies an advanced Imaging Wind and Rain Profiler (IWRAP), a C-band and K_u-band dual-polarized (vertical and horizontal polarization) profiling scatterometer radar system designed to measure the backscattered signal from precipitation and the ocean surface. A similar system, the High Altitude Imaging Wind and Rain Airborne Profiler (HIWRAP), flew on the NASA Global Hawk, an unmanned aircraft used during NASA hurricane-related field experiments.

16.2.3   Satellite radars used in hurricane research

The TRMM and GPM precipitation radars have been used as tools to pinpoint tropical cyclone locations and estimate the intensity and potential intensity change of these

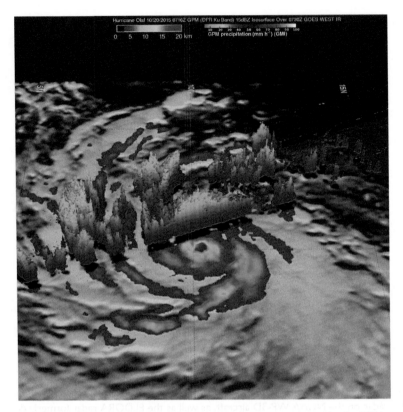

Figure 16.8   Three-dimensional depiction of the radar reflectivity measured by the GPM precipitation radar overlaid on a GOES-West infrared image of the hurricane (Courtesy NASA)

storms over the remote oceans during the time after each satellite was launched into orbit. As these satellites orbit the earth in low-earth orbits, they pass over any given feature on the earth at most twice per day. Because of this limitation, radar-based studies of tropical cyclone intensity change have focused on interpretation of the physical properties of radar echoes, rather than temporal changes in echo characteristics between the time of overpasses.

Figure 16.8, for example, shows a reconstruction of echoes and retrieved precipitation from GPM during a pass over Hurricane Olaf in October 2015. A key feature of the hurricane core region observed with satellites in many tropical cyclones is the presence of **convective bursts**, also referred to as **hot towers**, towering cumulus that extend well above the mean height of the eyewall clouds. Statistical studies of tropical cyclones developed from satellite-based radar suggest that the presence of these convective bursts is a precursor to rapid intensification. In the following section, we examine these and other structural features of tropical cyclones deduced from radar studies of tropical cyclones.

16.3   Tropical cyclone structure and kinematics

Much of what we have learned about tropical cyclone structure originates from radar measurements collected by airborne radars, ground-based radar measurements in

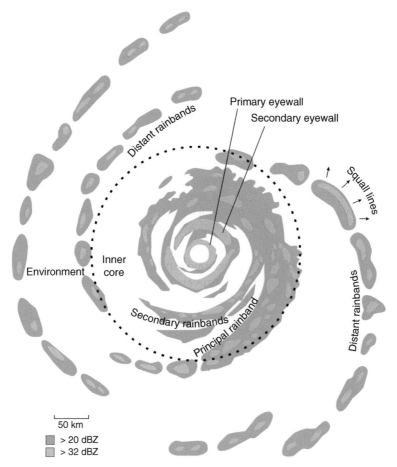

Figure 16.9 Schematic illustration of the radar reflectivity in a northern hemisphere tropical cyclone exhibiting double-eyewall structure (Adapted from Houze, R.A. Jr. (2010) Clouds in tropical cyclones. Mon. Weather Rev., **138**, 293–344. ©American Meteorological Society, used with permission)

hurricanes that made landfall along the US East and Gulf Coasts, and from satellite measurements made throughout the lifecycles of these storms. A tropical cyclone typically appears on satellite images as a large circular storm with spiral cloud bands extending away from the storm's central core (Figure 16.3). The corresponding archetypical structure of a well-developed tropical cyclone as viewed with radar is shown in Figure 16.9.

At the center of the circulation is the **eye**, a prominent, small region free of high clouds. The eye may be clear to the ocean, but more typically has low-level clouds near its base. In stronger storms, these clouds are sometimes organized into small-scale vortices that rotate within the eye. A ring of deep convective clouds called the **eyewall** surrounds the eye. Clouds within the eyewall extend from just above the ocean surface to the tropopause. Torrential precipitation occurs beneath the eyewall. Air in the eyewall rises upward and radially outward, so that the eyewall takes on the shape of a funnel (Figure 16.10), with clouds violently rotating around the eye in the lower troposphere and more slowly in the upper troposphere. Intense hurricanes, particularly while over the ocean, often develop a **double eyewall** (Figures 16.9 and

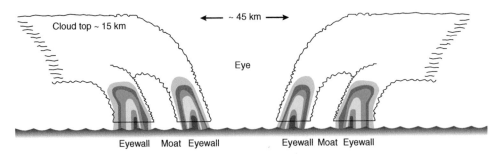

Figure 16.10   Idealized vertical cross section through a hurricane with double-eyewall structure. The scalloped region represents the cloud boundary of the convective features. The shading represents radar reflectivity at intervals of 25, 30, 35, 37.5, 40, and 45 dBZ (Adapted from Hence, D.A. and Houze, R.A. Jr (2012) Vertical structure of tropical cyclones with concentric eyewalls as seen by the TRMM precipitation radar. *J. Atmos. Sci.*, **69**, 1021–1036. © American Meteorological Society, used with permission)

16.10). The outer eyewall forms as outer rainbands contract into a closed ring about the inner eyewall. A wind maximum appears within each eyewall. Both the outer and inner eyewalls contract with time, with a hurricane typically reaching its peak intensity during the inner eyewall contraction period. The inner eyewall eventually weakens and dissipates, as the outer eyewall continues to contract and intensify. Once the inner eyewall disappears, the intensity of a hurricane will often decrease. This **eyewall replacement cycle** may occur more than once during a storm's lifetime.

From a kinematic perspective, the hurricane vortex can be divided into two regions, an inner core where the Rossby number, defined as

$$\mathrm{RO} = \frac{V}{fR} = \frac{V^2/R}{fV} = \frac{\text{centrifugal acceleration}}{\text{Coriolis acceleration}} \qquad (16.1)$$

where V is the rotational wind component, f is the Coriolis parameter, and R is the radius from the center of the eye, exceeds unity, and the region outside the core where the Rossby number is less than unity (Figure 16.11). In the inner core, air trajectories form nearly closed paths as air rotates within the hurricane gyre. In the outer region, the envelope of the storm, environmental air passes through the vortex and around the core. The primary band, called the **principal band**, typically extends outward in a long arc away from the central core and forms the boundary between the hurricane core and the envelope (Figures 16.9 and 16.11). The Rossby number is of order unity at this boundary. Dynamically, the principal band can be thought of as the boundary beyond which the Coriolis acceleration dominates the centrifugal acceleration. The principal band is part of what is termed the **stationary band complex**, a group of rainbands that extend outward from the hurricane core. The term "stationary" here is with respect to the hurricane center. In earth coordinates, the band complex moves with the hurricane, generally evolving slowly until the hurricane makes landfall. Figure 16.12 illustrates how the eyewall, core region, and stationary band complex appeared in both the reflectivity and the radial velocity fields of Hurricane Ike (2008) as it approached the shore of Texas. As can be seen from Figure 16.12, precipitation outside of the eyewall varies in intensity, although the winds can still be quite strong. The heaviest rain outside the eyewall occurs in the principal band that extends outward from the eyewall. Other secondary bands

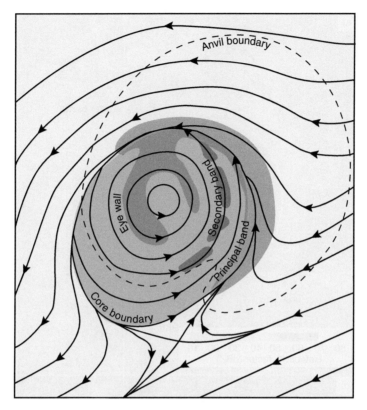

Figure 16.11 Schematic representation of the stationary band complex in a hurricane, the flows in which they are embedded, and the core and outer regions of a hurricane (Adapted From Willoughby, H.E., Marks, F.D. Jr, and Feinberg, R.J. (1984) Stationary and moving convective bands in hurricanes. *J. Atmos. Sci.*, **41**, 3189–3211. © American Meteorological Society, used with permission)

also wrap about hurricanes (Figure 16.9). These bands are typically, but not always, oriented near-parallel to the principal band orientation.

A tropical cyclone is a rotating vortex. An oversimplified description of the secondary circulation of airflow within this vortex would be that the air spirals inward just above the surface rises from the surface to the tropopause in the eyewall and then spirals outward from the eyewall at the tropopause. Hurricanes also have additional circulations associated with the eye, the spiral rainbands, and the regions between the spiral rainbands and the eyewall. Air in the upper and middle troposphere slowly descends within the eye of a hurricane, but not completely to the ocean surface. The base of the descending air is marked by an inversion at about the 1.5–2.0 km level. Weak convection typically occurs within the layer between the inversion and the ocean surface, producing scud clouds typically visible at the base of the eye. The other secondary circulations in hurricanes involve the spiral rainbands. Some of the air converging toward the center of the hurricane rises along preferred regions of convergence in the flow. These convergence zones lie along the principal band, and along other rainbands that spiral out from the core region. Except within the bands, air outside of the eyewall slowly descends to compensate for ascent within the eyewall and the rainbands.

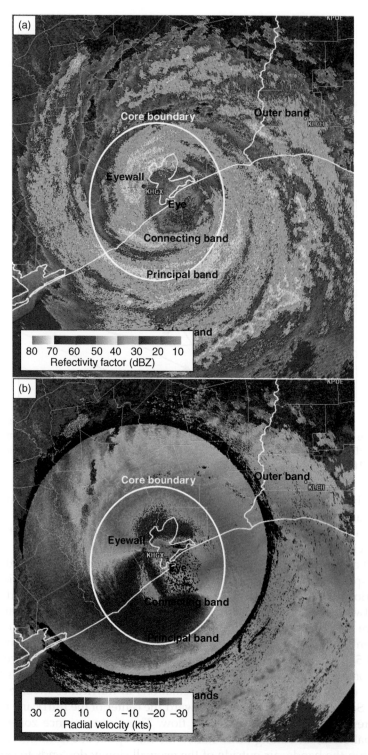

Figure 16.12   (a) Radar reflectivity factor and (b) radial velocity measured by the KHGX WSR-88D in Hurricane Ike (2008) at landfall on September 13, 2008 at 07:04 UTC illustrating the band features noted in the conceptual model in Figure 16.11. Note the folded radial velocities in the eyewall region

16.3.1   Eyewall and eye radar structure

The radar structure of the eyewall varies depending on the degree of organization and intensity of a hurricane. Figure 16.13 shows the radar reflectivity factor in Hurricane Andrew (1992), the last Category 5 storm to make landfall with that intensity in the USA through 2016. Andrew had a well-formed, symmetric circular eyewall. For comparison, Figure 16.14 shows the eyewall and outer bands of Hurricane Arthur (2014), which was a Category 2 hurricane with sustained winds of 100 mph at the time of landfall in North Carolina. The eyewall in this storm was less well defined and asymmetric. Although the rainfall beneath the eyewalls in Andrew and Arthur was torrential, the reflectivities barely reached 50 dBZ, a relatively low value compared to reflectivity values found in mid-latitude thunderstorms. Reflectivities on the west side of Hurricane Arthur barely reached 40 dBZ. The vertical structure of the mean reflectivity in tropical cyclones exhibiting single and concentric eyewalls, as measured by the spaceborne TRMM radar, appears in Figure 16.15. These data show that the mean reflectivity values are typically less than 40 dBZ. In general, the radar reflectivity falls in the range of 35–50 dBZ within the eyewall.

Several factors contribute to the lower values of reflectivity found in hurricane eyewalls. First, the melting level in hurricanes is quite high, typically near 5 km. In addition, the updrafts in the hurricane eyewall just above the freezing level are normally in the range of 5–8 m s^{-1}, although much higher vertical velocities, of the order of 10–20 m s^{-1}, have been observed in the upper regions of convective cells associated with convective bursts. Compare these values with mid-latitude thunderstorms, where the melting level is typical at 3–4 km elevation and updrafts range from 10

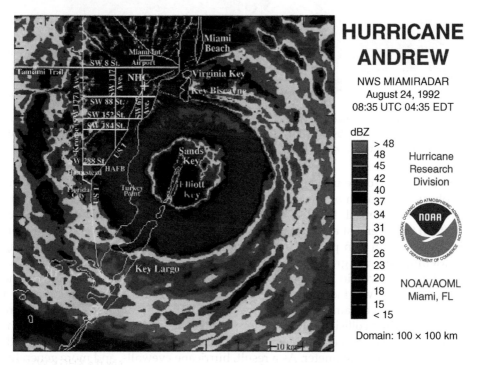

Figure 16.13   The eyewall of Hurricane Andrew (1992) (Courtesy of NOAA)

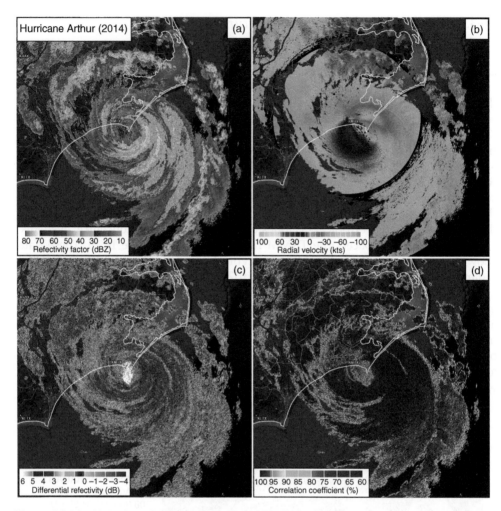

Figure 16.14 (a) Radar reflectivity factor, (b) radial velocity, (c) differential reflectivity, and (d) correlation coefficient measured by the KMHX radar on July 4, 2014 at 03:01 UTC as Hurricane Arthur made landfall on the outer banks of North Carolina

to $50\,\text{m s}^{-1}$ closer to the melting level. The implications of these differences on the microphysical evolution of precipitation are critical to understanding the reflectivity structure of the eyewall. Because of the high freezing levels and generally weaker updrafts in the layer just above the freezing level, hurricanes produce much of their rain through warm rain processes and produce little to no hail above the melting level. Melting hail and graupel that are largely responsible for the higher reflectivities that are observed in mid-latitude thunderstorms. A further factor limiting the value of the reflectivity factor in the eyewall is that in a hurricane eyewall, there is a large concentration of raindrops. The large concentration of raindrops leads to more frequently raindrop collisions and therefore droplet breakup, a factor that limits raindrop sizes. As we have seen in Chapter 5, the reflectivity factor depends on the sixth power of the raindrop diameter. As a result, hurricane eyewalls, and more generally, their outer bands are characterized by low values of Z_{DR}, typically less than $2\,\text{dB}$,

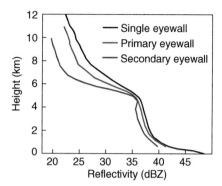

Figure 16.15   Vertical structure of the radar reflectivity factor in tropical cyclones with single and concentric (primary and secondary) eyewalls as measured by the TRMM precipitation radar (Adapted from Hence, D.A. and Houze, R.A. Jr (2012) Vertical structure of tropical cyclones with concentric eyewalls as seen by the TRMM precipitation radar. *J. Atmos. Sci.*, **69**, 1021–1036. © American Meteorological Society, used with permission)

indicative of small drops (Figure 16.14). Nevertheless, the rain is quite heavy, with K_{DP} values exceeding $1°km^{-1}$ indicative of large number concentrations of oblate droplets with rain rates exceeding an inch per hour, and values of ρ_{HV} approaching 100%. Values of ρ_{HV} do become lower when the beam crosses the melting level, as can be seen in the outer bands of Hurricane Arthur in Figure 16.14.

A curious characteristic of the radar echoes within the eye of hurricanes is the very high values of Z_{DR} that characterize the eye. In Hurricane Arthur, for example, the Z_{DR} values in the eye exceed 6 dB (Figure 16.14). This anomalously high Z_{DR} is believed to be caused by birds taking refuge within the eye (as discussed in Chapter 7).

As noted earlier, strong hurricanes, particularly while over the ocean, often exhibit a double eyewall, which is shown conceptually and in cross section in Figures 16.9 and 16.10. For example, the double eyewall of Hurricane Rita (2005) appears as concentric rings of higher reflectivity separated by a "moat" of lower reflectivity (Figure 16.16). The smoothness of the reflectivity field and radial velocity within Rita (Figure 16.17a,b) masks the actual nature of the vertical circulations. Figure 16.17c,d shows the vertical velocity and vertical vorticity derived from airborne dual-Doppler measurements made in Rita. These fields appear much more cellular, suggesting that the updrafts and downdrafts are organized on a much smaller scale than the primary circulation of the eyewall.

Convective bursts often occur within the eyewall circulation, particularly just before and during rapid intensification. As noted earlier, the cells associated with these bursts are commonly referred to as hot towers. The term "hot" is somewhat of a misnomer, as temperatures in the upper troposphere where these convective cells exhibit their strongest vertical velocities range from −20 °C to as cold as −80 °C. The term "hot" in this context implies that these regions possess positive buoyancy, the buoyancy resulting from latent heat release. Figure 16.18a shows an example of a hot tower as observed by the NASA EDOP radar. The corresponding vertical velocities in Figure 16.18b were corrected for particle fall velocity and represent actual vertical air motion, rather than vertical Doppler radial velocities. The tower can be seen extending upward from the eyewall with vertical velocities at the tropopause level of 15–20 m s^{-1}.

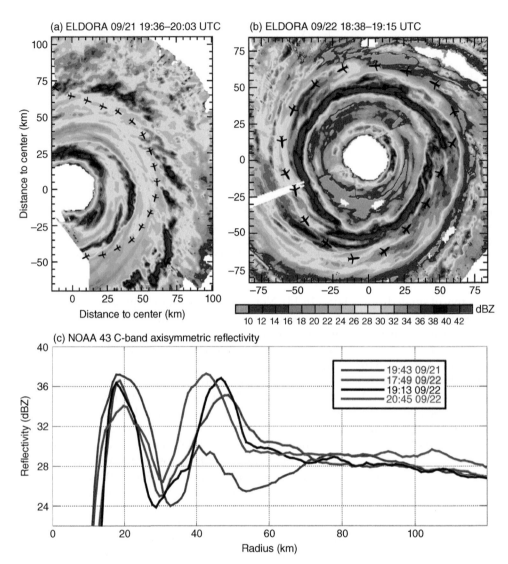

Figure 16.16   Radar reflectivity from ELDORA X-band radar at 3-km altitude during (a) 19:36–20:03 UTC September 21 and (b) 18:38–19:15 UTC September 22 and (c) axisymmetric reflectivity from NOAA C-band at ~3 km at four consecutive times shown in inset (From Bell, M.M., Montgomery, M.T., and Lee, W.-C. (2012) An axisymmetric view of concentric eyewall evolution in Hurricane Rita (2005). *J. Atmos. Sci.*, **69**, 2414–2432. © American Meteorological Society, used with permission)

In general, hot towers penetrate to about 16 km height and have maximum updrafts of 20 m s⁻¹ at about 12–14-km height. Studies have also shown that significant downdrafts of about 10–12 m s⁻¹ occur on the flanks of the updrafts with this downdraft air transported toward the eye (e.g., Figure 16.18b). Adiabatic heating associated with the descending air has been hypothesized to contribute to intensification of the warm core at the storm center, and therefore the overall intensity of a hurricane.

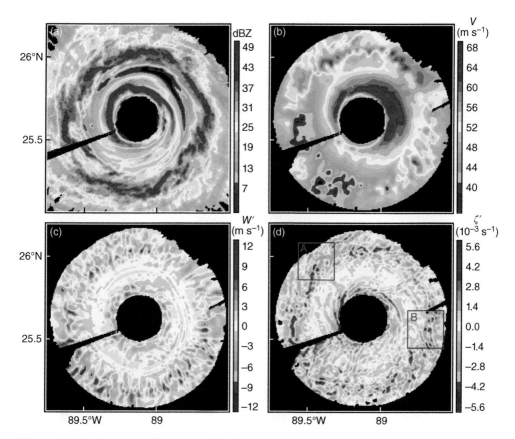

Figure 16.17 Plan view of Hurricane Rita's concentric eyewalls at 4-km altitude as observed by the ELDORA radar during 18:00–18:20 UTC September 22, 2005. The flight leg began and ended in the southwestern portion of the storm where there is a gap in the data. (a) Radar reflectivity. (b) Tangential velocity relative to the storm center. Positive values are cyclonic. (c) Vertical velocity perturbations, defined as velocity components from wavenumbers 2 and higher. (d) Vertical vorticity perturbations, defined as vorticity components from wavenumbers 2 and higher (From: Didlake, A.C. Jr and Houze, R.A. Jr (2011) Kinematics of the secondary eyewall observed in Hurricane Rita (2005). *J. Atmos. Sci.*, **68**, 1620–1636. © American Meteorological Society, used with permission)

16.3.2   Radar structure of principal band

Airborne radar analyses of the principal bands of northern hemisphere tropical cyclones have provided evidence that these bands undergo a transition from convective to stratiform structure as air flows from the southeast to northwest quadrant of hurricanes (Figure 16.19). This evolution, for example, is evident in both the radar image of the principal band of Hurricane Rita (2005), assembled from data collected by the NCAR ELDORA radar on the Naval Research Laboratory P-3 aircraft (Figure 16.20a) and the principal band of Hurricane Lili (2002), imaged at landfall by the KLCH WSR-88D (Figure 16.20b). An example of the stratiform region in the northwest quadrant of Rita appears in Figure 16.21.

A comparison of the circulations associated with these two regions is shown in Figure 16.22. Figure 16.22a corresponds to the cross section A and B in Figure 16.19,

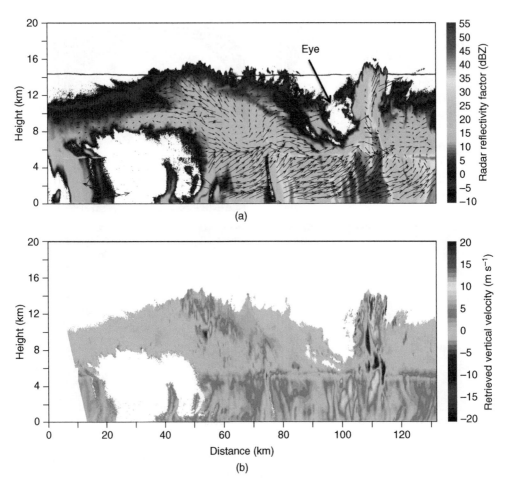

Figure 16.18 (a) Radar reflectivity factor and 2D wind field and (b) vertical velocity from the NASA EDOP radar in Hurricane Dennis illustrating a hot tower in the storm's eastern eyewall (From: Guimond, S.R., Heymsfield, G.M., and Turk, F.J. (2010) Multiscale observations of Hurricane Dennis (2005): the effects of hot towers on rapid intensification. *J. Atmos. Sci.*, **67**, 633–654. © American Meteorological Society, used with permission)

whereas Figure 16.22b corresponds to the cross section C and D. In the southeast sector, characterized by convective cells, air approaching the cyclone core rises in the convection and exhausts outward from the core (Figure 16.22a). Vertical momentum transport creates local jets in the primary circulation in the vicinity of the convective cells (V_1 and V_2 in Figure 16.22a). The convection in this region exhausts at a lower altitude than air rising through the eyewall. The circulation in the stratiform sector of the principal band is more complicated (Figure 16.22b) and is reminiscent of the stratiform sectors of mesoscale convective systems. Analyses indicate that a buoyancy gradient ($\partial B / \partial R$ in Figure 16.22b) develops between the inner and outer radius of the band at mid-levels due to differential latent heating and cooling associated with condensation and evaporation of particles. This gradient creates a pressure gradient and a flow akin to the rear-inflow jet of an MCS. The nose of the

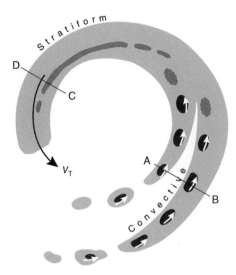

Figure 16.19 Plan view schematic of an organized rainband complex in a mature tropical cyclone. Reflectivity contours (20 and 35 dBZ) show embedded convective cells (red) that collapse (orange) and form stratiform precipitation traveling around the storm. The arrows represent tangential jets associated with each precipitation feature, with V_T indicating the jet within the stratiform sector (Adapted from Didlake, A.C. Jr and Houze, R.A. Jr (2013) Dynamics of the stratiform sector of a tropical cyclone rainband. *J. Atmos. Sci.*, **70**, 1891–1911. © American Meteorological Society, used with permission)

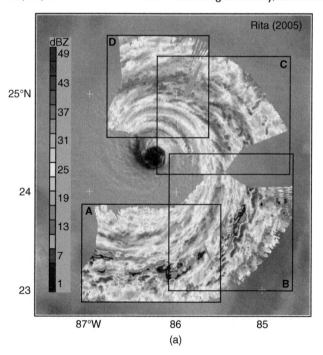

Figure 16.20 (a) Plan view of ELDORA reflectivity data at 2-km altitude observed during 18:45–19:27 UTC September 21, 2005. Visible imagery from Geostationary Operational Environmental Satellite-East is shown in the background. The boxes outline regions analyzed during individual flight legs (From Didlake, A.C. Jr. and Houze, R.A. Jr (2013) Convective-scale variations in the inner-core rainbands of a tropical cyclone. *J. Atmos. Sci.*, **70**, 504–523. © American Meteorological Society, used with permission.) (b) Convective bands in Hurricane Lili (2002) as observed with the KLCH WSR-88D at the time of landfall in Louisiana

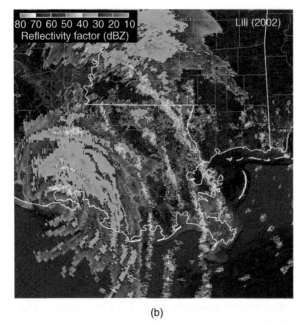

(b)

Figure 16.20   (*Continued*)

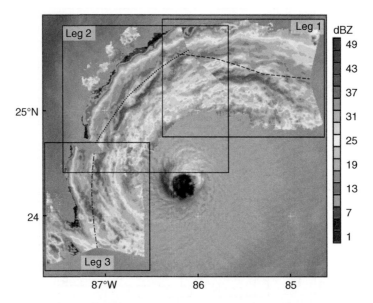

Figure 16.21   Plan view of ELDORA reflectivity data at 2-km altitude in Hurricane Rita (2005) observed on September 21, 2005. The boxes outline data sections from individual flight legs used to assemble the image. Flight tracks corresponding to each leg are drawn as dotted and/or dashed lines. Visible satellite imagery from the Geostationary Operational Environmental Satellite-East (GOES-E) is shown in the background (From Didlake, A.C. Jr and Houze, R.A. Jr (2013) Dynamics of the stratiform sector of a tropical cyclone rainband. *J. Atmos. Sci.*, **70**, 1891–1911. © American Meteorological Society, used with permission)

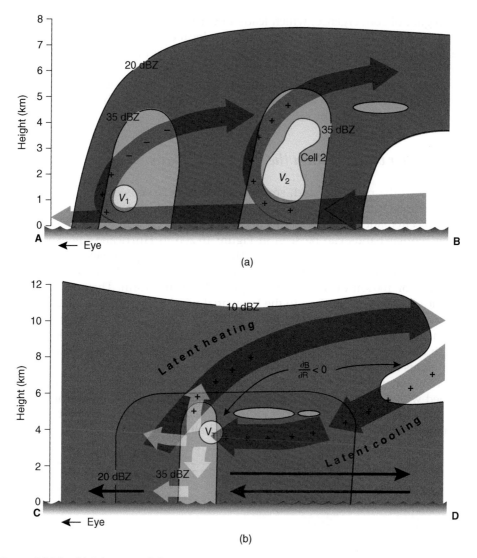

(a)

(b)

Figure 16.22 (a) Schematic of the convective motions associated with two mature convective cells at different radial distances from a storm center within the inner core (see line A and B in Figure 16.19). Reflectivity contours are drawn showing cell 1 at a smaller radius and cell 2 at a larger radius. The solid arrows represent the overturning secondary circulation within each cell, and the plus (minus) signs indicate regions of increasing (decreasing) tangential velocity; V_1 and V_2 represent the tangential velocity jets in each cell. (b) Schematic of the dynamics within a stratiform rainband (see line C and D in Figure 16.19). Reflectivity contours are drawn. The line arrows represent vortex-scale motions associated with the overall storm, and the broad arrows represent mesoscale motions associated with the stratiform rainband. The broad arrows of the descending inflow are driven by two regions of a radial buoyancy gradient $(\partial B/\partial R)$. The plus signs indicate regions of increasing tangential velocity by the secondary circulation. The circled region indicates the tangential jet (V_T). Latent cooling and latent heating occur in the indicated regions (Panel (a): From Didlake, A.C. Jr and Houze, R.A. Jr (2013) Convective-scale variations in the inner-core rainbands of a tropical cyclone. *J. Atmos. Sci.*, **70**, 504–523. © American Meteorological Society, used with permission); Panel (b): From Didlake, A.C. Jr and Houze, R.A. Jr (2013) Dynamics of the stratiform sector of a tropical cyclone rainband. *J. Atmos. Sci.*, **70**, 1891–1911. © American Meteorological Society, used with permission)

jet is a region of convergence, which leads to vertical circulations (indicated by the white arrows in Figure 16.22b). These lead to a local enhancement of reflectivity in the middle of the principal band, which can be seen in Figure 16.21 and is indicated schematically in Figure 16.19. These circulations and the structure of the principal band no doubt evolve substantially as hurricanes weaken, are sheared, or undergo landfall, but little is known about the mesoscale dynamics of these transitions as scant data exist to analyze the circulations.

16.3.3   Other bands within the hurricane vortex

In addition to the primary band, smaller scale bands have been observed extending approximately parallel to the principal band within the outer envelope of a hurricane's circulation. Figure 16.23 shows an example of such bands in Hurricane Andrew. Similar bands can be seen in Figure 16.20b in Hurricane Lili. They have been observed in many other hurricanes. The smaller scale spiral bands are typically within about 100 km of the hurricane center, have scales of 10 km across a band, and extend for 100 km or more as they spiral about the storm. They move around the hurricane with speeds close to the tangential wind at the level of the bands and are characterized by enhanced updrafts. They cause wind speed variations of at least $8 \, \text{m s}^{-1}$ across the bands. They appear visually similar to boundary layer rolls, although their depth is greater than typical boundary layer rolls observed in other environments over the ocean. Modeling studies suggest that they can form as a result of different dynamic processes, with the most likely as Kelvin–Helmholz instability combined with boundary layer shear. There is some evidence that horizontal shear associated with these bands might provide sufficient vorticity to support the spin-up of weak tornadoes, particularly in stronger landfalling storms.

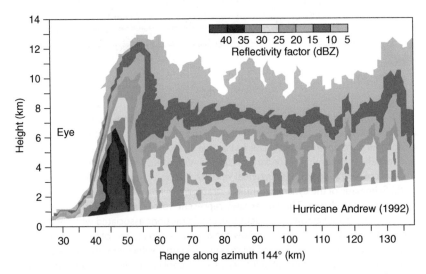

Figure 16.23   RHI through Hurricane Andrew (1992) near the time of landfall in Florida showing spiral rainbands (Adapted from Gall, R., Tuttle, J., and Hildebrand, P. (1998) Small-scale spiral bands observed in Hurricanes Andrew, Hugo, and Erin. *Mon. Weather Rev.*, **126**, 1749–1766. © American Meteorological Society, used with permission)

16.4   Operational use of radar to detect tropical cyclone hazards

The primary threats to life and property posed by tropical cyclones include destruction from hurricane winds, **storm surge** flooding at landfall, inland flooding, particularly along waterways, and damage from tornadoes produced within the storm's rainbands. These impacts can often be related to the structure and evolution of the radar reflectivity, radial velocity, and polarization variables measured by radars observing the storm. For this reason, radar systems are a critical tool to monitor tropical cyclone intensity and intensity change, estimate where and when flooding will be a serious threat, and provide warnings for locations where tornadoes may be occurring.

16.4.1   High winds and storm surge

The **sustained winds** (defined as the average 1-min wind speeds) within the core of a hurricane increase radially outward from the center of the eye to the **radius of maximum winds** and then decrease outward from the radius of maximum winds to the outer edge of the eyewall and beyond. Sustained winds are normally asymmetric about a hurricane, with the strongest winds found on the right side of a storm when looking in the direction of the storm motion vector. Winds in a hurricane increase with altitude from the surface to about 600 m and then weaken above that level. **Wind gusts**, short bursts of wind that may last only a few seconds but can be quite destructive, may occur anywhere within the hurricane and can exceed sustained winds by as much as 20–30%.

When analyzing the wind field of a hurricane with Doppler radar, the complex wind distribution described above must be interpreted within the context of the radar's viewing geometry. The fact that a Doppler radar only measures the wind component along the direction of the beam can make estimation of the maximum sustained winds in a hurricane challenging as the storm advances toward a coastline and the viewing angle changes with time. Surface wind gusts are even more difficult to estimate as at distances away from the radar, the beam is located at altitudes well above the earth's surface. The radar resolution is also generally too low to capture fluctuations in the radial velocity associated with gusts.

As an example, Figure 16.24 shows the radar reflectivity factor and radial velocity fields observed near landfall of two hurricanes, Katrina in 2005 (Figure 16.24a,b) and Ike in 2008 (Figure 16.24c,d). The position of the center of Katrina relative to the radar at the time of Figure 16.24a,b was ideal for the radar beam to be oriented directly along the wind within the northeast eyewall so that the approaching radial velocity value approximated the maximum wind speed at that location. The viewing geometry also permitted the strong winds directly along the southwest eyewall of the storm to be measured. In Hurricane Ike, the viewing geometry was less than ideal at the time of Figure 16.24c,d. The radar was near the center of circulation, and the beam was nearly perpendicular to most parts of the eyewall of the hurricane. In general, unless the radar beam is parallel to the storm propagation vector on the right side of the hurricane at the radius of maximum winds, the radar-measured radial velocity will be an underestimate of the true maximum wind in a hurricane.

Because of the destructive power of hurricanes, it is vitally important that the wind speed distribution and radius of maximum winds be estimated from single Doppler radar as these storms approach land. The greatest storm surge is typically centered

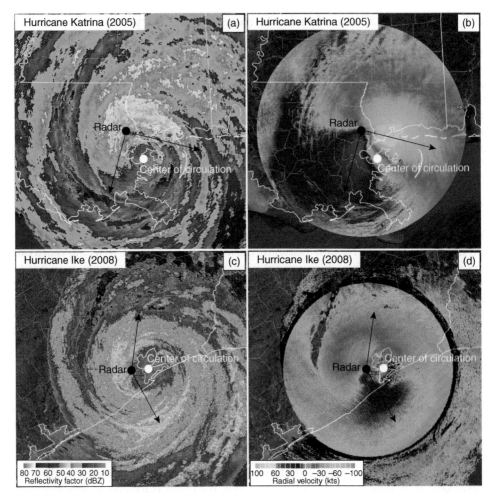

Figure 16.24   (a) Radar reflectivity factor and (b) radial velocity from the KLIX radar during Hurricane Katrina (2005) at landfall. (c) Radar reflectivity factor and (d) radial velocity from the KHGX radar during Hurricane Ike (2008)

near the location where the strongest winds blow toward the shore as a hurricane makes landfall, and the amplitude of the surge is closely related to the maximum wind speed. In the early years after deployment of the WSR-88D radars in the 1990s, a wind retrieval algorithm, termed the **Ground-Based Velocity Track Display (GBVTD)** technique, was developed to obtain these critical measurements. During the last two decades, the GBVTD algorithm has been improved, operationally implemented, and tested in a number of hurricane environments.[1] The GBVTD technique derives a tropical cyclone's tangential wind circulation from velocity data collected by a single ground-based Doppler radar. The GBVTD algorithm estimates the horizontal

[1]This algorithm is described in a series of papers over two decades, the first of which is Lee, W.-C., Jong-Dao Jou, B., Chang, P.-L., and Deng, S.-M. (1999) Tropical cyclone kinematic structure retrieved from single-Doppler radar observations. Part I: Interpretation of Doppler velocity patterns and the GBVTD technique. *Mon. Weather Rev.*, **127**, 2419–2439.

winds of a hurricane relative to the mean wind vector on rings concentric with the circulation center position. The accuracy of the GBVTD wind retrieval is closely related to the accuracy with which the center of circulation can be specified. Accompanying algorithms to GBVTD to determine the position of the circulation center have been developed.

Briefly, the mean wind around a GBVTD ring corresponds to the environmental wind (which is assumed to vary only with altitude across the inner core of the tropical cyclone). Harmonic analyses of the Doppler velocity data are performed around GBVTD rings, and the resulting Fourier coefficients are related to various wavenumber components of the tangential and radial wind that are intrinsic to the hurricane, one component of which is the environmental wind. Assumptions must be made to obtain the GBVTD solution because the system of equations in GBTVD is not closed. Different approaches to the closure assumptions have been developed and tested. The GBVTD approach continues to undergo testing and refinement and has proved to be a key tool in estimating the distribution and intensity of winds as tropical cyclones approach coastlines.

A second, independent approach termed **Tracking Radar Echoes by Correlation (TREC)** and in subsequent developments **TC circulation TREC (T-TREC)** has been used to estimate hurricane wind speeds from single Doppler radar.[2] The key assumption in TREC is that reflectivity features can be tracked. These features are assumed to act as passive, conservative tracers that are transported with the wind between the sampling time of successive scans at the same elevation angle. Because the TREC algorithm tracks features in the reflectivity pattern, its implementation requires that sufficient texture exists in the reflectivity field.

When employing TREC, the first scan is divided into a number of equal-sized Cartesian two-dimensional arrays of data spaced by a distance corresponding to the analysis grid spacing. Each array is statistically compared with all possible arrays of the same size developed from the next scan. The array that correlates highest to the first array is used and its location determines the endpoint of the estimated wind vector. An issue with the TREC technique is that it can produce poor results near the eyewall and outer rainbands of more intense tropical cyclones due to the fairly uniform reflectivity structure (in the direction of airflow) found in these regions (e.g., Figure 16.13). The T-TREC technique extends TREC to polar coordinates centered on the tropical cyclone and constrains the local echo-tracking region using the vortex rotation rate estimated from the radial data.

16.4.2   Heavy precipitation and flooding

Tropical cyclones, and the cloud systems that survive them after landfall, often produce flooding over large areas, sometimes extending hundreds to well over a thousand kilometers beyond the point of landfall. Extreme rainfall accumulation can occur in short time frames, with rainfall rates exceeding several centimeters per hour. Radar is an important tool in monitoring flash flood potential from these storms.

---

[2]These algorithms are described in Tuttle, J. and Gall, R. (1999) A single-radar technique for estimating the winds in tropical cyclones. *Bull. Am. Meteorol. Soc.*, **80**, 653–668, and Wang, M., Zhao, K., and Wu, D. (2011) The T-TREC technique for retrieving the winds of landfalling typhoons in China. *Acta Meteorol. Sin.* **25**, 91–103.

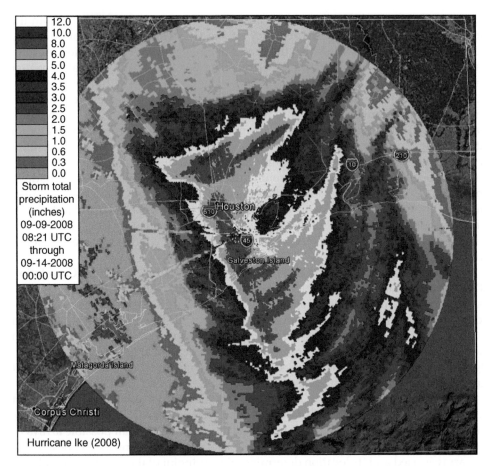

Figure 16.25 Storm total precipitation from Hurricane Ike (2008) as it make landfall and passed over Galveston and Houston, TX

The techniques to calculate rainfall are the same as those discussed in Chapter 13. For example, Figure 16.25 shows the total accumulated rainfall from Hurricane Ike (2008) right after landfall in Texas. The heaviest rain, exceeding 8 in., fell in two narrow bands, the first extending from Galveston Island northwestward through Houston, TX, and the second in a southwest–northeast swath north of Houston. The heavy rain in metropolitan Houston caused extensive flooding throughout the city. A sharp gradient in rainfall occurred with most of the rain falling from the storm's eastern eyewall. Much less rainfall occurred west of the track of the storm's center of circulation.

Extreme rainfall can occur well inland with tropical cyclones, and radar is essential in monitoring precipitation in these situations. An example is Hurricane Irene, which first made landfall along the Outer Banks of North Carolina, moved back over the ocean, and made a second landfall in New Jersey. The storm produced heavy rain across western New York and New England. Figure 16.26 shows the storm total precipitation across this region as measured by two WSR-88D radars. The regions of extreme rainfall straddling the western New York border are evident. Radar rainfall estimates during the storm led to flash flood warnings being issued throughout the region.

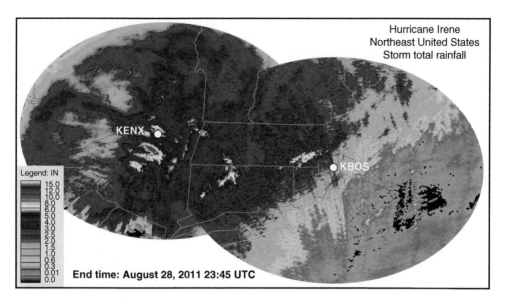

Figure 16.26   Storm total precipitation over southern New England and western New York state during the passage of the remnants of Hurricane Irene (2011)

16.4.3   Tornadoes

Tornadoes occurring within tropical cyclone circulations represent about 6% of all tornadoes occurring in the USA. When they occur, they are typically short-lived and weak, producing damage in the EF-0 to EF-1 range, although stronger tornadoes with intensity EF-3 or EF-4 have been documented. A single hurricane can produce a large number of tornadoes, the current record as of 2015 being Hurricane Ivan (2004), which produced 118 tornadoes. Tornadoes generated within hurricanes are typically found

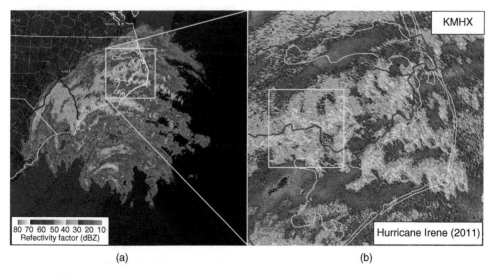

Figure 16.27   (a) Radar reflectivity from the KMHX radar as Hurricane Irene approached the outer banks of North Carolina. (b) Close-up of the echoes in the box in (a) showing a group of shallow supercells. The box in (b) is further expanded in Figure 16.28

within small low-topped supercells within outer hurricane rainbands on the right side of the hurricane. In this region, the speed and orientation of the winds create vertical shear profiles resembling Great Plains supercells, except that the shear is in a shallower layer and is often stronger. The tornadoes are short-lived, shallow, and often embedded in rain, but they can be detected with Doppler radars, particularly since polarization capabilities have been added so that debris signatures can be detected.

Figures 16.27 and 16.28 show an example from Hurricane Irene (2011). Figure 16.27a shows the hurricane from the KMHX WSR-88D in North Carolina. The storm's eye was still well offshore. The boxed region in Figure 16.27a is expanded in Figure 16.27b. This part of the hurricane's circulation is marked by a large group of cells, some of which are rotating and have supercellular characteristics. The area in the box in Figure 16.27b is further expanded in Figure 16.28. It is difficult in Figure 16.28a to identify any signatures that would definitively appear as a tornado

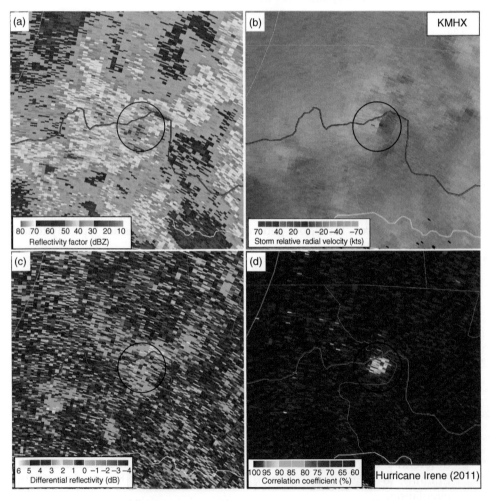

Figure 16.28 (a) Radar reflectivity factor, (b) storm-relative radial velocity, (c) differential reflectivity, and (d) correlation coefficient for echoes within the box in Figure 16.27b. A tornado occurred in the region marked by the black circle

in the reflectivity field (such as the debris signature in a hook echo). However, a velocity couplet is clearly evident in the storm-relative velocity field (Figure 16.28b), and coincident with it, low values of the correlation coefficient (Figure 16.28d) and Z_{DR} (Figure 16.28c), characteristic of debris. Unfortunately, not all tornadoes in hurricanes produce such clear signatures, but the example above demonstrates the potential of polarization Doppler radar to provide tornado identification and warning capability for tornadoes embedded within hurricane circulations.

Important terms

Convective bursts
Double eyewall
Eye
Eyewall
Eyewall replacement cycle
Global Precipitation Mission (GPM)
Ground-based velocity track display (GBVTD)
Hot towers
Principal band

Radius of maximum winds
Stationary band complex
Storm surge
Sustained winds
TC circulation TREC (T-TREC)
Tracking radar echoes by correlation (TREC)
Tropical Rainfall Measuring Mission (TRMM)
Wind gusts

Review questions

1. What are the key structural features of a tropical cyclone, and how do they appear on radar displays of the radar reflectivity factor, radial velocity, differential reflectivity, and correlation coefficient?

2. Why is the maximum radial velocity observed by a ground-based radar in a landfalling tropical cyclone often an underestimate of the maximum wind speed in the storm?

3. Why do reflectivity measurements in hurricane eyewalls typically range from 35 to 50 dBZ, while reflectivity measurements in convective regions of mid-latitude thunderstorms typically range from 45 to 65 dBZ?

4. Why does the differential reflectivity in the eyewall of a hurricane typically have a value of 2 dB, while the eye of a hurricane has a value of 6 dB?

5. How do the vertical circulations differ in the southeast versus northwest part of the principal band of a typical northern hemisphere tropical cyclone?

6. Where in a hurricane are convective bursts (hot towers) typically observed?

7. Where in a landfalling hurricane should a forecaster focus attention when trying to identify potential tornadoes? What features of the radial velocity, differential reflectivity, and correlation coefficient should the forecaster examine to pinpoint possible tornado locations within the hurricane circulation?

(continued)

8. A hurricane is making landfall on the west coast of Florida. How might a forecaster use the radial velocity field to identify where the greatest storm surge would occur during the time of landfall?

9. What are the current airborne and spaceborne radar systems used to operationally monitor tropical cyclones while over the ocean and outside the range of the ground-based radars?

10. Summarize the advantages and disadvantages of each type of radar system discussed in this chapter with regard to tropical cyclone forecasting and research.

Challenge problems

11. A hurricane moves directly over a Doppler radar so that the eye is centered on the radar site. Can the EVAD technique be used at this time to determine the maximum winds occurring in the hurricane? Can the EVAD technique be used to determine the forward motion of the storm? Explain your answers.

12. A hurricane is located over the eastern Atlantic at 30°N and is stationary. Assume that its core width is 120 km, identical to the swath width of the GPM K_a-band radar. The GPM satellite orbits the earth every 93 min. What is the minimum time period required to insure that the radar will sample the hurricane's core region?

13. A Category 5 hurricane over the Caribbean Sea at 20°N has double-eyewall structure. Winds measured by radar under the inner eyewall, at a radius of 20 km, are 75 m s^{-1}. Winds under the outer eyewall of 60 km are 45 m s^{-1}. What is the Rossby number at the radius of these two eyewalls? Allowing for an error of at most 10%, can we assume that the circulation under these eyewalls is in cyclostrophic balance (i.e., can the Coriolis acceleration be neglected relative to the centrifugal acceleration?).

14. A very slow moving hurricane makes landfall in Texas with the eyewall of the storm residing over Houston, TX, for 6 h. Assume that the radar reflectivity factor during this 6 h period was constant with a value of 43 dBZ. What is the difference in estimated radar rainfall if the rainfall retrieval algorithm uses the Marshall–Palmer Z–R relationship versus the Rosenfeld tropical Z–R relationship (see Table 13.1)?

15. Assume in Figure 16.18 that the melting layer is at 5 km altitude, that at this level in the eyewall, the pressure is 550 kPa, and that air rises through the eyewall from the melting level to 13 km, the level where the maximum vertical velocity was measured in the hot tower. What is the temperature in the "hot" tower at 13 km? (hint: use a Skew-T Log-P diagram).

17

Clouds and Vertical Motions

Objectives

By the end of this chapter, you should understand the following:

- Why certain radars are called cloud radars.
- Why cloud radars are typically used in vertical orientation.
- The various platforms on which cloud radars have been deployed.
- The advantages and disadvantages of using cloud radars.
- How cloud radars are affected by attenuation.
- How vertical air motion can be obtained with a cloud radar.
- Why cloud radars are particularly suited to determine statistical properties of clouds.
- How cloud radars have been applied to understand atmospheric structure under both cloudy and clear conditions.
- Applications of spaceborne cloud radar systems.

17.1 Introduction

Clouds blanket much of the earth, contribute substantially to our planet's albedo, trap infrared emissions from the earth's surface, and strongly influence our planet's radiation budget. Cirrus clouds shade vast areas of earth's tropics and middle latitudes, and large regions of the world's oceans are covered by stratocumulus. Other clouds blanket earth's landmasses, such as fields of fair weather cumulus in summer, or stratus sheets that overlie cold airmasses in winter. Most of these clouds are non-precipitating. They are composed of tiny droplets or ice crystals too small to be detected with centimeter wavelength radars such as those used in the WSR-88D network.

Studies of earth's climate, and future changes in earth's climate resulting from natural forces or human activities, require detailed knowledge of clouds—their coverage, vertical distribution, and particularly their microphysical properties. As all clouds except fog lie above the earth's surface, understanding of cloud processes can only be accomplished using aircraft or remote sensors. Aircraft *in situ* measurements, as important as they are, are expensive to obtain, limited in duration, and limited in

Radar Meteorology: A First Course, First Edition. Robert M. Rauber and Stephen W. Nesbitt.
© 2018 John Wiley & Sons Ltd. Published 2018 by John Wiley & Sons Ltd.
Companion website: www.wiley.com/go/Rauber/RadarMeteorology

Figure 17.1   The radar on the left is a K_a-band vertically pointing radar. The radars on the right are scanning K_a-band and W-band radars. These radars are located at the DOE Southern Great Plains ARM facility in Oklahoma (Courtesy of DOE/ARM)

coverage. Remote sensing of clouds is the only practical way to obtain the long-term statistics needed for climate studies. Ground-based remote sensors offer the opportunity to measure cloud properties continuously over long time periods at specific locations, whereas space-based remote sensors allow vast areas of the planet, particularly remote areas, to be investigated. Airborne remote sensors allow aircraft *in situ* measurements, collected along the aircraft path at a single altitude, to be placed into the context of the larger cloud system through which the aircraft is flying.

Two types of remote sensors have been employed extensively in cloud research: **cloud radars**, radars with millimeter wavelengths that are very sensitive and can detect small cloud particles, and **lidars**, remote sensors that use lasers either at visible or near infrared wavelengths. In this, the final chapter of this book, we focus specifically on cloud radars and their applications in weather and climate research. We concentrate here on the two radar frequencies most commonly used with cloud radars, 35 GHz (K_a band, $\lambda = 8.6$ mm) and 94 GHz (W band, $\lambda = 3.2$ mm). Two radars using K_a-band frequency and one using W-band frequency are pictured in Figure 17.1. These radars are part of a suite of radars studying cloud properties used at the U.S. Department of Energy Southern Great Plains Atmospheric Radiation Measurement (ARM) climate research facility.

17.2   Cloud radars

The earliest cloud radars were deployed in the 1950s and used in vertical pointing mode to monitor the depth and height of cloud decks near airports. It was not until the

1980s that vertically pointing cloud radars became a commonly used tool in research. The **Wyoming Cloud Radar** (WCR), the first airborne cloud radar, was installed on the University of Wyoming King Air in the 1990s. In the mid-1990s, the ARM program began to install cloud radars at locations around the world for climatic studies of clouds and their radiative properties. Since the 1990s, a large number of cloud radars have been deployed by government laboratories and universities for research applications. As of 2017, cloud radars fly on two NASA aircraft, and the NSF/NCAR G-V research aircraft, in addition to the Wyoming King Air. Cloud radars are also deployed in space on the **Global Precipitation Mission** (GPM) and **CloudSat** satellites. With rare exception, cloud radars are used in **vertically pointing mode** for reasons discussed in the next section. Many are Doppler radars, some with capability to record Doppler spectra. At the time of this writing, a few radars also have dual-polarization capability.

17.2.1   *Advantages and disadvantages of cloud radars*

Recall from Chapter 5 that the backscattering cross section for a spherical particle in the Rayleigh scattering regime is given by

$$\sigma = 4\pi r^2 \frac{S_r}{S_{inc}} = \frac{\pi^5 |K|^2 D^6}{\lambda^4} \tag{17.1}$$

The key parameter that makes short-wavelength radars so advantageous for studying clouds is the term in the denominator, λ^4. This implies that the backscattering cross section of a small spherical particle such as a cloud droplet for a radar at K_a-band is 15,848 times that of a WSR-88D S-band radar. The backscattering cross section is nearly six orders of magnitude larger than that of the WSR-88D for a radar at W-band wavelength. It is clear from these values that short-wavelength radars have the sensitivity to detect the very smallest particles composing clouds, hence the term "cloud radars."

Aside from increased sensitivity, another advantage of short-wavelength radars is their insensitivity to Bragg scattering (see Chapter 8). S-band and, to a lesser extent, C-band radars receive backscattered power from small-scale humidity fluctuations associated with turbulence in the vicinity of clouds. These echoes can be greater than those contributed by the cloud particles themselves, making interpretation of cloud properties difficult to impossible. Cloud radars are essentially immune from Bragg scattering, so the echoes can be interpreted as resulting from the particles composing the clouds. Finally, cloud radars require very small antennas and equipment, making them more easily transportable and lower cost than larger wavelength radars. But these enhanced capabilities all come at three significant costs.

The first is **attenuation**. Figure 17.2 shows the one-way attenuation due to molecular oxygen (blue line) and water vapor (red curves) at various wavelengths including those of cloud radars. These curves apply to a cloud-free atmosphere at a temperature of 294 K and a pressure of 1010 hPa. It is clear from these graphs that attenuation in a moist atmosphere, such as is common in the tropics, can be quite severe. Keep in mind that this graph shows attenuation *in the absence of clouds*. Signal attenuation induced by water droplets is significant. As an example, a cloud containing $1\,g\,m^{-3}$ of water distributed across a distribution of cloud droplets with diameters less than

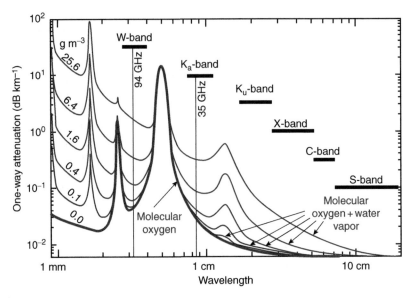

Figure 17.2 One-way attenuation versus wavelength for (a) a 1-km horizontal path at the surface with a pressure of 1010 mbar, a temperature of 294 K, and water vapor amounts varying from 0 to 26.6 g m^{-3} (Adapted From Kollias, P., Clothiaux, E.E., Miller, M.A., Albrecht, B.A., Stephens, G.L., and Ackerman, T.P. (2007) Millimeter-wavelength radars. New Frontier in atmospheric cloud and precipitation research. *Bull. Am. Meteorol. Soc.*, **88**, 1608–1624, © American Meteorological Society, used with permission)

100 μm will result in one-way attenuation near 0.4 dB km^{-1} at K$_a$-band (35 GHz) and 4.2 dB km^{-1} at W-band (94 GHz), about equivalent to that occurring in a very humid, tropical atmosphere with a mixing ratio of 25 g kg^{-1}. Fortunately, the attenuation of millimeter wavelength radars due to absorption in ice clouds is insignificant, only 0.03 dB km^{-1}, making cloud radars especially suitable for studying wintertime and higher altitude ice clouds. Attenuation by scattering is only significant in heavy snow at W-band, and can largely be ignored at K$_a$-band.

The second issue with cloud radars is that it is more difficult to satisfy the Rayleigh scattering criteria. For the Raleigh criteria to be satisfied, particles in the scattering volume must have diameters smaller than λ/π. For a 35 GHz, K$_a$-band radar, $D = 2.7$ mm, and for a 94 GHz, W-band radar, $D = 1$ mm. Many non-precipitating clouds have some larger particles that can exceed these limits, and precipitating clouds almost certainly have particles exceeding these limits, meaning that Mie scattering theory must be considered in determining the reflectivity for these conditions. For example, Figure 17.3 shows the reflectivity factor (Z) for spherical particles of different sizes at three wavelengths, 10 cm, 8.6 mm, and 2.7 mm. The point where Rayleigh scattering no longer applies is evident from the departure of the curves from the S-band line. In practice, clouds consist of countless particles of varied sizes (and shapes in the case of ice), so the nature of scattering in any random cloud cannot be predicted. Furthermore, non-Rayleigh scattering effects usually result in **saturation** of non-attenuated W-band reflectivities in rainfall at a level of 25–27 dBZ. Quantitative precipitation estimation from the radar reflectivity factor at millimeter wavelengths, for this reason, is much more problematic than at centimeter wavelengths.

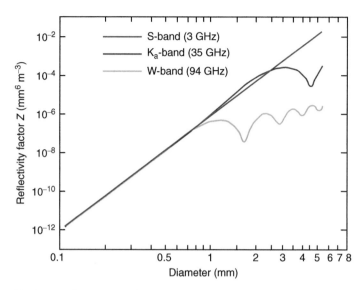

Figure 17.3 The radar reflectivity as a function of spherical particle diameter for three different radar frequencies: 3 GHz (red line), 35 GHz (blue line), and 94 GHz (green line). The particle concentration is 1 m^{-3}. Rayleigh backscattering is valid for all raindrop sizes at 3 GHz. At higher radar frequencies, deviation from the 3-GHz red line indicates the maximum raindrop size for which the Rayleigh approximation is valid. Note that as the raindrop size increases, there is a decrease in the radar reflectivities and mean Doppler velocities at higher frequencies (Adapted from Kollias, P., Clothiaux, E.E., Miller, M.A., Albrecht, B.A., Stephens, G.L., and Ackerman, T.P. (2007) Millimeter-wavelength radars. New Frontier in atmospheric cloud and precipitation research. *Bull. Am. Meteorol. Soc.*, **88**, 1608–1624. © American Meteorological Society, used with permission)

The third major issue is one of power. Millimeter wavelength radars require small waveguides, joints, and other components (see Figure 2.5). The power that can be transmitted through these components without causing arcing is much less than can be applied in a large-wavelength radar system. For example, an S-band (10 cm) waveguide might be rated for a peak power of 1.6 MW, whereas a W-band waveguide (3.2 mm) would be rated for a peak power of 0.003 MW. Less power decreases the overall radar sensitivity, although cloud radars are still much more sensitive to small particles than centimeter wavelength radars.

17.2.2 Examples of data from cloud radars

As noted earlier, cloud radars have been deployed on aircraft, in space, and on the ground. These varied platforms have permitted studies of individual clouds, cloud systems, and global cloud coverage. As examples, consider the following datasets, the first collected from the W-band WCR while flying through the comma-head of an "Alberta Clipper" winter cyclone, the second by the W-band CloudSat spaceborne radar flying over a marine extratropical cyclone, and the third by a ground-based radar viewing orographic clouds at Battle Pass near the crest of the Sierra Madre mountain range in southern Wyoming.

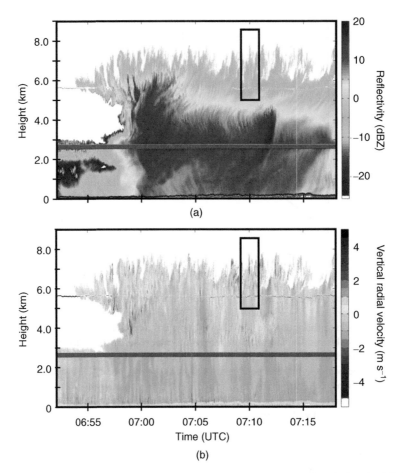

Figure 17.4   (a) Wyoming Cloud Radar (WCR) reflectivity (dBZ), for the 06:52–07:18 UTC, February 15, 2010 C-130 flight leg during a pass through the comma head region of an extratropical cyclone. The black box indicates the location for which data are shown in more detail in Figure 17.5. The thick gray line is the flight track. The thin line echo near 5.7 km is caused by ground reflection. (b) WCR vertical radial velocity for the same time period as in (a) (From Keeler, J.M., Jewett, B.F., Rauber, R.M., McFarquhar, G.M., Rasmussen, R.M., Xue, L., Liu, C., and Thompson, G. (2016) Dynamics of cloud-top generating cells in winter cyclones. Part I: idealized simulations in the context of field observations, *J. Atmos. Sci.*, **73**, 1507–1527. © American Meteorological Society, used with permission)

Figure 17.4, a broad view of the WCR reflectivity and vertical radial velocity of particles composing the comma head of the Alberta Clipper cyclone, shows a 7 km deep cloud system with cloud-top convective generating cells. These cells produce streamers of snow that descend into the broader stratiform echo. The nature of these cells can be understood by examining Figure 17.5, a zoomed in view of the boxed region in Figure 17.4. Here, we see that the cells contain 2–3 m s^{-1} updrafts and downdrafts, showing that the cloud top is highly turbulent, and a major source of ice particles seeding the stratiform cloud below.

The second example is a much broader view of clouds and precipitation associated with a cold front within a marine cyclone (Figure 17.6). The equivalent potential

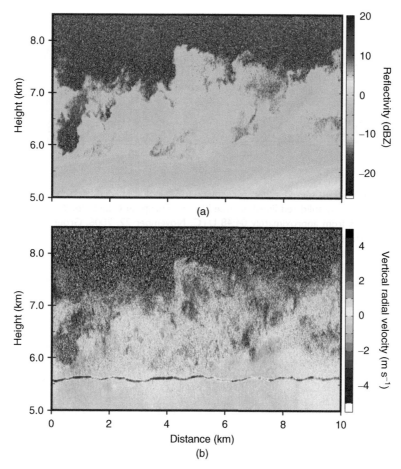

Figure 17.5 High-resolution view of (a) the reflectivity data and (b) the vertical radial velocity data in the box in Figure 17.5. The thin wavy line in the velocity field at 5.5 km is an artifact caused by ground reflection of the downward beam (From Keeler, J.M., Jewett, B.F., Rauber, R.M., McFarquhar, G.M., Rasmussen, R.M., Xue, L., Liu, C., and Thompson, G. (2016) Dynamics of cloud-top generating cells in winter cyclones. Part I: idealized simulations in the context of field observations, *J. Atmos. Sci.*, **73**, 1507–1527. © American Meteorological Society, used with permission)

temperature analyzed by the European Centre for Medium Range Forecasting is overlaid in the figure. Here, we can see a broad cloud region extending rearward over the cold front, a large stratiform cloud shield aloft with what appear to be cells near cloud top, and a large field of low-level shallow convective clouds in the cold air well to the cold side (left) of the deep frontal cloud system.

The final example shows reflectivity, vertical radial velocity, and spectral width data collected by a ground-based vertically pointing K_a-band radar located in Wyoming's Sierra Madre Mountains during a cloud seeding study (Figure 17.7). Temperatures in Figure 17.7 were determined from a collocated microwave radiometer, and snow was falling at the time of the measurements. The data show several tall convective cells passing over the radar site. Updrafts of 1–2 m s^{-1} were present in the upper parts of the cells, which were also marked by increased turbulence. Subsidence

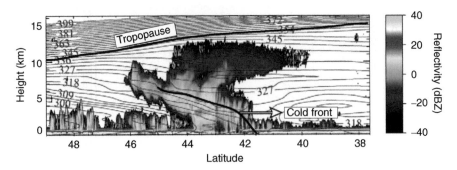

Figure 17.6 CloudSat observed radar reflectivity (dBZ, color shaded), overlaid with equivalent potential temperature (K, solid red lines) analyzed by the European Center for Medium Range Forecasting. The data are from approximately 04:48 UTC, November 22, 2006. Ground echo is present near the surface. The positions of a cold front and tropopause are marked with heavy black lines, and the direction of movement of the front is indicated with a white arrow (Adapted from Posselt, D.J., Stephens, G.L., and Miller, M. (2008) CLOUDSAT: adding a new dimension to a classical view of extratropical cyclones. *Bull. Am. Meteorol. Soc.*, **89**, 599–609. © American Meteorological Society, used with permission)

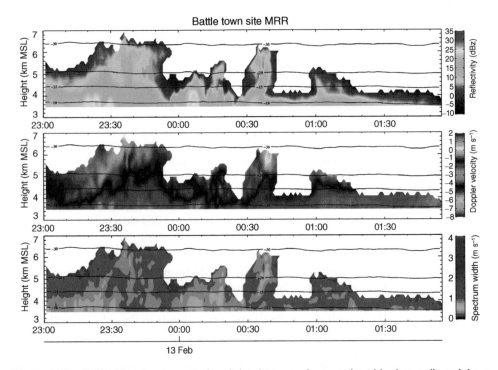

Figure 17.7 Reflectivity factor, vertical radial velocity, and spectral width data collected by a ground-based vertically pointing K_a-band radar located in Wyoming's Sierra Madre Mountains (Courtesy Katja Friedrich)

at low levels was due to orographic flow, rather than changes in particle fall speed, because the site was located just downwind of the mountain crest, where low-level subsidence is typical. Wintertime airmasses over the high desert of Wyoming were thought to be stable, but measurements from this radar showed that embedded and emergent convective cells occur frequently in the orographic flow, a surprising result that would not have been possible to determine with conventional scanning radars.

17.3 Application of cloud radars

Cloud radars are used in a wide variety of scientific applications. Their use has been growing rapidly as more and more of these radars become available within the meteorological community. In this section, we limit our review to a few key examples. Other applications are easily found in the scientific literature, for those readers interested in further exploring how these radars contribute to scientific studies.

17.3.1 Determining vertical motions in clouds

Vertical air motion is one of the most important quantities in atmospheric science, as it directly relates to both the kinematic forcing and the microphysical evolution of clouds. Retrieval of vertical air motions with Doppler radars has been a primary goal of atmospheric research with Doppler radars since these radars were first deployed to study atmospheric processes, as precipitation formation in clouds is largely controlled by updraft speed. Vertically pointing Doppler radars have the advantage of directly measuring vertical particle motion, from which vertical air motion can be estimated. In contrast, vertical velocity estimation from volume-scanning radars requires dual-Doppler synthesis and the application of airmass continuity. Both steps have a significant error margin, implying more uncertainty and poorer resolution.

A vertically pointing Doppler radar measures the reflectivity-weighted total fall velocity of particles, W, within a pulse volume, which includes the reflectivity-weighted **terminal velocity** of the particles ($\overline{w}_t$) plus any vertical air motion (w). Cloud radars are sufficiently sensitive to detect essentially all the particles in a cloud, but the sixth power reflectivity-weighting biases both W and $\overline{w}_t$ toward the larger particles in the particle size distribution. Note that all radars have finite beamwidths, so there is also a slight contribution to the radial velocity from horizontal motion of particles toward the edge of the beam. This is normally not a significant issue for narrow (e.g., ~1°) beam radars but can become a concern for wind profilers that have beamwidths of several degrees. To retrieve w using

$$w = W - \overline{w}_t \qquad (17.2)$$

it is obvious that one must estimate $\overline{w}_t$. The terminal velocity of a single particle, w_t, is a function of its size, shape, and density and the density of air through which the particle is falling. The terminal velocity of water droplets as a function of their size has been well established from both laboratory measurements and theoretical calculations (Figure 17.8). The terminal velocities of ice particles are much more complicated. Measurements of ice particle terminal velocities have been made during field experiments and in laboratory settings and have been estimated from theory using

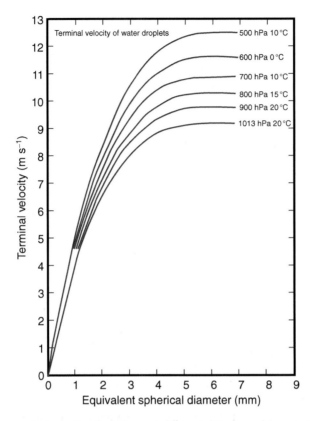

Figure 17.8   The terminal velocity of raindrops at different altitudes and temperatures (From Beard, K.V. (1976) The terminal velocity and shape of cloud and precipitation drops aloft. *J. Atmos. Sci.*, **33**, 851–867. © American Meteorological Society, used with permission)

idealized shapes. The results of these studies show that ice particles can have a range of terminal velocities that depend on their shape, size, density, and degree of riming (Figure 17.9a,b).

Several approaches to retrieve w from measurements of W using vertically pointing radar measurements have been developed.[1] These have been applied to cloud types other than deep convection. In stronger convection, the production of particles with large w_t (e.g., graupel or hail) is correlated with the occurrence of stronger updrafts, making application of these methods less accurate. Here, we limit our discussion to two approaches used in past studies of cloud structure.

The first involves calculation of the running mean of W at a given altitude, $\overline{W}$, over a time period sufficiently long that the expected magnitude and number of updrafts and downdrafts sampled are approximately equal in stratiform clouds. To the degree that this is true, the average value of w, $\overline{w}$, over the averaging interval will approach

---

[1] A review and comparison of methods to retrieve w from W can be found in Protat, A. and Williams, C. (2011) The accuracy of radar estimates of ice terminal fall speed from vertically pointing Doppler radar measurements. *J. Clim. Appl. Meteorol.*, **30**, 2120–2137.

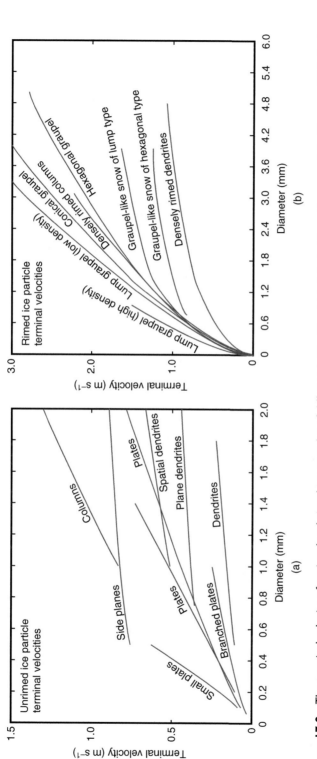

Figure 17.9 The terminal velocity of unrimed and rimed ice particles of different habits (Panel (a): Adapted from Rosenow, A.A., Plummer, D.M., Rauber, R.M., McFarquhar, G.M., Jewett, B.F., and Leon, D. (2014) Vertical velocity and physical structure of generating cells and convection in the comma head region of continental winter cyclones. *J. Atmos. Sci.*, **71**, 1538–1558. © American Meteorological Society, used with permission. Panel (b): Adapted from Locatelli, J.D. and Hobbs, P.V. (1974) Fall speeds and masses of solid precipitation particles. *J. Geophys. Res.*, **79**, 2185–2197. © American Geophysical Union, used with permission)

zero as the updrafts and downdrafts will cancel out in the average. Under these conditions, the value of $\overline{W}$ should approach the value of $\overline{w}_t$. Studies suggest that a 20 min running mean is sufficiently long in most cases. The value of w is then retrieved by subtracting $\overline{w}_t$ from individual measurements of W at the center of the averaging interval.

The second approach is to define a series of reflectivity bins (ΔZ in dBZ) that encompass the data in a cloud system. This method is especially attractive for liquid water clouds as $\overline{w}_t$ strongly depends on droplet diameter (Figure 17.8) and thus reflectivity. All W measurements within each reflectivity bin at each altitude are then averaged. Provided the number of measurements in each bin is sufficiently large, the magnitude and number of updrafts and downdrafts sampled should be approximately equal. Again, to the degree that this is true, the value of $\overline{w}$ over the averaging interval will approach zero as the updrafts and downdrafts will cancel out in the average. Under these conditions, the value of $\overline{W}$ should equal a unique value of $\overline{w}_t$ *for each reflectivity bin value.* The value of w is retrieved by first examining the value of Z and then subtracting the appropriate value of $\overline{w}_t$ from individual measurements of W at the center of the averaging interval.

For any given ice cloud, all Doppler velocities measured for each (Z_e, ΔH) pair, where ΔH is a height interval, are averaged and are assumed to be equal to $\overline{w}_t$ corresponding to this (Z_e, H) pair. The result obtained with this simple method is illustrated in Figure 17.10a–d, a cirrus cloud shield which, over a 12 h period, passed over central Oklahoma and the Atmospheric Radiation Program (ARM) Southern Great Plains site. The original measurements of reflectivity and vertical radial velocity are shown in Figure 17.10a,b. The particle terminal velocity was determined as a function of reflectivity in 20 min averaging intervals, the results of which appear in Figure 17.10c. The derived vertical motion field appears in Figure 17.10d. One can see from this field that the motions in the cirrus are quite turbulent, with fine-scale updrafts and downdrafts as large as $0.5 \, \text{m s}^{-1}$.

17.3.2   Determining statistical cloud properties

Fixed vertically pointing cloud radars can be run in unattended mode for long time periods, so that the data collected can be used to deduce climatological characteristics of clouds over a radar site. In fact, obtaining such statistics was the basis for establishment of the Department of Energy's **Atmospheric Radiation Program** (ARM), which since 1992 has monitored cloud systems using radars and other instruments in key climate-sensitive regions of the globe. In many locations, data have been collected for sufficiently long times so that cloud statistics useful in climate modeling have been obtained. Here, we provide two examples that illustrate the types of analyses that can be performed with climatological radar datasets such as those obtained by the ARM program.

As a first example, Figure 17.11 shows a 1 year statistical analysis of the height, depth, and thickness of cirrus clouds occurring over the ARM Southern Great Plains Site in Oklahoma in 1997, as determined from analysis of vertically pointing K_a-band radar data. Cirrus clouds in this study were defined based on temperature criteria and that the clouds were isolated from any other clouds at lower altitudes. The data show that the modal altitude at which cirrus clouds occur over the site throughout

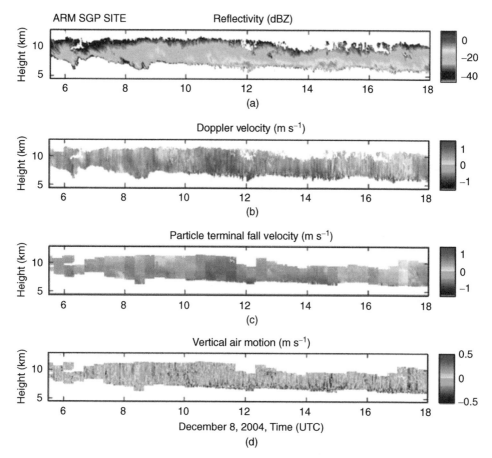

Figure 17.10 Example of Doppler vertical radial velocity decomposition into reflectivity-weighted terminal fall velocity, $\overline{w}_t$, and vertical air motion, w, on December 8, 2004 at the ARM SGP site. Positive velocity values indicate downward motion. (a) Radar reflectivity at 10-s and 45-m resolution, (b) Doppler vertical radial velocity at 10-s and 45-m resolution, (c) $\overline{w}_t$ at 10-s and 135-m resolution, and (d) w at 10-s and 135-m resolution (Adapted from: Kalesse, H. and Kollias, P. (2013) Climatology of high cloud dynamics using profiling ARM Doppler radar observations. *J. Clim.*, **26**, 6340–6359, © American Meteorological Society, used with permission)

the year was 10 km. Cirrus clouds were typically located just below the tropopause, with nearly all cirrus within 1–2 km of the tropopause altitude. Although cirrus clouds occurred as low as 5 km, and as high as 16 km over the site, they were most likely to occur in the 8.5–10.0 km range. Most cirrus clouds were of the order of 1 km thick. The analyses showed that cirrus clouds were present 22% of the time over the site. Obviously, although not shown here, monthly or seasonal statistics can be compiled to determine how these values change throughout the annual cycle, and comparisons can be made with other years to determine any long-term changes that may be occurring in high cloud coverage due to changes in the climate system.

As a second example, consider the data collected by an ARM K_a-band radar located in the equatorial western Pacific on the small island of Manus, Papua

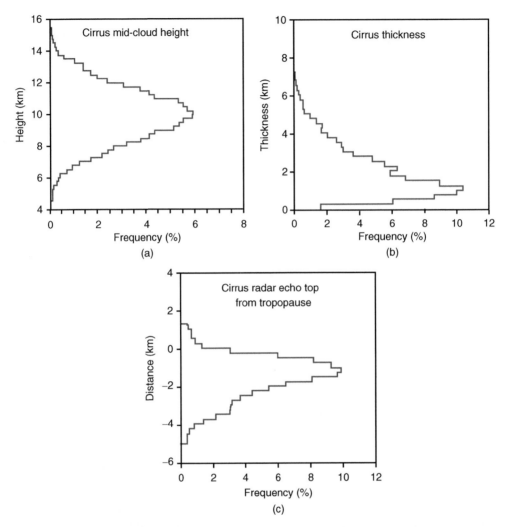

Figure 17.11   Properties of mid-latitude cirrus clouds obtained from a continuous vertically point-ing millimeter-wave cloud radar. Data collected at the Atmospheric Radiation Measurement Program Southern Great Plains site in Oklahoma. Frequency distribution for the annual dataset of (a) mid-cloud height, (b) layer thickness, and (c) radar echo top from the tropopause. Negative values denote that the radar echo top was below the tropopause (Adapted from Mace, G.G., Clothiaux, E.E., and Acker-man, T.P. (2001) The composite characteristics of cirrus clouds: bulk properties revealed by one year of continuous cloud radar data. *J. Clim.*, 14, 2185–2203, © American Meteorological Society, used with permission)

New Guinea. Manus is the northern island of Papua New Guinea, which itself is located just north of Australia's Cape York Peninsula. Analysis of data collected in high-altitude clouds at Manus over a period of 11.5 years showed that two classes of high ice clouds occurred, the first class associated with gravity waves and the second where gravity waves were absent. A statistical comparison of the properties of these clouds is shown in Figure 17.12. The data show that although cloud heights were generally similar, high ice clouds associated with gravity waves were somewhat

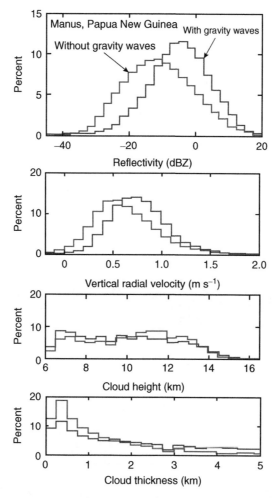

Figure 17.12 Statistics of reflectivity factor, vertical radial velocity, cloud height, and cloud thickness for high clouds with and without gravity waves over Manus, Papua New Guinea (Adapted from: Kalesse, H. and Kollias, P. (2013) Climatology of high cloud dynamics using profiling ARM Doppler radar observations. *J. Clim.*, **26**, 6340–6359. © American Meteorological Society, used with permission)

thicker and had higher reflectivity and corresponding larger downward radial velocities, implying that ice particles were larger and fell faster as a result of the waves.

These examples illustrate but a small fraction of the scientific analyses underway with cloud radars around the globe. Cloud statistics studies often combine cloud radar and lidar data, as, to a first order, radar reflectivity provides an estimate of a characteristic particle diameter, whereas a backscatter lidar provides an estimate of particle number concentration. Both diameter and number concentration are essential in radiative transfer and thus important to climate studies. As clouds play an important role in modulating climate through their effects on radiative forcing and precipitation, cloud radars are a critical tool in investigations of climate processes and anthropogenic and naturally forced climate change.

17.3.3   Understanding atmospheric and storm structure

Cloud radars have been used in numerous meteorological investigations, such as studies of convective initiation along Great Plains dryline, cloud seeding effects in winter orographic clouds, lake-effect storm structure, precipitation formation in marine stratocumulus, and snow generation in winter cyclones. Here, we limit the discussion to three examples to illustrate the power these radars have to elucidate the fine-scale structure of weather systems.

The first example consists of 1 min of data collected by the High-performance Instrumented Airborne Platform for Environmental Research (**HIAPER**) **cloud radar** (HCR) (Figure 17.13). The radar was mounted on the NSF/NCAR Gulfstream V aircraft during a flight over the comma-head region of a winter cyclone along the northeast coast of the USA. The storm produced nearly a foot of snow in the Boston, Massachusetts area. The radar was pointed at nadir from a flight altitude

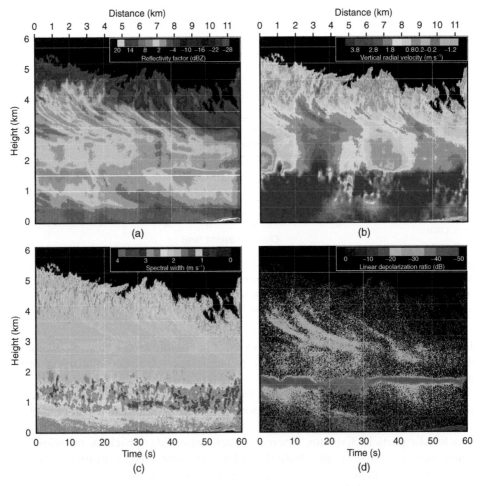

Figure 17.13   HIAPER cloud radar (HCR) observations of structural features of clouds in a 12 km region of a "Nor'Easter" winter cyclone along the East Coast of the USA. The panels show (a) reflectivity factor, (b) vertical radial velocity, (c) spectral width, and (d) linear depolarization ratio

of 12,800 m (42,000 ft). Streamers of precipitation can be seen descending from the cloud-top region in the reflectivity field (Figure 17.13a) and merging into a broader region of more uniform precipitation near 3 km. The cloud-top region is marked by updrafts and downdrafts of 1–2 m s^{-1} (Figure 17.13b) and turbulence as indicated by large spectral width values (Figure 17.13c). The melting layer appears clearly in L_{DR} signal (Figure 17.13d). Particle fall velocities increase below the melting layer as ice particles melt to raindrops. The region below the melting layer is marked by a marked increase in turbulence. The melting is not uniform—pockets of slower falling particles are interspersed with faster falling particles. The fine-scale non-uniformity in the precipitation features is in itself a remarkable feature of this data.

Our second example comes from data collected over the Oklahoma Panhandle. The Great Plains dryline, an extremely important meteorological feature, is a common locus of convective initiation and severe storm development in this region. Before cloud formation, the position of the dryline is often evident as a fine line in the reflectivity field, the echo associated with insects accumulating within converging airflow. In 2002, University of Wyoming King Air flights with the WCR were carried out across the dryline in clear air conditions as part of the International H_2O Project (IHOP). Figure 17.14 shows an example of the WCR equivalent reflectivity factor and vertical radial velocity collected during a pass across a double dryline boundary, as well as measurements of potential and virtual potential temperature, mixing ratio and along-track wind at the aircraft level. The radar echo is entirely due to insects. Updrafts of over 5 m s^{-1} are apparent along each dryline boundary, as well as enhancements in reflectivity associated with increased insect density due to convergence. Solenoidal circulations are evident at each boundary. Data such as these allowed scientists to relate the horizontal density difference across the drylines to the intensity of the resulting solenoidal circulations and show that the density contrast can maintain the sustained convergence and the radar fine line appearing on the WSR-88D radar. These analyses help clarify why the dryline is a trigger point for development of severe storms.

As a final example, we consider a very different type of weather hazard, lake-effect snow. Lake-effect snow is common in winter along all the downwind shores of the Laurentian Great Lakes, including Lake Ontario. Under the right wind regime, a single intense band will often form down the axis of Lake Ontario, which can persist for days and dump snow on the Tug Hill Plateau beyond the Lake's eastern shore. Snowfall totals for the Tug Hill region average more than 200 in. (16.7 ft.; 5.1 m) per winter, and as much as 140 in. have fallen during a single event. Figure 17.15 shows an example of such a band on January 7, 2014. A flight with the WCR on the King Air was carried out along the path indicated on the figure. The fine scale W-band measurements of the band's structure are shown in Figure 17.16. The reflectivity in Figure 17.16a shows clear evidence of attenuation at low levels, a sign that supercooled water is present in the cloud. Strong updrafts with core values exceeding 5 m s^{-1} are present at the band center in Figure 17.16b. Note the strong downward motions as well. Recalling that we are observing particle motion, not air motion, these strong downward radial motions are likely due to fast falling particles, as might be expected for heavily rimed ice particles that form in the presence of supercooled water. The band can be seen to be composed of narrowly spaced updrafts, probably organized in lines along the wind. These data, together with other

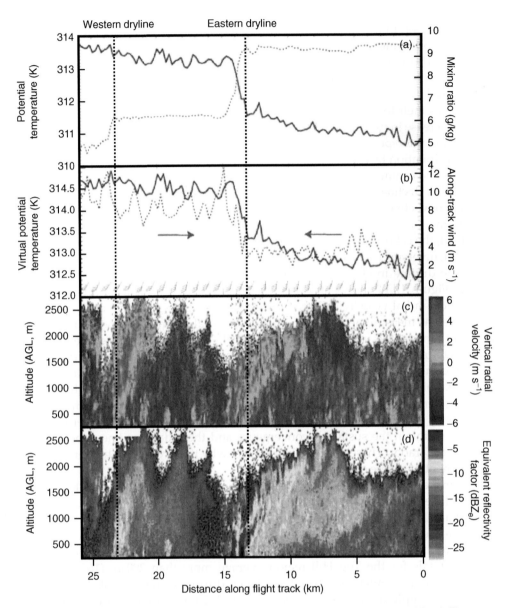

Figure 17.14 WKA observations of a dryline at a flight level 160 m above ground level. The moist side is on the right. Flight-level data are shown on top. This includes potential and virtual potential temperature, mixing ratio, and the along-track (dryline normal) wind. The main convergence belt is highlighted by the opposing blue arrows. Bottom panels show the vertical radial velocity (positive values indicate ascent) and reflectivity in the vertical plane above the aircraft measured by the Wyoming Cloud Radar (a millimeter wavelength cloud radar) (From Miao, Q. and Geerts, B. (2007) Finescale vertical structure and dynamics of some dryline boundaries observed in IHOP. *Mon. Weather Rev.*, **135**, 4161–4184, © American Meteorological Society, used with permission)

Figure 17.15   Lake-effect precipitation band over the eastern part of Lake Ontario. The yellow line denotes the location of the data presented in Figure 17.16 (Courtesy of Christopher Johnston)

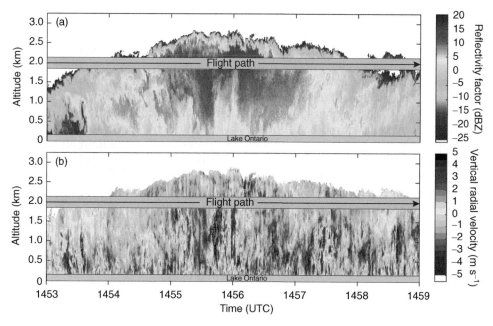

Figure 17.16   Reflectivity factor (a) and vertical radial velocity (b) measured by the Wyoming Cloud Radar as it flew across the lake-effect precipitation band in Figure 17.15 (Courtesy Christopher Johnston)

project measurements, are helping clarify how convection organizes over the lake and the microphysical processes that lead to sustained heavy snow.

17.3.4   Understanding global cloud properties

Cloud radars have been placed in orbit both to study global cloud properties and measure precipitation. As of 2015, two cloud radars are flying on satellites in low-earth orbits. The first of these, the Cloud Profiling Radar (CPR), is a W-band (94 GHz) radar deployed on the CloudSat satellite. The CloudSat mission, launched in 2006, was specifically designed to advance our understanding of cloud abundance, distribution, structure, and radiative properties. CloudSat's radar is designed to detect the much smaller particles of liquid water and ice that constitute the large cloud masses that cover much of the earth. CloudSat is one of several satellites that are part of the **Afternoon Constellation**, more commonly called the **A-Train**, that fly in a common orbit. The train of satellites flies in a nearly circular, sun-synchronous orbit with an equatorial altitude of approximately 705 km. The constellation of satellites together monitors global atmospheric cloud properties with the primary purpose of understanding earth's climate and climate change. Sophisticated algorithms have been developed to combine data from a cloud radar, a cloud lidar, and passive sensors in the microwave, infrared, and visible spectra on these satellites to produce specific products that characterize cloud properties, such as ice water content, liquid water content, and column integrated precipitation. The products are then used to characterize the statistical properties of clouds around the globe.

The second cloud radar currently in space, flown on the GPM satellite, is one component of a dual-frequency radar system consisting of a K_a-band and K_u-band radar. GPM was launched in 2014, a joint project of between NASA and the Japan Aerospace Exploration Agency. The Dual-frequency Precipitation Radar (DPR) aboard GPM provides three-dimensional views of storm structure across an narrow observation swath as the satellite orbits the earth. The K_u-band radar covers a 245 km swath, while at the time of writing, the K_a-band radar covers a 120 km swath nested in the center of the K_u-band swath. Algorithms for the DPR are designed to estimate precipitation rate and characteristics, or moments, of the size distribution of precipitation particles (see Chapter 13). The GPM mission follows the successful Tropical Rainfall Measuring Mission (TRMM), which flew a single-frequency (K_u-band) radar between the years 1997 and 2015. Unlike TRMM, GPM provides global coverage and will provide the first estimates of precipitation in remote, cold regions of the earth where ground measurements are sparse.

Important terms

A-Train	HIAPER cloud radar
Afternoon Constellation	Lidars
Atmospheric Radiation Program	Saturation
Attenuation	Terminal velocity
Cloud radars	Vertical air motion
CloudSat	Vertically pointing mode
Global Precipitation mission	Wyoming Cloud Radar

Review questions

1. What are three primary advantages of using cloud radars compared to those with longer wavelengths?
2. What are three primary disadvantages of using cloud radars compared to those with longer wavelengths?
3. Why are 3.2 and 8.6 mm preferred wavelengths for cloud radars (hint: examine Figure 17.2).
4. Why are cloud radars rarely used in a scanning mode similar to the WSR-88D radars?
5. Why are W-band radars generally more suited to studies of high altitude and wintertime clouds, compared to low-altitude summertime clouds and storms?
6. Is it likely that a W-band radar will measure a reflectivity factor of 40 dBZ in a cloud? Explain. (Hint: examine Figure 17.3).
7. Explain qualitatively how vertical air motion, w, can be retrieved from measurements of particle vertical radial velocity, W, in a stratiform cloud system.
8. Why do the methods described in this chapter for retrieving vertical air motion in stratiform clouds not work well in convective clouds?
9. Contrast some advantages and disadvantages of using spaceborne versus ground-based cloud radars to monitor global cloud properties.
10. Where is signal attenuation most likely to be occurring in Figure 17.13a? What evidence exists on the figure to support your answer?

Challenge problems

11. Assume that you have just been funded to use an airborne W-band radar to study the warm rain process in trade-wind cumulus clouds. The clouds are typically 1–3 km deep and contain no ice. Your flight plans call for overflights of the clouds with the radar at nadir, and flights just above the ocean surface with the radar pointed at the zenith. What issues might you encounter, and how might you attempt to deal with them?
12. In the text, we referred to "solenoidal circulations" along the drylines appearing in Figure 17.14c. Make a copy of the figure and sketch these circulations based on the radial velocity field.
13. Vertically pointing W-band Doppler radial velocities are rarely if ever folded. Why might this be so (hint: Consider Eq. (6.14), and Figures 17.8 and 17.9)?

(continued)

14. Cirrus cloud coverage may be enhanced by the presence of contrails that frequently appear in the sky behind commercial aircraft. Suppose you were just funded by the Federal Aviation Administration to do a 5 year study of the effect of commercial jet aircraft on the climatological coverage of cirrus clouds over the USA. You have at your disposal 10 vertically pointing mobile W-band radar systems. How might you design the study to answer the question?

15. Spaceborne cloud radars must correct for attenuation to make quantitative measurements of precipitation. Examine the scientific literature and write a short qualitative summary describing how attenuation correction is accomplished for these radars, and how differences in attenuation-corrected reflectivity between two radar frequencies can be used as a measure of particle size and precipitation rate.

APPENDIX A

List of Variables (and Chapters)

$\mathbf{S}$	Scattering matrix	(7)
$\mathbf{T}$	Transmission matrix	(7)
$\vec{A}$	Amplitude of electric field	(1, 7)
$\vec{B}$	Magnetic induction	(1)
$\vec{D}$	Electric displacement vector	(1)
$\vec{E}$	Electric field intensity	(1)
$\vec{H}$	Magnetic field intensity	(1)
$\vec{I}$	Electrical current	(1)
$\vec{V}$	Charge velocity	(1)
A	Axis ratio	(7)
A_{DP}	Specific differential attenuation	(9)
A_h	Specific attenuation	(9)
A_e	Effective area of antenna	(5)
A_M	Mass-weighted mean axis ratio	(7)
A_Z	Reflectivity-weighted mean axis ratio	(7)
B	Buoyancy	(16)
C	Autocorrelation function	(6)
C_e	Earth curvature	(4)
C_r	Radar constant	(5)
C_{ray}	Ray curvature	(4)
$C_{r,earth}$	Ray curvature relative to earth	(4)
C_1	Constant in equation for index of refraction	(4)
C_2	Constant in equation for index of refraction	(4)
C_3	Constant in equation for index of refraction	(4)
CC	Correlation coefficient	(14)
D	Diameter of spherical particle (e.g., raindrop)	(5, 13, 17)
D_0	Median volume diameter	(7)
D	Determinant	(12)

Radar Meteorology: A First Course, First Edition. Robert M. Rauber and Stephen W. Nesbitt.
© 2018 John Wiley & Sons Ltd. Published 2018 by John Wiley & Sons Ltd.
Companion website: www.wiley.com/go/Rauber/RadarMeteorology

D_v	Divergence	(11)
DWR	Dual-wavelength ratio	(9)
E_0	Electric field amplitude	(6)
E_r	Electric field of received wave	(6)
E_t	Electric field of transmitted wave	(6)
E_x	Electric field in x direction	(1)
E_{xm}	Maximum electric field in x direction	(1)
E_y	Electric field in y direction	(1)
E_{ym}	Maximum electric field in y direction	(1)
F	Amplitude of frequency components	(6)
$G_{\theta,\phi}$	Gain function	(2, 5)
G	Gain on beam axis	(5)
H	Scale height	(11)
K	Dielectric factor	(5, 13, 17)
K	Rainfall attenuation ratio	(9)
K_{DP}	Specific differential phase	(9, 13)
L_x	Power loss in hardware downstream of directional coupler	(3)
L_y	Power loss between the directional coupler and power meter	(3)
L_{DR}	Linear depolarization ratio	(7,17)
LWC	Liquid water content	(9)
M	Number of discrete samples	(6)
N	Radio refractivity	(4, 9)
N	Total droplet concentration	(13)
P	Total air pressure	(4)
P_{avg}	Average transmitted power	(3)
P_d	Dry air pressure	(4)
P_r	Received power	(5, 9)
P_{r0}	Received power in the absence of attenuation	(5, 9)
P_m	Power measured at the directional coupler	(3)
P_w	Power	(3, 9)
P_t	Transmitted power	(5)
PRF	Pulse repetition frequency	(3,6)
R	Radius of influence	(12)
R	Rainfall rate	(13)
R	Radius from center of tropical cyclone	(16)
R_0	Reference rainfall rate	(13)
R_n	Signal from nth pulse	(6)
RO	Rossby number	(16)
R_Z	Reference rainfall rate	(13)
S	Power flux density	(6)
S	Snowfall rate	(13)
S^{HH}	Scattering matrix element, incident horizontal, scattered horizontal	(7)
S^{HV}	Scattering matrix element, incident horizontal, scattered vertical	(7)

f_t	Transmitted frequency	(2,3,6,9)
f_v	Forward scattering amplitude, vertical polarization	(7)
f_L	STALO frequency	(2)
f_C	COHO frequency	(2)
g_u	Geometric terms in multiple Doppler synthesis	(12)
g_v	Geometric terms in multiple Doppler synthesis	(12)
g_w	Geometric terms in multiple Doppler synthesis	(12)
h	Height of a ray above the altitude of the radar antenna	(4)
h'	Height of a ray above the altitude of the radar antenna in National Weather Service equation	(4)
h_0	Height of a radar antenna above the Earth's surface	(4)
h_t	Height of terrain	(12)
h_{top}	Height of cloud top	(12)
h_u	Geometric terms in multiple Doppler synthesis	(12)
h_v	Geometric terms in multiple Doppler synthesis	(12)
i	Square root of negative 1	(1,4)
l	Pulse length	(3, 4,5)
k	Used as a constant or variable	(4)
k	Attenuation coefficient	(9)
k_a	Absorptive component of index of refraction	(4)
k_L	Attenuation coefficient	(9)
k_r	Rain attenuation coefficient	(9)
m	Complex index of refraction	(4)
n	Refractive index of a medium	(1,4,9)
n_0	Concentration parameter in exponential size distribution	(13)
n_1	Concentration parameter in gamma size distribution	(13)
$n(D)$	Particle size distribution function	(13)
q	Charge	(1)
r	Range from radar	(1, 3, 4, 5, 6, 17)
r_p	Distance between points in Cartesian and spherical coordinates	(12)
r_{max}	Maximum unambiguous range	(3, 6, 12)
s	Distance along the earth's surface	(4)
t	Time	(1, 3, 5, 6, 7, 9)
u	West–east component of wind	(6, 11,12)
u_0	Mean value of west–east component of wind	(11)
v	South–north component of wind	(11,12)
v_0	Mean value south–north component of wind	(11)
v_c	Speed of light in a medium	(1,4)
v_{max}	Nyquist velocity	(6)
v_r	Radial velocity	(6, 12)
v_{re}	Radial velocity when beam points east	(11)
v_{rn}	Radial velocity when beam points north	(11)
v_{rz}	Radial velocity when beam points upward	(11)
w	Vertical wind component	(6,11,12, 13)
$\overline{w}$	w averaged over many samples	(17)

ϕ	Angular coordinate relative to radar beam	(2, 5, 6)
ϕ	Phase	(6, 7, 9)
ϕ_{HH}	Phase of horizontal polarized wave	(7)
ϕ_{VV}	Phase of vertical polarized wave	(7)
ϕ_{DP}	Intrinsic differential phase	(7)
χ	Susceptibility of a medium	(1, 9)
χ	Angle relative to center of the earth	(4)
ω	Angular frequency	(1, 6, 7)
ω_d	Doppler angular frequency	(6)

APPENDIX B

Derivation of the Exact Equation for a Ray Path through a Spherically Stratified Atmosphere[1]

If dl is an infinitesimal segment along a ray path, then S, the optical path length between two points A and B, is given by

$$S = \int_A^B n\,dl \tag{B.1}$$

Fermat's principle states that

$$\delta S = \delta \int_A^B n\,dl = 0 \tag{B.2}$$

Formally, the principle states that the optical length of the path followed by light between two fixed points, A and B, is an extremum, or more simply that light follows the path of shortest optical length connecting two points, where the optical length depends on the refractive index of the medium.

A path segment through the atmosphere on a spherical earth with radius a, in terms of "flat earth" coordinates (s, h), in which the index of refraction is only a function of height, h, is given by

$$dl^2 = dh^2 + \left(\frac{a_e + h}{a_e}\right)^2 ds^2 \tag{B.3}$$

where s is distance along the earth's surface. Substituting y for dl/ds in (B.3), we can write

$$y^2 = \left(\frac{dl}{ds}\right)^2 = \left(\frac{dh}{ds}\right)^2 + \left(\frac{a_e + h}{a_e}\right)^2 \tag{B.4}$$

Let δ be a variation from a point on a ray to another nearby path of integration, conveniently chosen to be at the same value of s. From Fermat's principle, we can write

$$\delta S = \delta \int_A^B n\,dl = \delta \int_A^B [ny]ds = 0 \tag{B.5}$$

[1]The derivation of the ray equation in Appendix B follows Hartree, D.R., Michel, J.G.L., and Nicolson, P. (1946) Practical methods for the solution of the equations of tropospheric refraction, in *Meteorological Factors in Radio Wave Propagation*, Physical Society, London, pp. 127–168.

Radar Meteorology: A First Course, First Edition. Robert M. Rauber and Stephen W. Nesbitt.
© 2018 John Wiley & Sons Ltd. Published 2018 by John Wiley & Sons Ltd.
Companion website: www.wiley.com/go/Rauber/RadarMeteorology

An expansion of the integrand gives

$$\delta S = \int_A^B \left[\left(y\frac{dn}{dh} + n\frac{\partial y}{\partial h} \right) \delta h + n\frac{\partial y}{\partial \left(\frac{\partial h}{\partial s} \right)} \delta \left(\frac{dh}{ds} \right) + y\frac{\partial n}{\partial \left(\frac{\partial h}{\partial s} \right)} \delta \left(\frac{dh}{ds} \right) \right] ds \qquad (B.6)$$

As n is only a function of h, the last term equals zero, and this reduces to

$$\delta S = \int_A^B \left[\left(y\frac{dn}{dh} + n\frac{\partial y}{\partial h} \right) \delta h + n\frac{\partial y}{\partial \left(\frac{\partial h}{\partial s} \right)} \delta \left(\frac{dh}{ds} \right) \right] ds \qquad (B.7)$$

Taking a derivative of (B.4) with respect to dh/ds yields

$$\frac{\partial y}{\partial \left(\frac{dh}{ds} \right)} = \frac{1}{y}\frac{\partial h}{\partial s} = \frac{dh}{dl} = \sin \theta \qquad (B.8)$$

where θ is the inclination of the ray to the horizontal. Also, as δ represents variation at constant s, the term $\delta(dh/ds)$ can be written as

$$\delta \left(\frac{\partial h}{\partial s} \right) = \frac{d}{ds}(\delta h) \qquad (B.9)$$

Substituting (B.8) and (B.9) into the last term in (B.6), integrating the last term in (B.6) by parts and gathering terms yields

$$\delta S = [(n \ \sin \ \theta)\delta h]_A^B + \int_A^B \left[\left(y\frac{dn}{dh} + n\frac{\partial y}{\partial h} \right) - \frac{d}{ds}\left(\frac{n}{y}\frac{dh}{ds} \right) \right] \delta h ds \qquad (B.10)$$

For Fermat's law (B.2) to be satisfied for a ray between points A and B, the integrand in (B.7) must equal zero at both limits, so that, for all δh, it must be true that

$$\frac{d}{ds}\left(\frac{n}{y}\frac{dh}{ds} \right) - \left(y\frac{dn}{dh} + n\frac{\partial y}{\partial h} \right) = 0 \qquad (B.11)$$

Taking a derivative of (B.4) with respect to h gives

$$y\frac{dy}{dh} = \frac{1}{a_e}\left(\frac{a_e + h}{a_e} \right) = 0 \qquad (B.12)$$

Taking a derivative of (B.4) with respect to s gives

$$y\frac{dy}{dx} = \frac{dh}{ds}\left[\frac{d^2h}{ds^2} + \frac{1}{a_e}\left(\frac{a_e + h}{a_e} \right) \right] = 0 \qquad (B.13)$$

Substituting (B.12) and (B.13) into (B.11) after expansion of derivatives and some algebraic manipulation yields

$$\frac{d^2h}{ds^2} - \left(\frac{2}{a_e + h} + \frac{1}{n}\frac{dn}{dh} \right)\left(\frac{dh}{ds} \right)^2 - \left(\frac{a_e + h}{a_e} \right)^2 \left(\frac{1}{a_e + h} + \frac{1}{n}\frac{dn}{dh} \right) = 0 \qquad (B.14)$$

the exact equation for a ray in an atmosphere where n is only a function of height.

Index

above ground level (AGL), 161
absolute humidity, 207
absorption
  atmospheric, 208
  of electromagnetic energy, 12, 18
  propagation and, 8
  water vapor, 210
absorption bands, 207
absorption lines, 207
accumulated rainfall, 325
active remote sensors, 49
Afternoon Constellation, 432
AGL. *See* above ground level
air density, 302
airborne radars
  in atmospheric research, 224
  C-band radar on, 386–388
  in hurricane research, 388–389
  NOAA WP-3D, 386–388
  as specialized radar, 43–44
  tropical cyclones studied with, 384–385
  vertically pointing, 58, 60, 208, 415
  WCR as, 224–225
aircraft echo, 196–198
Alberta Clipper, 417–418
algorithms
  legacy unfolding, 239
  multi-PRF dealiasing, 239–240
  for precipitation, 327
  radar, 324
  Sachidananda–Zrnic, 115
alternate transmission and reception, 130
altitude diagram, 373, 375
ambiguity, 52
amplitude, 128
anafront, 368–370
analog to digital converter, 37
angular coordinates, 84
angular frequency, 121

anisotropic turbulence, 201
annual average precipitation, 333
anode, 27
anomalous propagation, 74–77, 184, 188–189
antenna beam pattern, 33, 289–290
antenna section, 32–34
ARM. *See* Atmospheric Radiation
     Measurement
A-scope display, 59
ASOS. *See* Automated Surface Observing
     System
Atlantic basin, 385
atmosphere, 11, 441–442
atmospheric absorption, 208
atmospheric attenuation, 208–209
atmospheric density, 274
atmospheric gases, 207–212
Atmospheric Radiation Measurement
     (ARM), 224, 424
atmospheric refraction
  radar ray and, 71
  of radar ray path, 74
  standard, 69
atmospheric research, 224
atmospheric structure, 428–432
A-Train, 432
attenuation, 83
  atmospheric, 208–209
  by atmospheric gases, 207–212
  Beer-Lambert law on, 206
  of C-band radar, 215–216
  by cloud droplets, 212–214
  coefficient, 206
  convective updrafts causing, 225
  differential, 133, 139–141, 205–206
  by hail, 219–224
  large-wavelength radars and signal, 24
  liquid coefficient and, 212
  normalized specific, 221

Radar Meteorology: A First Course, First Edition. Robert M. Rauber and Stephen W. Nesbitt.
© 2018 John Wiley & Sons Ltd. Published 2018 by John Wiley & Sons Ltd.
Companion website: www.wiley.com/go/Rauber/RadarMeteorology

Printed and bound by CPI Group (UK) Ltd, Croydon, CR0 4YY

23/10/2024

14578384-0002